The Arboviruses: Epidemiology and Ecology

Volume III

Editor

Thomas P. Monath, M.D.

Director
Division of Vector-Borne Viral Diseases
Centers for Disease Control
Public Health Service
U.S. Department of Health and Human Services
Fort Collins, Colorado

CRC Press
Taylor & Francis Group
Boca Raton London New York

CRC Press is an imprint of the
Taylor & Francis Group, an **informa** business

T0174916

CRC Press
Taylor & Francis Group
6000 Broken Sound Parkway NW, Suite 300
Boca Raton, FL 33487-2742

Reissued 2019 by CRC Press

© 1988 by Taylor & Francis Group, LLC
CRC Press is an imprint of Taylor & Francis Group, an Informa business

No claim to original U.S. Government works

A Library of Congress record exists under LC control number:

Publisher's Note
The publisher has gone to great lengths to ensure the quality of this reprint but points out that some imperfections in the original copies may be apparent.

Disclaimer
The publisher has made every effort to trace copyright holders and welcomes correspondence from those they have been unable to contact.

ISBN 13: 978-0-367-23540-6 (hbk)
ISBN 13: 978-0-367-23542-0 (pbk)
ISBN 13: 978-0-429-28027-6 (ebk)

Visit the Taylor & Francis Web site at http://www.taylorandfrancis.com and the CRC Press Web site at http://www.crcpress.com

FOREWORD

The term "arbovirus" is used to describe a diverse array of viruses which share a common feature, namely transmission by arthropod vectors. This ecological grouping now includes over 500 viruses, most belonging to five families — the Togaviridae, Flaviviridae, Bunyaviridae, Reoviridae, and Rhabdoviridae. Over 100 of these agents have been associated with naturally acquired disease in humans and/or domestic animals, and among these approximately 50 of the most important pathogenic viruses have been selected for detailed review under this title.

The complexity of arbovirus ecology requires a fundamental understanding of the influence of each of the multiple components (virus, vector, viremic host, clinical host, and environment) on infection and transmission cycles. The first volume, devoted to the variables which affect arbovirus transmission, provides this background and also contains guidelines for the design of future epidemiological investigations. Armed with these general principles, the student, teacher, or research worker will be able to structure specific knowledge about individual arbovirus infections found in Volumes II through V.

Recent textbooks are available which provide comprehensive coverage of the clinical aspects, pathogenesis, virological characteristics, and molecular biology of arboviruses. The intent of this book is different, for it focuses on the epidemiology and ecology of the arboviruses, the risk factors underlying the appearance of disease in the community, and the roles of arthropod vector and vertebrate hosts in virus transmission. Emphasis is placed on the field and laboratory evidence for involvement of vector and host species and on the ecological dynamics which determine their ability to spread infection. Elements of transmission cycles which are susceptible to surveillance, field investigation, prevention, and control are elucidated.

A number of arboviruses which have caused human disease only on rare occasions are not included in the book or are mentioned in passing within chapters on related diseases. Although these viruses (for example, Spondweni, Ilheus, Rio Bravo, Usutu, Orungo, Wanowrie) are inherently interesting and may, with changing ecologic conditions, turn out to be medically important, little or no information about their epidemiology (insofar as it relates to clinical hosts) is available. Finally, the scope of this book has been limited strictly to arthropod-borne infections and other viruses sometimes considered under the aegis of arbovirology (e.g., rodent-borne viral hemorrhagic fevers) are not included.

The compilation of a book of this scope required sacrifices in time and energy by a large number of contributors, all of whom faced multiple other commitments. This sacrifice will, I expect, be partially compensated by the availability of a useful compendium of collective knowledge.

Thomas P. Monath
October 1986

THE EDITOR

Thomas P. Monath is Director of the Division of Vector-Borne Viral Diseases, Centers for Disease Control, and is an affiliate faculty member of the Department of Microbiology, College of Veterinary Medicine, Colorado State University.

He received his undergraduate and M.D. degrees from Harvard University and his clinical training in Internal Medicine at the Peter Bent Brigham Hospital, Boston. In 1968 he joined the U.S. Public Health Service, serving as Medical Officer in the Arbovirology Unit, Centers for Disease Control, Atlanta, and later as Chief of the Arbovirus Section. Between 1970 and 1972, he was assigned to the Virus Research Laboratory of the Rockefeller Foundation, University of Ibadan, Nigeria, where he conducted field research on the epidemiology of yellow fever and Lassa fever. Since 1974, Dr. Monath has been Director of the Division of Vector-Borne Viral Diseases, Fort Collins, Colorado. In 1984 — 1985 he spent a sabbatical year in the Gastroenterology Unit of the Massachusetts General Hospital.

Dr. Monath is a Fellow of the American College of Physicians, the Infectious Disease Society, and the Royal Society of Tropical Medicine and Hygiene. He is a member of the American Society of Virologists, the American Society of Tropical Medicine and Hygiene, and the Association of Military Surgeons. He serves on the Editorial Boards of the *American Journal of Tropical Medicine and Hygiene, Acta Tropica,* and the *Journal of Virological Methods.* Dr. Monath is a member of the Committee on Research Grants, Board of Science and Technology for International Development, National Research Council, and is currently Chairman of the AIBS Infectious Diseases and Immunology Peer Review Panel to the U. S. Army Medical Research and Development Command. He is a member of the World Health Organization Expert Committee on Virus Diseases and the Pan American Health Organization Scientific Advisory Committee on Dengue, Yellow Fever, and *Aedes aegypti.* He has served as Chairman of the Executive Council of the American Committee on Arthropod-Borne Viruses, as a Councilor of the American Society of Tropical Medicine and Hygiene, and as a member of the Directory Board, International Comparative Virology Organization and the U. S.-Japan Cooperative Medical Research Program Panel on Virus Diseases.

Dr. Monath has authored or coauthored over 140 scientific publications in the field of virology and is editor of the book, *St. Louis Encephalitis,* published by the American Public Health Association. His main research interests are the ecology, epidemiology, and pathogenesis of arbovirus infections.

TO THE MEMORY OF MY PARENTS

CONTRIBUTORS

Kalyan Banerjee, M.B., B.S., Ph.D.
Head
Department of Virology
National Institute of Virology
Pune, India

Donald S. Burke, M.D.
Chief
Department of Virus Diseases
Walter Reed Army Institute of Research
Washington, D. C.

F. Glyn Davies
Head
ODA Virology Research Project
Veterinary Research Laboratories
Kabete, Kenya

Bruce K. Johnson, Ph.D.
Deputy Director
Arbovirus Division
Virus Research Center
Nairobi, Kenya

Yuji Kono, D.V.M., Ph.D.
Second Research Division
National Institute of Animal Health
Kannondai, Tsukuba, Ibaraki, Japan

Colin J. Leake, Ph.D.
Lecturer
Department of Entomology
London School of Hygiene and Tropical
 Medicine
London, England

James W. LeDuc, Ph.D.
Chief
Department of Epidemiology
Disease Assessment Division
USAMRIID, Fort Detrick
Frederick, Maryland

Dimitri K. Lvov, M.D.
Academician Acad. Med. Sci. U.S.S.R.
Laboratory of Viral Ecology
The D. I. Ivanovsky Institute of Virology
Academy of Medical Sciences
Moscow, U.S.S.R.

Ian D. Marshall, Ph.D.
Department of Microbiology
John Curtin School of Medical Research
Australian National University
Canberra, Australia

Charles D. Morris, Ph.D.
Mosquito Research Coordinator
Environmental Services of Polk County
Bartow, Florida

Francisco P. Pinheiro, M.D.
Regional Advisor in Viral Diseases
Communicable Diseases Program
Pan American Health Organization
Washington, D.C.

Hugh W. Reid, Ph.D.
Principal Veterinary Research Officer
Department of Microbiology
Animal Diseases Research Association
Edinburgh, Scotland

Robert E. Shope, M.D.
Professor of Epidemiology
Department of Epidemiology and Public
 Health
Yale Arbovirus Research Unit
Yale University
New Haven, Connecticut

Amelia Travassos da Rosa
Instituto Evandro Chagas
Fundaçao Servico Especial de Saude
 Publica
Belém, Brazil

John P. Woodall, Ph.D.
World Health Organization
Geneva, Switzerland

TABLE OF CONTENTS

Volume III

Chapter 24

EASTERN EQUINE ENCEPHALOMYELITIS

C. D. Morris

TABLE OF CONTENTS

I. HISTORICAL BACKGROUND

A. Discovery of Epidemics, the Agent, and History

Eastern equine encephalomyelitis virus (EEEV) has been the cause of epizootics of eastern equine encephalitis (EEE) in North American horses as far back as 1831, but undoubtedly the virus itself was present in its endemic cycle long before that.[1,2] The virus, however, was not isolated until a major outbreak occurred in horses in coastal areas of Delaware, Maryland, New Jersey, and Virginia in 1933.[3,4] There were additional outbreaks in 1934 in Virginia and 1935 in North Carolina. It was during the 1935 outbreak that birds were considered to be a possible reservoir host,[5] but it was not until 1950 that the first virus isolation was made from a wild bird.[6] Subsequent studies have shown that many birds, including virtually all passerine species, are susceptible to EEEV infection.[7,8]

The first human cases of EEE were confirmed in 1938 by virus isolation from brain tissue in New England.[9,10] Between 1938 and 1961 there was a total of 112 recorded human cases during outbreaks in Massachusetts in 1938,[11] Mississippi, Arkansas, and Texas in 1941,[12,13] Michigan in 1942—1943,[14] Louisiana in 1946 to 1949,[15-17] and New Jersey in 1959.[18,19] Between 1961 and 1985 there have been 99 human cases, at the rate of fewer than 15 a year nationwide.[20,21]

The fact that the virus could cause encephalitis in wild and domestic pheasants was first documented in Connecticut in 1938.[22] There were major epiornitics in pheasants in New Jersey between 1936 and 1946,[23-25] and outbreaks in pheasants, Pekin ducks, chukars, and pigeons in many states along the eastern seaboard and elsewhere.[26-28] Other vertebrates besides human beings, horses, and birds have been found naturally infected with EEEV. None of these are considered to play any role in the epidemiology of the virus nor is the virus considered a major disease of other animals.

Mosquitoes were first incriminated as potential vectors of EEEV in 1934, and a number of studies have shown species of *Aedes, Culex,* and *Coquillettidia* could become infected with and transmit EEEV from one vertebrate to another.[29-32] The first isolation from arthropods in nature were from mites and lice in 1947;[33] the first from a mosquito (*Coquillettidia perturbans*) was in 1949,[34] and the first from *Culiseta melanura* in 1951.[35] Since then numerous virus strains have been isolated in nature from mosquitoes, the vast majority from *Cs. melanura.*

B. Social and Economic Impact

Outbreaks of EEE are infrequent, and perhaps because of their infrequency, the disease has a significant economic and social impact once an endemic focus has been identified. The first time the disease occurs in an area, there is a loss of horses and/or poultry, perhaps human morbidity and mortality, and the human fear of the unknown. This was particularly evident in central New York State in 1971, when EEE was first discovered in the area. There was panic in the local population and the economy suffered as visitors removed their children from camps and refused to use the area for camping, fishing, and hunting. The outbreak occurred just prior to the New York State Fair which was held near the outbreak area, and there was a 12% reduction in fair attendance in 1971 as compared to 1970 and 1972.

The social and economic impact of mosquito control for EEE prevention is considerable once the potential for disease is recognized in an area. Economic losses have been documented for the 1959 outbreak in New Jersey.[36] The economic impact entails costs for surveillance, prevention, and control. In many cases these costs, especially for mosquito control, are not trivial. Besides the financial costs, the effects of environmental alterations from heavy use of pesticide and water management projects, which have occurred over the past 40 years, are substantial.

II. THE VIRUS

A. Antigenic Relationships

EEEV is a member of the *Alphavirus* genus, family Togoviridae. It is related to but antigenically distinct from a sympatric member of the western equine encephalomyelitis (WEE) virus complex, Highlands J (HJ) virus. Epidemiologically, EEEV has many similarities to WEEV in that both viruses cause encephalitis in horses and man, have wild avian hosts, and are transmitted from birds to mammals by mosquitoes. EEEV and HJ virus are most closely related epidemiologically in that they share geographic distributions, are transmitted by the mosquito *Cs. melanura*,[37-39] and infect a wide spectrum of wild avians, especially passerines.[40-43] HJ is often more prevalent in mosquitoes than EEEV, and HJ is an ecological precursor of EEEV, its presence preceding EEEV in mosquitoes. While EEEV causes frequent and serious encephalitis in horses, HJ has infrequently been the cause of mild equine encephalitis.[44,45] Isolations of HJ during both routine surveillance and outbreaks of EEEV are common.[42,43,46]

While unrelated serologically, Hart Park and Flanders, two viruses in the Hart Park group of Rhabdoviruses, are also intimately associated with *Cs. melanura* and passerine birds.[47,48] Neither virus has been shown to cause disease in man, birds, horses, or other animals.

B. Host Range

The known host range of EEEV is extensive if experimental infection data are included. EEEV has been isolated from, or antibodies to EEEV have been found in, naturally infected ringnecked pheasants, pigeons, chukar partridge, Pekin ducks, turkeys, and a multitude of wild birds, particularly passerines, but also including owls, whooping cranes, and shore birds.[26,28,49-54] Virus and antibodies have been found less frequently in naturally infected mammals, even though most small mammals tested are highly susceptible.[55-62] Turtles and snakes have also been infected naturally and experimentally, as have swine, bovines, hamsters, and fish.[57,63-67]

The isolation of EEEV from mosquitoes is very common during epizootics, but much less common during interepizootic periods. EEEV isolations have been reported in the literature from 23 species in six genera. More than 80% of these have been from *Cs. melanura*, but there have been isolations made from (in roughly decreasing frequency) *Cx. nigripalpus*, *Cq. perturbans*, *Cs. morsitans*, *Ae. sollicitans*, *Anopheles quadrimaculatus*, *Cx. salinarius*, *Cx. pipiens*, *Cx. restuans*, *Cx.* sp., *Ae. canadensis*, *Ae. atlanticus-tormentor*, *Ae. cantator*, *Ae. infirmatus*, *Aedes* sp., *An. crucians*, *Cx. (Melaniconium)* sp., *Ae. vexans*, *Cs. minnesotae*, *Cx. territans*, *Uranotaenia sapphirina*, *Cx. quinquefasciatus*, *Ae. triseriatus*, *Ae. mitchellae*, and *An. punctipennis*.

C. Strain Variation and Assay Methods

Prior to 1961 all EEEV strains were considered similar. In 1961, Casals[68] used hemagglutination-inhibition (HI) tests to compare several strains of EEEV from different geographic locations and concluded that there were three groups: those in New Jersey, those near Belém, Brazil, and those from Central America. In 1962, Casals[69] concluded that 12 EEE strains fell into four groups: northeastern U.S., Central America, Brazil, and Argentina. Finally, in 1964, Casals[70] compared 19 strains of EEEV by kinetic HI test and concluded that there were only two groups, which he designated North American and South American subtypes. The North American subtype included all isolates from all hosts from the area between Massachusetts and the Caribbean. The South American subtype was limited to the South American continent.

For 20 years this dichotomy of EEEV strains has guided epidemiologic thought. In 1981, Walder et al.,[71] using polyacrylamide gel electrophoresis (PAGE), examined 24 strains isolated from many parts of the range of the virus. They concluded that while the core proteins were similar for all strains, there were seven groups which could be distinguished by the mobility of the two envelope glycoproteins. Three of the four groups which contained more than one strain also contained strains from both of Casals' North American and South American groups. These results demonstrate that EEEV of different epidemiological, biological, and geographic origins may also differ structurally. In contrast, strains from a single subpopulation of EEEV have similar glycoproteins although they may differ in antigenic, surface charge, or virulence properties. The host cell had little if any influence on the glycoprotein patterns. These structural differences are apparently stable genetic markers that can be considered indicative of strain differences.[72]

There is evidence that there are differences among epidemic and endemic strains of EEE and differences in virulence between strains from different areas.[49,73] WEE viruses isolated in different times of the year also differ in virulence.[74] There is considerable antigenic variation among strains of WEE, particularly among recent isolates of HJ from the eastern U.S.[75] Recently isolated strains of WEE from the same area seem to vary among themselves and suggest that WEE is more subject to antigenic change than is commonly believed.[75] Based on these findings, it may be possible that EEEV is also prone to significant antigenic change.

The factors which influence virus antigenicity and virulence have been examined.[76] Variations in virus types in nature are known to occur due to the species of host infected and to the antigenic characteristics which may change during the passage of the virus through different hosts or through adaptation to a given system. Virologists are familiar with the egg- and mouse-adapted strains of the same virus which differ antigenically. Phenotypic mixing does occur in *Alphavirus*,[77] but thus far genome reassortment has not been demonstrated.[78]

It is sometimes necessary to pass an agent through an assay system many times to get a cytopathic effect (CPE).[79] The occasional failure of EEEV to propagate and the common disappearance of CPE in the second to fourth passage is a recognized and unexplained phenomenon.[79]

The concept of strain differences among EEEV isolates has historically reflected the ability of available techniques to distinguish differences among viruses. No doubt, the application of monoclonal antibody reagents and the development of new techniques will eventually allow us to detect even more differences between isolates and subpopulations of viruses. One ultimate conclusion could be that most virus strains are allopatric. Another likely conclusion is that there is a continuum of change in the virus from one end of its endemic range to the other. The concept of one or two virus strains for all of the New World is simply not tenable in light of current research on EEEV and other *Alphavirus*. The concept of species or strains often is for the convenience of scientists rather than an accurate description of nature.

For more than 20 years the HI, neutralization (N), and to a lesser degree complement-fixation (CF) tests have been the mainstay of laboratory techniques for the assay of EEEV antibodies.[51,80] Virus isolation has historically been, and still is, accomplished primarily by detecting disease in suckling mice and wet chicks or CPE in tissue cultures such as HeLa, BHK, and duck or chick embryo. More recently, the fluorescent antibody (FA), indirect FA (IFA), enzyme-linked immunosorbent assay (ELISA), plaque reduction, N test, and electrophoretic techniques have been developed to detect and identify EEEV and antibodies to EEEV.[72,81-84]

Current widely used procedures are probably inadequate to detect EEEV in all its forms. Amino acid sequencing techniques indicate a high degree of conservation in repeated sequences among closely related *Alphavirus*, including EEEV.[85] RNA probes keyed to these sequences should be able to detect both virulent and avirulent virions. It is, however, the degree of variation in the heterologous, nonrepeated sequences which will determine if EEEV epidemiologic types constitute a continuous or discrete distribution.

III. DISEASE ASSOCIATIONS AND EXPERIMENTAL INFECTION STUDIES OF ANIMALS

A. Humans

The course of infection of EEEV in the human is dependent on the age of the person and the presence of neural infection.[61,86-88] There are two types of illness, systemic and encephalitic. A systemic infection is abrupt and characterized by malaise, arthralgia, and myalgia. In a few hours the patient is chilly and experiences severe muscular shaking which lasts for a few days. Maximum temperatures reach 100 to 104°F. The illness lasts 1 to 2 weeks, there is no CNS involvement, and recovery is complete.[51]

The encephalitic form of the disease in infants is characterized by an abrupt onset, whereas in older children and adults onset of active encephalitis typically occurs after a few days of indisposition.[89] Symptoms include fever of 102 to 106.4°F, irritability, restlessness, drowsiness, anorexia, vomiting, diarrhea, headache, cyanosis, convulsions, and coma.[15,51,89] Tremors and muscular twitching are usually accompanied by neck rigidity which is continuous. Cerebrospinal fluid usually has increased pressure with 200 to 2000 cells, 60 to 90% neutrophils.

Death typically occurs in 2 to 10 days but there is a report of an infant that died 54 days after onset.[15] Death is due to encephalitis, sometimes with evidence of myocardial insufficiency and pulmonary involvement.[89] Brain pathology shows encephalomyelitis characterized by intense vascular engorgement, perivascular and parenchymous in the cortex, midbrain, and brain stem.[9,89] The spinal cord is little involved.

During epidemics the initial mortality approximates 70%. Nearly all who recover have disabling mental and physical sequelae which are progressive.[15] In one study the initial mortality was 74%, but 9 years postinfection total mortality associated with EEE sequelae reached 90%, with only a 3% complete recovery rate.[89]

Studies with rhesus monkeys showed that the outcome of the disease was associated with the route of inoculation.[61,90] Intracerebral injection, as well as intranasal and intralingual inoculations, produced fatal illness in most young monkeys. Intravenous and subcutaneous injections did not produce encephalitis, although antibodies were produced. Ocular infusion or ingestion of the virus did not cause infection, and no healthy animals contracted the illness when caged with sick and dying animals.

Unpassed isolates of EEEV from mosquitoes and a fatal human case both produced a fatal encephalitis in monkeys similar to that seen in humans. Apparently wild strains of EEEV have a very high primate neurovirulence regardless of the natural host.[90] The apparent to inapparent case ratio in the 1959 outbreak in New Jersey was 1:23.[91] This probably represents the frequency with which virus invades the CNS rather than variation in the neurovirulence of the strains.[61] Most of the nonencephalitic cases are subclinical, if not inapparent.

B. Domestic Animals

Four major patterns of infection in horses with EEEV have been observed, both by subcutaneous inoculation and mosquito bite.[80] The first is characterized by a biphasic febrile reaction. The outcome can be fatal or the animals may recover with or without CNS sequelae. The maximum viremia of 19 inoculated equines in one study was 5.8 dex.[92] Viremia was not related to dosage given. The second type of infection is characterized by a single temperature rise, the horses showing no signs of CNS involvement although circulating virus was demonstrated. The third pattern of infection is the presence of small amounts of virus without febrile illness and the fourth type is the absence of both a febrile response and demonstrable circulating virus even though there is development of specific antibodies.

When illness occurs, the signs of infection are typically in the order of depression, progressive incoordination, rocking motion, convulsions, and prostration.[80] Harrison[93] recognized two types of clinical disease, the ataxic form and the paralytic form. In the ataxic form, incoordination in the anterior quarters led to the loss of equilibrium. The animal remained upright by leaning its flank on a vertical wall and keeping its nostrils directed toward a corner. The hind legs were often crossed. There are often visual problems which may result in at least a partial blindness. Fever reached 103 to 106°F.

In the paralytic form the horse showed an almost total depressive state. The ataxia of locomotion was difficult to observe. The eyelids were lightly tumefied with eyes half closed. The animal was in a state of somnolence or sleep with an accelerated heart beat. The jaws locked and death occurred in 1 to 3 days.

A survey of 67 equine deaths due to EEE in the U.S. in 1971 showed that the most common signs of illness were depression (49% of cases), fever above 103°F (30%), ataxia (25%), paralysis (25%), anorexia (20%), and stupor (20%).[94] Other symptoms of note include irregular gait, grinding of teeth, incoordination, circling, staggering, recumbency, and hyperexcitability. The pathology of EEEV in horses involves primarily the gray matter of the brain.[80] Suckling pigs and calves are also known to be susceptible to disease.[63,65,95]

Clinical signs of EEEV infection in domestic fowl (pheasants, chickens, chukars, ducks, turkeys) and wild birds have been described in detail.[22,26,52,96-100] Most species show signs referable to CNS involvement consisting primarily of leg paralysis, tremors, and somnolence, followed by prostration and death. EEEV infection in chickens and whooping cranes are viserotrophic rather than neurotrophic.[22,96,97] Chickens typically exhibit diarrhea (bloody at times), whereas only three of seven cranes showed any clinical signs (lethargy and ataxia) prior to death. It has been suggested that the primary sites of infection are the liver, spleen, and kidney.[80,97]

Younger animals, including people, birds, and rodents, are typically more susceptible to CNS disease.[22,61,73,88,96] This may be attributed to a greater susceptibility of cerebral tissue and the facts that virus multiplies more and circulates longer in the blood of young.[61]

C. Wildlife

EEEV infection in wild birds may result in disease in introduced species such as the English sparrow,[101] ring-necked pheasant, and domestic pigeon.[9] Die-offs of native birds, especially small species, have been noted during EEEV epizootics, but the caues of death was not conclusively identified as EEEV.[41,43] In 1984, 7 of 39 captive whooping cranes died from EEE, but no morbidity or mortality was observed in cohabiting sandhill cranes.[97] Inoculated passerines often die whereas most shorebirds do not.[8,102]

Bats are susceptible to natural infection and by inoculation and mosquito bite in the laboratory, but not by ingestion of virus.[60,103] Bats could not transmit the virus through bite, urine, or feces.[103] Natural EEE infections have not been demonstrated in white-tailed deer,[55,56] and only 1 of 276 small mammals in Massachusetts had antibody-positive sera; none were viremic.[57] In contrast, 3% of 1172 Florida raccoon and opossum sera had antibodies to EEEV.[58] Voles, woodchucks, and cottontail rabbits are highly susceptible to and invariably die from experimental infections.[59] The opossum survived inoculation without signs of disease.

Reptiles and amphibians have also been found to be naturally infected with EEEV,[57,64,104] to be susceptible to experimental infection,[57,64] to maintain high viremia over several months,[57,105] and to experimentally carry the virus through hibernation.[57] Inoculated snakes were especially susceptible to infection and often died in captivity.[64] As with many laboratory studies on wild animals, it is uncertain whether the cause of these deaths was from infection or the effects of captivity. Turtles and lizards, on the other hand, were susceptible to infection but did not die.

D. Adverse Effects on Arthropods

Insects other than mosquitoes can be infected with EEEV and support virus replication.[106] A grasshopper, bedbug, milkweed bug, carpet beetle, Indian meal moth, and soft tick all supported virus replication, whereas the cockroach and housefly did not. Transstadial transmission was demonstrated in the milkweed bug and carpet beetle. There was no transovarial transmission in the milkweed bug. No adverse affects of infection were seen in any of the six arthropod species.

Little work has been done on the effect of EEEV on mosquitoes, but no CPE was observed in EEEV-infected salivary glands of *Ae. triseriatus*.[107] There have been no reports of differences in behavior or longevity of laboratory-infected mosquitoes.

IV. EPIDEMIOLOGY

A. Geographic Distribution and Incidence

On a gross geographic scale, outbreaks of EEEV in North America have been documented in the Canadian province of Quebec and in essentially all of the states of the U.S. east of the Mississippi River. West of the river, the virus has been isolated from Minnesota, South Dakota, and Texas. The incidence of EEE in humans since 1955 has ranged from 0 to 36 (mean of 7) human cases per year.[20,21] The number of inapparent cases per overt case was estimated at 23 for the 1959 outbreak in New Jersey, giving an attack rate of approximately 1/1000.[91] Between 1956 and 1972 there were 684 reported equine cases with a range from 1 to 132/year.[20] The equine attack rate in New Jersey varied from 0.4 to 80.5/10,000 between 1933 and 1959.[18]

B. Seasonal Distribution

In the northern part of the virus range, both human and equine cases occur between July and October. Human cases occur year-round in Florida, but are concentrated between May and August.[108] The virus or disease is often recognized first in either wild birds, penned pheasants, or equines prior to human involvement.[18,109-111] Epizootic activity in a specific endemic area occurs about every 5 to 10 years. Peaks of equine cases occur approximately every 5 years, whereas peaks of human cases occur about every 10 years.[18,20,112,113]

C. Risk Factors

People at greatest risk to EEE are children under 15 and adults over 55. Those age groups make up to 70 to 90% of the cases in a given outbreak.[18,114] Both sexes are equally affected, although typically there are more males than females.[108,114] Inapparent infections are the same for all age groups and for both sexes.[91] The disease is rural in distribution and most cases are associated with wooded areas adjacent to swamps and marshes. Horses of all breeds are susceptible.[115]

V. TRANSMISSION CYCLES AND OVERWINTERING

The classical concept of EEEV epidemiology is based on the assumptions that: (1) EEEV activity in geographically isolated locations is interrelated epidemiologically, if not interdependent, and (2) there is one and only one EEE virus (with perhaps a latent phase) that is epidemiologically distinct from and independent of other arboviruses and microorganisms that share the same mosquito and avian hosts. Prior condensations of EEEV epidemiology typically conclude that: (1) *Cs. melanura* is the principal endemic vector, (2) wild birds, primarily passerines, serve as the primary amplifying hosts, (3) epizootics begin in swamps and move locally outward through viremic birds, (4) in some cases, migrating birds introduce the virus into distant "clean" ecosystems, (5) humans and equines are dead-end hosts, (6) there is a single most important epidemic or epizootic vector, (7) nonavian vertebrates are potential but unproven maintenance hosts, and (8) the overwintering mechanism is unknown or an overwintering mechanism is not required.

Some of these conclusions are logically untenable; others seem to be attempts to simplify and unify for convenience. Alternative interpretations of published data are possible, given a different set of *a priori* conditions. New hypotheses stated within a scenario based on one such an alternative framework will now be presented.

The first *a priori* concept is that there are many distinct, geographically isolated foci of EEEV subpopulations, each with its own isolated communities of invertebrate vectors and vertebrate hosts. In the past, EEEV was probably distributed over a much larger and more contiguous area. Natural change and human development disrupted this continuity and developed islands of EEEV. These isolated foci are operative at the level of individual swamps or swamp complexes, not just between states or geographic areas of the continent. The basic unit of EEEV is a natural wetland community. The ecologic principles which govern island forests and wetlands also govern the organisms which inhabit these forests and wetlands, including EEEV, its invertebrate vectors, and vertebrate hosts.

The fact that each year EEEV epizootic activity in North America is typically in a different location or in several distant locations, rather than being widespread, is presented as support. Even within a state or county, virus activity will be associated with one region or swamp complex one year, and another region or swamp complex the next.[108,112,116]

Associated with the formation of wetland islands was the development of subpopulations of EEEV within each unique habitat. In all likelihood, the subpopulation of EEEV from different areas of the U.S. have different infection rates and transmission rates in different subpopulations of mosquitoes, similar to what has been observed for La Crosse virus and *Ae. triseriatus*.[117] Behavioral differences have already been observed between subpopulations of *Cs. melanura*.[118-127]

The second *a priori* concept is that within a focus, the EEEV subpopulation is multiphased or genotypically unstable, with the capability of changing from one phase to another. This is not to say that there are two or three or more strains of virus, each replicating itself, but rather that a virus subpopulation can take on and replicate numerous and diverse morpho-logic/physiologic phases, the highly virulent form of which we call EEEV.

Because EEEV has been isolated every month of the year in Florida, it has been proposed that there is active virus circulation year-round in the state, with no overwintering mechanism required.[108] This may be true in the southern tip of Florida, but in those parts of the state where EEEV is most prevalent there are long periods during the winter without adult mosquitoes. Thus, a winter reservoir seems a prerequisite for virtually all North American endemic foci.

Each focus of EEEV is here considered allopatric, and while differences in the epidemiology of the virus should be expected among the foci, there remains the constant fact that *Cs. melanura* is present in all foci. Since the virus in any form does not appear to harm arthropods, it has been suggested that the virus is of insect origin.[106] Also, since over 80% of the EEEV and 90% of the HJ isolates have come from *Cs. melanura*, this species is considered here and by others to be the prime candidate as the overwintering host for EEEV.[58,128]

Based on these concepts, the following is proposed as a typical transmission cycle of EEEV in North America. The virus, in an avirulent phase, resides in the overwintering invertebrate host, *Cs. melanura*. The spring brood, or perhaps only the summer brood, of *Cs. melanura* transmits the virus to its predominant hosts, passerine birds, and the virus is continuously circulated between these two host species. Sometime during this cycle the avirulent phase mutates to a phase which may cause CPE in atypical hosts. This phase change occurs in the transfer from *Cs. melanura* to passerine birds and is thus essentially limited to the swamp and associated habitats where *Cs. melanura* is most abundant. The detection of HJ antibody in birds 3 weeks prior to virus isolation from mosquitoes suggests that transformation occurs in birds rather than in mosquitoes.[53]

Following breeding, passerine birds undergo severe reproductive system regression and a complete feather molt.[129] This typically begins in mid-July for most passerines,[130] and during this period birds are relatively quiet and inactive.[131] As any avid birder knows, this is a bad time for bird watching. Molting is followed by metabolic and behavioral changes associated with postbreeding random wanderings, and then fall migration.[129] These major avian physiologic changes provide the virus with a substantially different set of host parameters. If the virus is going to change to the highly virulent phase we call EEEV, this would seem to be the best time and location.

The timing of these relatively sudden changes in avian physiology and proposed virus virulence could explain the seemingly sudden appearance of EEEV in mosquitoes and wild birds in endemic foci in July. This sudden appearance also frequently coincides with the peak emergence of the summer brood of *Cs. melanura* in northern latitudes.

The consequences of this highly virulent virus in wild birds are the production of high viremias and homologous antibodies, and perhaps some mortality. At this stage other hematophagous arthropod species in the swamp habitat become involved as secondary vectors. Based on the feeding habits of these vectors, small mammals, domestic fowl, equines, and man may become infected.[132] There is no single epizootic vector species for all foci or for all hosts within a focus.

The post-nesting random wanderings of avians, particularly young birds, may serve as a mechanism of dissemination of the virulent phase within the limits of the premigratory range, which can be north or south from the breeding area.[131] Migratory birds which settle in an outbreak area should be as susceptible to infection as resident birds, the probability of becoming so being largely dependent upon the length of stay in the area. Any influx of susceptible migrants would tend to amplify the epiornitic, and in that sense migratory birds could contribute to EEEV epidemics.[46] Viremic migratory birds may leave endemic foci, but it is highly problematic that they serve as a source of new outbreaks. If they did, it would be apparent by a spread of EEEV activity from north to south concurrent with fall bird migrations; this has not been demonstrated. What has been demonstrated is that EEEV outbreaks occur in the same foci repeatedly and only where there is *Cs. melanura*. If *Cs. melanura* were not essential in epidemic areas, there should be outbreaks throughout the migratory range of infected passerines; this also does not happen.

EEE in man and equines is closely associated with the swamp breeding grounds of *Cs. melanura*.[133,134] Disease in either vertebrate host seldom occurs beyond 5 mi from these foci[43,112,113] and prevalence of antibodies in birds decreases with distance from the swamp.[41,43] This distance is within the flight range of *Cs. melanura* and the mosquito species which probably serve as epidemic vectors, such as *Cq. perturbans*, *Ae. vexans*, *Ae. canadensis*, *Ae. sollicitans*, *Cs. nigripalpus*, and others.[134-139] Thus, both mosquitoes and birds are capable of disseminating the virus from the swamp foci.

It is generally believed that equines are dead-end hosts, yet experimental studies have shown viremias high enough to infect mosquitoes,[32,37,92] and thus horses may enter into the epidemic maintenance of EEEV.[50]

While a number of laboratory studies have evaluated the vector potential of many mosquito species, results of these studies should be viewed with caution since both the virus and mosquitoes have been laboratory adapted.[37,140] A case in point is that laboratory infection and transmission studies classified *Cs. melanura* as only a fair to poor vector.[32]

EEEV is frequently isolated from different *Aedes* and *Culex* species in different parts of the continent. However, EEEV has been isolated from a single species, *Cq. perturbans* in many parts of the virus range, including Georgia, Florida, New York, Michigan, Massachusetts, and New Jersey.[34,42,141-144]

Cq. perturbans is an opportunistic feeder and feeds equally on birds and mammals, avidly on man and equines, thus making it an ideal potential epizootic vector.[145-147]

While mosquitoes have classically been incriminated as the vector, there has been essentially no examination of other hematophagous Diptera during an outbreak, including midges (Ceratopogonidae:*Culicoides*) and black flies (Simuliidae). Experimental infection and transmission studies of nonculicine groups have been done only on mites and lice, even though EEEV has been isolated from black flies and *Culicoides*.[140,148-151]

Adult mosquito population densities are closely linked to environmental conditions, and densities are highest during years with high rainfall. Associations between EEEV outbreaks and excessive rainfall have been observed.[111,128,152] It is important to reemphasize that mosquito population density is not, in itself, a presage of either detectable endemic or epidemic EEEV activity.

EEEV was isolated from overwintering larval *Cs. melanura*,[153] and there is laboratory evidence of transovarial transmission of EEEV by *Cq. perturbans*.[154,155] One isolate of EEEV has also been made from male *Cs. melanura*, but concurrent isolations of EEEV from females in the same collection suggests the possibility of contamination.[156] More recent laboratory studies indicate, however, that EEEV in *Cs. melanura* does not infect ovarian tissue and that the virus is not transovarially transmitted.[157,158] Although species other than *Cs. melanura* have been suggested as possible overwintering hosts, recent field studies on populations of *Cs. melanura*, *Cq. perturbans*, *Cs. morsitans*, and *Ae. cantator* all failed to demonstrate virus in males, first brood females, or overwintering larvae.[142,157,159-163] Results of those studies that tested late instar larvae for virus may be invalid since Ksiazek et al.[164] demonstrated an inhibitory effect of larval components on the detection of alphaviruses by plaque assay, a phenomenon confirmed for EEEV in *Cs. melanura*.[165] However, if the concept of an avirulent phase of EEEV is accepted, then one may not detect EEEV in mosquito larvae using current isolation techniques. The absence of virulent EEEV in male mosquitoes in these studies is consistent with the hypothesis presented earlier that virus transformation occurs in the bird, and thus the virulent form is accessible only to female mosquitoes.

Virus latency in birds, as a possible reservoir mechanism of EEEV, cannot be disproven by current knowledge. It is difficult to justify latency in birds as a reservoir when the interval between epiornitics in a given focus usually exceeds the 1- to 2-year average life span of most passerines.[131] If latency in nonculicine hosts does serve as a reservoir, new research initiatives should emphasize long-lived reptilians (especially snakes and turtles) from which avians evolved, and which seem to be less susceptible to disease when infected with EEEV.

VI. ECOLOGY

A. Environment

The primary habitats for EEEV are lowland areas in the eastern half of North America. The microenvironments in these foci are quite similar and specific, and the sizes of the foci quite small. Swamps and environs in Maryland, New York, New Jersey, Massachusetts, Florida, and Michigan have been described as being associated with muck-peat soil associations dominated by hardwoods.[43,46,116,160,165-167] In the northern part of the virus range the indicator tree species are red maple and hornbeam. In New Jersey, Maryland, and Florida the key species are red maple, cedar, and loblolly bay.[116,165,167,168] All of these trees can develop in a wet, mucky habitat and provide the type of root system favorable for *Cs. melanura* oviposition and larval development.

B. Oviposition

The bionomics of *Cs. melanura* larvae requires dark, highly organic water.[166,168-170] Breeding is usually concentrated on the edge of the swamps but also occurs in the swamp interior.[168,169] Overwintering larvae pupate in early spring and the first adults emerge in late May in the North, earlier in the South. In late June to early July in the North (earlier in the South) there is typically a second emergence in summer. While it is assumed that the second brood is the offspring of the first, it has been suggested that the two broods are physiologically, if not genetically, different.[166] Since EEEV is not normally found in mosquitoes until after the summer brood emerges, there may be some significant difference in the physiology of the two broods which influences vector potential.

C. Biting Activity and Host Preferences

The blood feeding habits of *Cs. melanura* are universally similar and ideally suited for EEEV transmission. All subpopulations of *Cs. melanura* examined so far are almost exclusively ornithophilic, and typically over 90% feed on passeriformes.[118,126,147,171] They do occasionally feed on reptiles and less frequently on mammals and man.[118,122,172,173]

VII. SURVEILLANCE

Monitoring for endemic, epizootic, or epidemic EEEV activity in an area can be accomplished by one of several methods, depending on available resources and objectives of the monitoring system.

Most states where EEEV is endemic have a statewide diagnostic service for veterinarians and physicians to submit blood and organ samples from suspected EEE cases. Reporting is required in some states, voluntary in others. These systems are inappropriate as an early warning system since the turnaround time of most is too long. They do, however, monitor nonendemic activity for the historical record.

Where veterinarians and epidemiologists are familiar with the parochial signs and symptoms of equine EEE, field diagnosis becomes an accurate and more rapid system which to base increased mosquito control efforts. By the time EEEV has become epizootic, however, prevention of additional cases requires the undesirable, widespread, indiscriminant spraying of mosquito adulticides. With or without spraying, human and equine cases typically persist, although perhaps at a reduced rate, until the advent of cold weather which interrupts the man- or horse-mosquito interactions.

The use of horse cases as a monitoring system is biased by the levels of vaccination in the population at risk. The higher the level of vaccination in the equine population, the less reliable the EEE monitoring system.[128] Valuable animals are usually vaccinated as prescribed, while animals belonging to poor and/or uncaring owners may be vaccinated only during times of confirmed EEEV activity, if at all.

Sentinel chickens are used for monitoring EEEV activity, but substantial evidence indicates they are inappropriate as an early warning system for epidemic activity.[50,108,174,175] In some foci, passerine birds and mosquitoes are monitored for endemic EEEV activity.[46,176] Virus isolations from mosquitoes, especially from other than *Cs. melanura*, signal possible epidemic activity and elude appropriate warnings and recommendations for increased mosquito control activity.[46,128,176] Presently, these techniques are time consuming and costly. Again, often the turnaround time from field to laboratory results is too long for implementing effective EEE-preventive mosquito control. The key to a better system lies in the development of rapid field-applicable techniques for detecting virus in birds and invertebrates. The ELISA test is currently being developed for this purpose.[177]

VIII. INVESTIGATION OF OUTBREAKS

Much of the literature on EEE is the result of investigations of outbreaks of disease in equines and man. Future investigations of outbreaks made under the prevailing concept that EEEV is autonomous in the ecosystem and limited to mosquitoes and birds will add little to elucidate the epidemiology of this virus. Sufficient information is now available to predict the fundamental results of any such investigation of an EEEV outbreak. If investigators take a more open mind to interactions of additional viruses, arthropods, and vertebrates, there is a great deal that could be examined during an outbreak. It would, of course, be preferable to establish longitudinal studies in endemic foci that would address the questions during interepidemic periods as well. The knowledge that *Cs. melanura* is a quintessential factor for EEEV means that mosquito surveys can easily delimit areas with a potential for EEEV; the higher the density of *Cs. melanura*, the higher the potential. The higher the populations of other man-biting mosquitoes, the higher the potential.

The prevention of outbreaks in man is predicated upon effective surveillance and appropriate mosquito control. Techniques for both are currently available, and refinements and new methods are being developed. Prevention of EEE in equines and domestic fowl by immunization has been available for decades and is routinely practiced by responsible owners. The recurrence of human and equine EEE in known endemic foci results from educational deficiencies and/or complacency by owners or public authorities, not from a lack of scientific knowledge of the epidemiology and control of the disease.

IX. PREVENTION AND CONTROL

Changing land use has significant impact on the occurrence of EEE in two ways. The initial epidemics of the 1930s in the Northeast resulted from close contact between mosquitoes and people who lived near endemic foci. As this area was developed and the significance of wetlands in disease epidemiology was recognized, many of these breeding areas were filled or drained and the incidence of EEE dropped. This process continues in many locations today.

The corollary, which is exemplified in central New York State and central Florida, is that as cultural development occurs, man and his domestic animals move into areas near previously unknown endemic foci and both contract the disease. EEE is an established veterinary problem in most of Florida.[108,115] It is becoming a medical problem because development is spreading inland, closer to the freshwater marshes and swamps which are endemic foci of EEEV.

Vaccines for humans are presently used to immunize virus laboratory technicians. Vaccines have proven to be very effective for preventing equine EEE, but recent evidence suggests that there are windows of susceptibility of foals in current vaccination practices.[115] Reared pheasants are often vaccinated routinely, although there is some question of efficacy of these vaccines.[25,49,179,180] Both of the problems with vaccines and vaccination protocols appear to be correctable.

Prevention and control of EEE has historically been addressed by the use of mosquito control programs. The efficacy of these programs is difficult to evaluate in most cases because mosquito control is often undertaken in response to an ongoing epizootic. Evidence that ongoing programs are effective is seen in the absence of epidemics of EEE since the advent of modern mosquito control in the 1940s. Two difficulties with mosquito control as a means of EEE control are that EEEV foci are in rural/suburban areas, and the mosquito breeding areas are extensive and difficult to reach and treat by means other than aircraft.

Mosquito control decisions for EEE control in some states are linked to the abundance of certain mosquito species, with or without concurrent virus isolations.[46,128,176] The indicator species in New York and Massachusetts is *Cs. melanura* and in New Jersey *Cs. melanura* and *Ae. sollicitans*. Diurnal resting shelters have proven to be an excellent monitoring technique for all physiologic states and both sexes of *Cs. melanura*.[46,118,165,182] CDC miniature light traps are used to monitor host-seeking *Cs. melanura* and man-biting species.[128,177,183] Human-bait sweep collections are used to monitor *Ae. sollicitans*.[46]

One well-timed aerial adulticide over endemic swamps in early to mid-July in central New York has been effective in preventing EEE cases in adjacent uplands. This control effort is aimed at *Cs. melanura* and is accomplished within 2 weeks after peak emergence of large summer broods. The size of the summer brood tends clearly to be either large or small, and the decision to spray is, therefore, usually clear-cut. Spraying can often be limited to the perimeter of large swamps where mosquito and bird densities are greatest.[41,119,120]

The current preferred methods for mosquito control are the application of conventional insecticides and through source reduction or habitat management. Societal demands for preserving natural wetlands means that source reduction will not likely be available in the 21st century. Natural wetlands are essential and must continue to be used for preserving watersheds, improving the quality of potable water, preserving natural history, and filtering sewage effluent.[185] There is a decided trend in the U.S. to create or restore destroyed wetlands for these purposes.[185-187] Mosquito control must be, and is being, integrated into increasingly complicated wetlands management plans.[187-189] Also, societal pressure for greater controls on the use of, or for eliminating, agrichemicals vital to current mosquito control programs continues to increase. Tomorrow's challenge in EEE prevention via mosquito control will be substantial under these greater restrictions.

ACKNOWLEDGMENTS

The author gratefully recognizes the significant contributions of Janine Callahan and John Howard to this report.

REFERENCES

1. **Hanson, R. P.,** An epizootic of equine encephalomyelitis that occurred in Massachusetts in 1831, *Am. J. Trop. Med. Hyg.,* 6, 858, 1957.
2. **Simms, B. T.,** Report of Infectious Equine Encephalomyelitis in the United States in 1949, Bulletin, Bureau of Animal Industry, U.S. Department of Agriculture, Washington, D.C., 1950, 4.
3. **Giltner, L. T. and Shahan, M. S.,** The 1933 outbreak of infectious equine encephalomyelitis in the eastern states, *North Am. Vet.,* 14, 25, 1933.
4. **Ten Broeck, C., Hurst, E. W., and Traub, E.,** Epidemiology of equine encephalomyelitis in the eastern United States, *J. Exp. Med.,* 62, 677, 1935.
5. **Ten Broeck, C.,** Birds as possible carriers of the virus of equine encephalomyelitis, *Arch. Pathol.,* 25, 759, 1938.
6. **Kissling, R. E., Rubin, H., Chamberlain, R. W., and Eidson, M. E.,** Recovery of virus of eastern equine encephalomyelitis from blood of a purple grackle, *Proc. Soc. Exp. Biol. Med.,* 77, 398, 1951.
7. **Kissling, R. E., Chamberlain, R. W., Sikes, R. K., and Eidson, M. E.,** Studies on the North American arthropod-borne encephalitides. III. Eastern equine encephalitis in wild birds, *Am. J. Hyg.,* 60, 251, 1954.
8. **Kissling, R. E., Chamberlain, R. W., Nelson, D. B., and Stamm, D. D.,** Studies on the North American arthropod-borne encephalitides. VIII. Equine encephalitis studies in Louisiana, *Am. J. Hyg.,* 62, 233, 1955.
9. **Fothergill, L. D., Dingle, J. H., Farber, S., and Connerley, M. L.,** Human encephalitis caused by the virus of the eastern variety of equine encephalomyelitis, *N. Engl. J. Med.,* 219, 411, 1938.
10. **Howitt, B.,** Recovery of the virus of equine encephalomyelitis from the brain of a child, *Science,* 88, 455, 1938.
11. **Feemster, R. F.,** Outbreak of encephalitis in man due to the eastern virus of equine encephalomyelitis, *Am. J. Public Health,* 28, 1403, 1938.

12. **Randall, R.,** Westward spread of eastern type equine encephalomyelitis virus, *Science*, 93, 595, 1941.
13. **Shahan, M. S. and Giltner, L. T.,** A review of the epizootiology of equine encephalomyelitis in the United States, *J. Am. Vet. Med. Assoc.*, 107, 279, 1945.
14. **Mohler, J. R.,** Report of Infectious Equine Encephalomyelitis in the United States, Bulletin, Bureau of Animal Industry, U. S. Department of Agriculture, Washington, D.C., 1943.
15. **Hauser, G. H.,** Human equine encephalomyelitis, eastern type, in Louisiana, *New Orleans Med. Surg. J.*, 100, 551, 1948.
16. **Howitt, B. F., Bishop, L. K., Gorrie, R. H., Kissling, R. E., Hauser, G. H., and Trueting, W. L.,** An outbreak of equine encephalomyelitis, eastern type, in southwestern Louisiana, *Proc. Soc. Exp. Biol. Med.*, 68, 70, 1948.
17. **Oglesby, W. T.,** 1947 outbreak of infectious equine encephalomyelitis in Louisiana, *J. Am. Vet. Med. Assoc.*, 113, 267, 1947.
18. **Goldfield, M. and Sussman, O.,** The 1959 outbreak of eastern encephalitis in New Jersey. I. Introduction and description of outbreak, *Am. J. Epidemiol.*, 87, 1, 1968.
19. **Rowan, D. F., Goldfield, M., Welsch, J. N., Taylor, B. F., and Sussman, O.,** The 1959 outbreak of eastern encephalitis in New Jersey. II. Isolation and indentification of viruses, *Am. J. Epidemiol.*, 78, 11, 1968.
20. **Monath, T. P.,** Arthropod-borne encephalitides in the Americas, *Bull. WHO*, 53, 513, 1979.
21. **Francy, D. B.,** personal communication, 1985.
22. **Tyzzer, E. E., Sellards, A. W., and Bennett, B. L.,** The occurrence in nature of equine encephalomyelitis in the ring-necked pheasant, *Science*, 83, 505, 1938.
23. **Beaudette, F. R. and Hudson, C. B.,** Additional outbreaks of equine encephalomyelitis in New Jersey pheasants, *J. Am. Vet. Med. Assoc.*, 107, 384, 1945.
24. **Beaudette, F. R. and Black, J. J.,** Equine encephalomyelitis in New Jersey pheasants, *J. Am. Vet. Med. Assoc.*, 112, 140, 1948.
25. **Beaudette, F. R., Black, J. J., Hudson, C. B., and Bivins, J. A.,** Equine encephalomyelitis in pheasants from 1947 to 1951, *J. Am. Vet. Med. Assoc.*, 121, 478, 1952.
26. **Moulthrop, I. M. and Gordy, B. A.,** Eastern viral encephalomyelitis in chukar *(Alectoris graeca)*, *Avian Dis.*, 4, 4, 1960.
27. **Parikh, G. C., Colburn, Z. D., and Larson, D. R.,** 1967 eastern equine encephalitis outbreak on a South Dakota pheasant farm, *Bacteriol. Proc.*, 69, 159, 1969.
28. **Dougherty, E., III and Price, J. I.,** Eastern encephalitis in white Peking ducklings on Long Island, *Avian Dis.*, 4, 247, 1960.
29. **Merrill, M. H., Lacaillade, C. W., Jr., and Ten Broeck, C.,** Mosquito transmission of equine encephalomyelitis, *Science*, 80, 251, 1934.
30. **Davis, W. A.,** A study of birds and mosquitoes as hosts for the virus of eastern equine encephalomyelitis, *Am. J. Hyg.*, 32, 45, 1940.
31. **Ten Broeck, C. and Merrill, M. H.,** Transmission of equine encephalomyelitis by mosquitoes, *Am. J. Pathol.*, 11, 847, 1935.
32. **Chamberlain, R. W., Sikes, R. K., Nelson, D. B., and Sudia, W. D.,** Studies on the North American arthropod-borne encephalitides. VI. Quantitative determinations of virus-vector relationships, *Am. J. Hyg.*, 60, 278, 1954.
33. **Howitt, B. F., Dodge, H. R., Bishop, L. K., and Gorrie, R. H.,** Virus of eastern equine encephalomyelitis isolated from chicken mites *(Dermanyssus gallinae)* and chicken lice *(Eomenacanthus stramineus)*, *Proc. Soc. Exp. Biol. Med.*, 68, 622, 1948.
34. **Howitt, B. F., Dodge, H. R., Bishop, L. K., and Gorrie, R. H.,** Recovery of the virus of eastern equine encephalomyelitis from mosquitoes, *(Mansonia perturbans)* collected in Georgia, *Science*, 110, 141, 1949.
35. **Chamberlain, R. W., Rubin, H., Kissling, R. E., and Eidson, M. E.,** Recovery of eastern encephalomyelitis from a mosquito, *Culiseta melanura* (Coquillett), *Proc. Soc. Exp. Biol. Med.*, 77, 396, 1951.
36. **Bontempo, S. A.,** Economic consequences of the 1959 outbreak of eastern encephalitis among humans, *Public Health News*, 41, 103, 1960.
37. **Howard, J. J. and Wallis, R. C.,** Infection and transmission of eastern equine encephalomyelitis virus with colonized *Culiseta melanura* (Coquillett), *Am. J. Trop. Med. Hyg.*, 23, 522, 1974.
38. **Hayes, C. G. and Wallis, R. C.,** The ecology of western equine encephalomyelitis in the eastern United States, *Adv. Virus Res.*, 21, 37, 1977.
39. **Wayes, C. G.,** Vector competence of colonized *Culiseta melanura* (Diptera:Culicidae) for western equine encephalomyelitis virus, *J. Med. Entomol.*, 15, 253, 1979.
40. **Kissling, R. W., Chamberlain, R. W., Sudia, W. D., and Stamm, D. D.,** Western equine encephalitis in wild birds, *Am. J. Hyg.*, 66, 48, 1957.
41. **Emord, D. E. and Morris, C. D.,** Epizootiology of eastern equine encephalomyelitis virus in upstate New York, USA. VI. Antibody prevalence in wild birds during an interepizootic period, *J. Med. Entomol.*, 4, 395, 1984.

42. **Maxfield, H. K.,** unpublished data, 1977.
43. **McLean, R. G., Frier, G., Parham, G. L., Francy, D. B., Monath, T. P., Campos, E. G., Therrien, A., Kerschner, J., and Calisher, C. H.,** Investigations of the vertebrate hosts of eastern equine encephalitis during an epizootic in Michigan, 1980, *Am. J. Trop. Med. Hyg.,* 34, 1190, 1985.
44. **Jennings, W. L., Allen, R. H., and Lewis, A. L.,** Western equine encephalomyelitis in a Florida horse, *Am. J. Trop. Med. Hyg.,* 15, 96, 1966.
45. **Hoff, G. L., Bigler, W. J., Buff, E. E., and Beck, E.,** Occurrence and distribution of western equine encephalomyelitis in Florida, *J. Am. Vet. Med. Assoc.,* 172, 351, 1978.
46. **Crans, W. J.,** unpublished data, 1985.
47. **Main, A. J., Hildreth, S. W., Wallis, R. C., and Elston, J.,** Arbovirus surveillance in Connecticut. III. Flanders virus, *Mosq. News,* 39, 560, 1979.
48. **Whitney, E.,** Flanders strain, an arbovirus newly isolated from mosquitoes and birds of New York State, *Am. J. Trop. Med. Hyg.,* 13, 123, 1964.
49. **Kissling, R. E.,** Eastern equine encephalomyelitis in pheasants, *J. Am. Vet. Med. Assoc.,* 132, 466, 1958.
50. **Kissling, R. E.,** Host relationships of the arthropod-borne encephalitides, *Ann. N.Y. Acad. Sci.,* 70, 320, 1958.
51. **Clark, D. H.,** Two nonfatal human infections with virus of eastern encephalitis, *Am. J. Trop. Med. Hyg.,* 10, 67, 1961.
52. **Spalatin, J., Karstad, L., Anderson, J. R., Lauerman, L., and Hanson, R. P.,** Natural and experimental infections in Wisconsin turkeys with the virus of eastern encephalitis, *Zoonoses Res.,* 1, 29, 1961.
53. **Emord, D. E., Morris, C. D., Howard, J. J., and Srihongse, S.,** unpublished data, 1984.
54. **Karstad, L., Vadlamudi, S., Hanson, R. P., Trainer, D. O., Jr., and Lee, V. H.,** Eastern equine encephalitis studies in Wisconsin, *J. Infect. Dis.,* 106, 53, 1960.
55. **Trainer, D. O. and Hanson, R. P.,** Serologic evidence of arbovirus infections in wild ruminants, *Am. J. Epidemiol.,* 90, 354, 1969.
56. **Trainer, D. O.,** The use of wildlife to monitor zoonoses, *J. Wildl. Dis.,* 6, 397, 1970.
57. **Hayes, R. O., Daniels, J. B., Maxfield, H. K., and Wheeler, R. E.,** Field and laboratory studies on eastern encephalitis in warm- and cold-blooded vertebrates, *Am. J. Trop. Med. Hyg.,* 13, 595, 1964.
58. **Wellings, F. M.,** personal communication, 1985.
59. **Syverton, J. T. and Berry, G. P.,** Host range of equine encephalomyelitis. Susceptibility of the North American cottontail rabbit, jack rabbit, field vole, woodchuck and opossum to experimental infection, *Am. J. Hyg.,* 32, 19, 1940.
60. **Main, A. J.,** Eastern equine encephalomyelitis virus in experimentally infected bats, *J. Wildl. Dis.,* 15, 467, 1979.
61. **Hurst, E. W.,** Some observations on the pathology of eastern equine encephalomyelitis and louping-ill in young and old animals, with special reference to routes of entry of the viruses into the nervous system, *J. Comp. Pathol.,* 60, 237, 1950.
62. **Sabin, A. B. and Olitsky, P. K.,** Variations in pathways by which equine encephalomyelitis viruses invade the CNS of mice and guinea-pigs, *Proc. Soc. Exp. Biol. Med.,* 38, 595, 1938.
63. **Karstad, L. and Hanson, R. P.,** Natural and experimental infections in swine with the virus of eastern equine encephalitis, *J. Infect. Dis.,* 105, 293, 1959.
64. **Karstad, L.,** Reptiles as possible reservoir hosts for eastern encephalitis virus, *Trans. North Am. Wildl. Conf.,* 26, 1961.
65. **Purcell, A. R., Peckham, J. C., Cole, J. R., Stewart, W. C., and Mitchell, F. E.,** Naturally occurring and artificially induced eastern encephalomyelitis in pigs, *J. Am. Vet. Med. Assoc.,* 161, 1143, 1972.
66. **Soret, M. G. and Sanders, M.,** In vitro method for cultivating eastern equine encephalomyelitis virus in teleost embryos, *Proc. Soc. Exp. Biol. Med.,* 87, 526, 1954.
67. **Satriano, S. F., Luginbuhl, R. E., Wallic, R. C., Jungherr, E. L., and Williamson, L. A.,** Investigation of eastern equine encephalomyelitis. IV. Susceptibility and transmission studies with virus of pheasant origin, *Am. J. Hyg.,* 67, 21, 1957.
68. **Casals, J.,** Antigenic variants in arthropod-borne viruses, paper presented at 10th Pacific Sci. Congr., 1961, 458.
69. **Casals, J.,** Antigenic variants of an arthropod-borne virus, eastern equine encephalitis, paper presented at 8th Int. Congr. Microbiol., Montreal, 1962, 100.
70. **Casals, J.,** Antigenic variants of eastern equine encephalitis virus, *J. Exp. Med.,* 119, 547, 1964.
71. **Walder, R., Rosato, R. R., and Eddy, G. A.,** Virion polypeptide heterogeneity among virulent and avirulent strains of eastern equine encephalitis (EEE) virus, *Arch. Virol.,* 68, 237, 1981.
72. **Walder, R., Jahrling, P. B., and Eddy, G. A.,** Differentiation markers of eastern equine encephalitis (EEE) viruses and virulence, *Zentrabl. Bakteriol. Microbiol. Hyg.,* 9(Suppl.), 237, 1980.
73. **Hanson, R. P., Vadlamudi, S., Trainer, D. O., and Anslow, R.,** Comparison of the resistance of different-aged pheasants to eastern encephalitis virus from different sources, *Am. J. Vet. Res.,* 29, 723, 1968.

74. **Reeves, W. C., Bellamy, R. E., and Scrivani, R. P.,** Relationships of mosquito vectors to winter survival of encephalitis viruses, *Am. J. Hyg.,* 67, 78, 1958.
75. **Karabatsos, N., Bourke, A. T. C., and Henderson, J. R.,** Antigenic variation among strains of western equine encephalomyelitis virus, *Am. J. Trop. Med. Hyg.,* 12, 408, 1963.
76. **Smith, C. E. G.,** Factors in the past and future evolution of the arboviruses, *Trans. R. Soc. Trop. Med. Hyg.,* 54, 113, 1960.
77. **Porterfield, J. S.,** Antigenic characteristics and classification of togaviridae, in *The Togaviruses, Biology, Structure, Replication,* Schlesinger, R. W., Ed., Academic Press, New York, 1980, chap. 2.
78. **Beaty, B. J., Sundin, D. R., Chandler, L. J., and Bishop, D. H. L.,** Evolution of bunyaviruses by genome reassortment in dually infected mosquitoes *(Aedes triseriatus), Science,* 230, 548, 1985.
79. **Eiring, A. G. and Scherer, W. F.,** Appearance of persistently cytopathic eastern and western encephalitis viruses after "blind" passage in cultures of L mouse fibroblasts, *J. Immunol.,* 87, 96, 1961.
80. **Kissling, R. E., Chamberlain, R. W., Eidson, M. E., Sikes, R. K., and Bucca, M. A.,** Studies on the North American arthropod-borne encephalitides. II. Eastern equine encephalitis in horses, *Am. J. Hyg.,* 60, 237, 1954.
81. **Monath, T. P., McLean, R. G., Cropp, C. B., Parham, G. L., Lazuick, J. S., and Calisher, C. H.,** Diagnosis of eastern equine encephalomyelitis by immunofluorescent staining of brain tissue, *Am. J. Vet. Res.,* 42, 1418, 1981.
82. **Frazier, C. L. and Shope, R. F.,** Detection of antibodies to alphaviruses by enzyme-linked immunosorbent assay, *J. Clin. Microbiol.,* 10, 583, 1979.
83. **Hildreth, S. W. and Beaty, B. J.,** Detection of eastern equine encephalomyelitis virus and highlands J virus antigens within mosquito pools by enzyme immunoassay (EIA). I. Laboratory study, *Am. J. Trop. Med. Hyg.,* 33, 965, 1984.
84. **Hildreth, S. W., Beaty, B. J., Maxfield, H. K., Gilfillan, R. F., and Rosenau, B. J.,** Detection of eastern equine encephalomyelitis virus and highlands J virus antigens within mosquito pools by enzyme immunoassay (EIA). II. Retrospective field test of the EIA, *Am. J. Trop. Med. Hyg.,* 33, 973, 1984.
85. **Ou, J. H., Trent, D. W., and Strauss, J. H.,** The 3'-non-coding regions of alphavirus RNAs contain repeating sequences, *J. Mol. Biol.,* 156, 719, 1982.
86. **Wyckoff, R. W. G. and Tesar, W. C.,** Equine encephalomyelitis in monkeys, *Immunology,* 37, 329, 1939.
87. **Sabin, A. B.,** Progression of different nasally instilled viruses along different nervous pathways in the same host, *Proc. Soc. Exp. Biol. Med.,* 38, 270, 1938.
88. **Sabin, A. B. and Olitsky, P. K.,** Age of host and capacity of equine encephalomyelitis viruses to invade the CNS, *Proc. Soc. Exp. Biol. Med.,* 38, 597, 1938.
89. **Ayres, J. C. and Feemster, R. F.,** The sequelae of eastern equine encephalomyelitis, *N. Engl. J. Med.,* 240, 960, 1949.
90. **Nathanson, N., Stolley, P. D., and Boolukos, P. J.,** Eastern equine encephalitis. Distribution of central nervous system lesions in man and rhesus monkey, *J. Comp. Pathol.,* 79, 109, 1969.
91. **Goldfield, M., Welsh, J. N., and Taylor, B. F.,** The 1959 outbreak of eastern encephalitis in New Jersey. V. The inapparent infection:disease ratio, *Am. J. Epidemiol.,* 87, 32, 1968.
92. **Byrne, R. J., French, G. R., Yancey, F. S., Gochenour, W. S., Russell, P. K., Ramsburg, H. H., Brand, O. A., Scheider, F. G., and Buescher, E. L.,** Clinical and immunologic interrelationship among Venezuelan, eastern, and western equine encephalomyelitis viruses in burros, *Am. J. Vet. Res.,* 25, 24, 1964.
93. **Harrison, R. J.,** The first Canadian case of eastern equine encephalitis (EEE) observed in the eastern Townships, Quebec, *Ann. Entomol. Soc. Quebec,* 20, 27, 1975.
94. **Maness, K. S. C. and Calisher, C. H.,** Eastern equine encephalitis in the United States, 1971: past and prologue, *Curr. Microbiol.,* 5, 311, 1981.
95. **Spradbrow, P.,** Arbovirus infections of domestic animals, *Vet. Bull. (London),* 36, 55, 1966.
96. **Coleman, P. H. and Kissling, R. E.,** Arbovirus infections, in *Diseases of Poultry,* 6th ed., Hofstad, M. S., Ed., Iowa State University Press, Ames, 1972, 21.
97. **Dein, F. J., Carpenter, J. W., Clark, G. G., Montali, R. J., Crabbs, C. L., Tsai, T. F., and Docherty, D. E.,** unpublished data, 1986.
98. **Jungherr, E. L., Helmboldt, C. F., Satriano, S. F., and Luginbuhl, R. E.,** Investigation of eastern equine encephalomyelitis. III. Pathology in pheasants and incidental observations in feral birds, *Am. J. Hyg.,* 67, 10, 1958.
99. **Faddoul, G. P. and Fellows, G. W.,** Clinical manifestations of eastern equine encephalomyelitis in pheasants, *Avian Dis.,* 9, 530, 1965.
100. **Ranck, F. M., Jr., Gainer, J. H., Hanley, J. E., and Nelson, S. L.,** Natural outbreaks of eastern and western encephalitis in pen-raised chukars in Florida, *Avian Dis.,* 9, 8, 1965.
101. **Stamm, D. D. and Kissling, R. E.,** Influence of season on EEE infection in English sparrows, *Proc. Soc. Exp. Biol. Med.,* 92, 374, 1956.

102. **Turtinen, L. W.**, Studies on Eastern Equine Encephalitis Virus Infection in Wild Birds, Ph. D. thesis, University of Wisconsin, Madison, 1978.
103. **Main, A. J.**, Virologic and serologic survey for eastern equine encephalomyelitis and certain other viruses in colonial bats of New England, *J. Wildl. Dis.*, 15, 455, 1979.
104. **Goldfield, M. and Sussman, O.**, EE and WE in New Jersey's non-avian vertebrates, paper presented at Wildl. Dis. Assoc., Urbana, Ill., June 15, 1967, 1.
105. **Smith, A. L. and Anderson, C. R.**, Susceptibility of two turtle species to eastern equine encephalitis virus, *J. Wildl. Dis.*, 16, 615, 1980.
106. **Hurlbut, H. S. and Thomas, J. I.**, The experimental host range of the arthropod-borne animal viruses in arthropods, *Virology*, 12, 391, 1960.
107. **Whitfield, S. G., Murphy, F. A., and Sudia, W. D.**, Eastern equine encephalomyelitis virus: an electron microscopic study of *Aedes triseriatus* (Say) salivary gland infection, *Virology*, 43, 110, 1971.
108. **Bigler, W. J., Lassing, E. B., Buff, E. E., Prather, E. C., Beck, E. C., and Hoff, G. L.**, Endemic eastern equine encephalomyelitis in Florida: a twenty-year analysis, 1955—1974, *Am. J. Trop. Med. Hyg.*, 25, 884, 1976.
109. **Beadle, L. D.**, Eastern equine encephalitis in the United States, *Mosq. News*, 12, 102, 1952.
110. **Morris, C. D., Whitney, E., Bast, T. F., and Deibel, R.**, An outbreak of eastern equine encephalomyelitis in upstate New York during 1971, *Am. J. Trop. Med. Hyg.*, 22, 561, 1973.
111. **Howard, J. J.**, unpublished data, 1985.
112. **Morris, C. D.**, unpublished data, 1986.
113. **Shope, R. E.**, Understanding EEE emergencies and their control, in *Eastern Equine Encephalomyelitis (EEE) and Public Health in Southwestern Michigan*, Engemann, J. G., Ed., Science for Citizens Center of Southwestern Michigan, Western Michigan University, Kalamazoo, 1982, 8.
114. **McGowan, J. E., Jr., Bryan, J. A., and Gregg, M. B.**, Surveillance of arboviral encephalitis in the United States, 1955—1971, *Am. J. Epidemiol.*, 97, 199, 1973.
115. **Gibbs, P. E.**, personal communication, 1985.
116. **Gamble, J.**, personal communication, 1985.
117. **Grimstad, P. R., Craig, G. B., Ross, Q. E., and Yuill, T. M.**, *Aedes triseriatus* and La Crosse virus: geographic variation in vector susceptibility and ability to transmit, *Am. J. Trop. Med. Hyg.*, 26, 990, 1977.
118. **Morris, C. D., Zimmerman, R. H., and Edman, J. D.**, Epizootiology of eastern equine encephalomyelitis virus in upstate New York, U. S. A. II. Population dynamics and vector potential of adult *Culiseta melanura* (Diptera:Culicidae) in relation to distance from breeding site, *J. Med. Entomol.*, 17, 453, 1980.
119. **Morris, C. D.**, Phenology of trophic and gonobiologic states in *Culiseta morsitans* and *Culiseta melanura* (Diptera:Culicidae), *J. Med. Entomol.*, 21, 38, 1984.
120. **Howard, J. J., Emord, D. E., and Morris, C. D.**, Epizootiology of eastern equine encephalomyelitis virus in upstate New York, USA. V. Habitat preference of host-seeking mosquitoes (Diptera:Culicidae), *J. Med. Entomol.*, 20, 62, 1983.
121. **Hayes, R. O.**, Observations on the swarming of *Culiseta melanura* (Coquillett), *Mosq. News*, 18, 70, 1958.
122. **Hayes, R. O.**, Host preferences of *Culiseta melanura* and allied mosquitoes, *Mosq. News*, 21, 179, 1961.
123. **Hayes, R. O.**, The diel activity cycle of *Culiseta melanura* (Coquillett) and allied mosquitoes, *Mosq. News*, 22, 352, 1962.
124. **Burbutis, P. P., and Lake, R. W.**, The biology of *Culiseta melanura* (Coquillett) in New Jersey, paper presented at N.J. Mosq. Extermination Assoc., Atlantic City, March 7 to 9, 1956.
125. **Main, A. J., Tonn, R. J., Randall, E. J., and Anderson, K. S.**, Mosquito densities at heights of five and twenty-five feet in southeastern Massachusetts, *Mosq. News*, 26, 243, 1966.
126. **Nasci, R. S. and Edman, J. D.**, Blood-feeding patterns of *Culiseta melanura* (Diptera:Culicidae) and associated sylvan mosquitoes in southeastern Massachusetts eastern equine encephalitis enzootic foci, *J. Med. Entomol.*, 18, 493, 1981.
127. **Nasci, R. S. and Edman, J. D.**, Vertical and temporal flight activity of the mosquito *Culiseta melanura* (Diptera:Culicidae) in southeastern Massachusetts, *J. Med. Entomol.*, 18, 501, 1981.
128. **Maxfield, H. K.**, Field studies of EEE in Massachusetts, in *Eastern Equine Encephalomyelitis (EEE) and Public Health in Southwestern Michigan*, Engemann, J. G., Ed., Science for Citizens Center of Southwestern Michigan, West Michigan University, Kalamazoo, 1982, 6.
129. **Meier, A. H. and Russo, A. C.**, Circadian organization of the avian annual cycle, in *Current Ornithology*, Vol. 2, Johnson, R. F., Ed., Plenum Press, New York, 1985, 9.
130. **Stokes, D. W.**, *A Guide to the Behavior of Common Birds*, Little, Brown, Boston, 1979, 336.
131. **Wallace, G. J. and Mahan, H. D.**, *An Introduction to Ornithology*, 3rd ed., Macmillan, New York, 1975, 546.
132. **Dalrymple, J. M., Young, O. P., Eldridge, B. F., and Russell, P. K.**, Ecology of arboviruses in a Maryland freshwater swamp. III. Vertebrate hosts, *Am. J. Epidemiol.*, 96, 129, 1972.

133. **Morris, C. D., Caines, A. R., Woodall, J. P., and Bast, T. F.,** Eastern equine encephalomyelitis in upstate New York 1972—1974, *Am. J. Trop. Med. Hyg.,* 24, 986, 1975.

134. **Sirhongse, S., Grayson, M. A., Morris, C. D., Deibel, R., and Duncan, C. S.,** Eastern equine encephalomyelitis in upstate New York: studies of a 1976 epizootic by a modified serologic technique, hemagglutination reduction, for rapid detection of virus infections, *Am. J. Trop. Med. Hyg.,* 27, 1240, 1978.

135. **Horsfall, W. R., Fowler, H. W., Jr., Moretti, L. J., and Larsen, J. R.,** *Bionomics and Embryology of the Inland Floodwater Mosquito Aedes vexans,* University of Illinois Press, Urbana, 1973, 211.

136. **Crans, W. J., Downing, J. D., and Slaff, M. E.,** Behavioral changes in the salt marsh mosquito, *Aedes sollicitans,* as a result of increased physiological age, *Mosq. News,* 36, 437, 1976.

137. **Crans, W. J.,** The status of *Aedes sollicitans* as a vector of eastern equine encephalitis in New Jersey, *Mosq. News,* 37, 85, 1977.

138. **Nayar, J. K.,** Bionomics and physiology of *Culex nigripalpus* (Diptera:Culicidae) of Florida: an important vector of diseases, *Fla. Agric. Exp. Stn., Tech. Bull.,* 827, 73, 1982.

139. **MacCreary, D. and Stearns, L. A.,** Mosquito migration across Delaware Bay, paper presented at N.J. Mosq. Extermination Assoc., Atlantic City, March 17 to 19, 1937, 188.

140. **Chamberlain, R. W. and Sudia, W. D.,** The effects of temperature upon the extrinsic incubation of eastern equine encephalitis in mosquitoes, *Am. J. Hyg.,* 62, 295, 1955.

141. **Wellings, F. M., Lewis, A. L., and Pierce, L. V.,** Agents encountered during arboviral ecological studies: Tampa Bay area, Florida, 1963 to 1970, *Am. J. Trop. Med. Hyg.,* 21, 201, 1972.

142. **Morris, C. D. and Srihongse, S.,** An evaluation of the hypothesis of transovarial transmission of eastern equine encephalomyelitis virus by *Culiseta melanura, Am. J. Trop. Med. Hyg.,* 27, 1246, 1978.

143. **Francy, D. B.,** Eastern equine encephalomyelitis (EEE) and consideration of some alternatives for dealing with EEE as a public health problem, in *Eastern Equine Encephalomyelitis (EEE) and Public Health in Southwestern Michigan,* Engemann, J. G., Ed., Science for Citizens Center of Southwestern Michigan, Western Michigan University, Kalamazoo, 1982, 15.

144. **Veazey, J., Adam, D., and Gusciora, W.,** Arbovirus surveillance for eastern encephalitis in New Jersey mosquitoes during the years 1960—1969. I. An overview with specific reference to *Culex pipiens, Cx. restuans, Cx. salinarius* and *Cx. territans,* paper presented at N.J. Mosq. Control Assoc., Cherry Hill, March 12 to 14, 1980, 160.

145. **Means, R. G.,** Host preference of mosquitoes (Diptera:Culicidae) in Suffolk County, New York, *Ann. Entomol. Soc. Am.,* 61, 116, 1968.

146. **Edman, J. D.,** Host-feeding patterns of Florida mosquitoes. I. *Aedes, Anopheles, Coquillettidia, Mansonia* and *Psorophora, J. Med. Entomol.,* 8, 687, 1971.

147. **Tempelis, C. H.,** Host-feeding patterns of mosquitoes, with a review of advances in analysis of blood meals by serology, *J. Med. Entomol.,* 6, 635, 1975.

148. **Clark, G. M., Lutz, A. E., and Fadness, L.,** Observations on the ability of *Haemogamasus liponyssoides* Ewing and *Ornithonyssus bacoti* (Hirst) (Acarina, Gamasina) to retain eastern equine encephalitis virus: preliminary report, *Am. J. Trop. Med. Hyg.,* 15, 107, 1966.

149. **Chamberlain, R. W. and Sikes, R. K.,** Laboratory investigations on the role of bird mites in the transmission of eastern and western equine encephalitis, *Am. J. Trop. Med. Hyg.,* 4, 106, 1955.

150. **Anderson, J. R., Lee, V. H., Vadlamudi, S., Hanson, R. P., and DeFoliart, G. R.,** Isolation of eastern encephalitis virus from Diptera in Wisconsin, *Mosq. News,* 21, 244, 1961.

151. **Karstad, L. H., Fletcher, O. K., Spalatin, J., Roberts, R., and Hanson, R. P.,** Eastern equine encephalomyelitis virus isolated from three species of Diptera from Georgia, *Science,* 125, 395, 1957.

152. **Hayes, R. O. and Hess, A. D.,** Climatological conditions associated with outbreaks of eastern encephalitis, *Am. J. Trop. Med. Hyg.,* 13, 851, 1964.

153. **Philbrook, F. R., Hayes, R. O., MacCready, R. A., Daniels, J., Wheeler, R. E., Parsons, M. A., and Anderson, K. S.,** *Special Report of the Department of Public Health on Continued Investigations and Study of Equine Encephalitis,* Wright & Potter, Boston, 1958, 7.

154. **Chamberlain, R. W., Sikes, R. K., and Nelson, D. B.,** Infection of *Mansonia perturbans* and *Psorophora ferox* mosquitoes with Venezuelan equine encephalomyelitis virus, *Proc. Soc. Exp. Biol. Med.,* 91, 215, 1956.

155. **Schaeffer, M. and Arnold, E. H.,** Studies on the North American arthropod-borne encephalitides. I. Introduction. Contributions of newer field-laboratory approaches, *Am. J. Hyg.,* 60, 231, 1954.

156. **Chamberlain, R. W. and Sudia, W. D.,** Mechanism of transmission of viruses by mosquitoes, in *Annual Review of Entomology,* Vol. 6, Steinhaus, E. A. and Smith, R. F., Eds., Annual Reviews, Palo Alto, Calif., 1961, 371.

157. **Sprance, H. E.,** Experimental evidence against the transovarial transmission of eastern equine encephalitis virus in *Culiseta melanura, Mosq. News,* 41, 168, 1981.

158. **Scott, T. W., Hildreth, S. W., and Beaty, B. J.,** The distribution and development of eastern equine encephalitis virus in its enzootic mosquito vector, *Culiseta melanura, Am. J. Trop. Med. Hyg.,* 33, 300, 1984.

159. **Watts, D. M., Clark, G. G., Crabbs, C. L., Rossi, C. A., Olin, T. R., and Bailey, C. L.,** unpublished data, 1986.

160. **Muul, I., Johnson, B. K., and Harrison, B. A.,** Ecological studies of *Culiseta melanura* (Diptera:Culicidae), *J. Med. Entomol.,* 11, 739, 1975.

161. **Hayes, R. O., Beadle, L. D., Hess, A. D., Sussman, O., and Bonese, M. J.,** Entomological aspects of the 1959 outbreak of eastern encephalitis in New Jersey, *Am. J. Trop. Med. Hyg.,* 11, 115, 1962.

162. **Howard, J. J. and Francy, D. B.,** unpublished data, 1985.

163. **Clark, G. G., Crans, W. J., and Crabbs, C. L.,** Absence of eastern equine encephalitis (EEE) virus in immature *Coquillettidia perturbans* associated with equine cases of EEE, *J. Am. Mosq. Control Assoc.,* 1, 540, 1985.

164. **Ksiazek, T. G., Hardy, J. L., and Reeves, W. C.,** Effect of normal mosquito extracts upon arbovirus recoveries from mosquito pools, *Am. J. Trop. Med. Hyg.,* 34, 578, 1985.

165. **Moussa, M. A., Gould, D. J., Nolan, M. P., Jr., and Hayes, D. E.,** Observations on *Culiseta melanura* (Coquillett) in relation to encephalitis in southern Maryland, *Mosq. News,* 26, 385, 1966.

166. **Joseph, S. R. and Bickley, W. E.,** *Culiseta melanura (Coquillett) on the Eastern Shore of Maryland (Diptera:* Culicidae), Bull. A-161, University of Maryland Agricultural Experiment Station, College Park, 1969, 1.

167. **Morris, C. D., Corey, M. E., Emord, D. E., and Howard, J. J.,** Epizootiology of eastern equine encephalomyelitis virus in upstate New York, U. S. A. I. Introduction, demography and natural environment of an endemic focus, *J. Med. Entomol.,* 17, 442, 1980.

168. **Pierson, J. W. and Morris, C. D.,** Epizootiology of eastern equine encephalomyelitis virus in upstate New York, U. S. A. IV. Distribution of *Culiseta* (Diptera:Culicidae) larvae in a freshwater swamp, *J. Med. Entomol.,* 19, 423, 1982.

169. **Williams, J. E., Watts, D. M., and Reed, T. J.,** Distribution of culicine mosquitoes within Pocomoke cypress swamp, Maryland, *Mosq. News,* 31, 371, 1971.

170. **Siverly, R. E. and Schoof, H. F.,** Biology of *Culiseta melanura* (Coquillett) in southeast Georgia, *Mosq. News,* 22, 274, 1962.

171. **Edman, J. D., Webber, L. A., and Kale, H. W., II,** Host-feeding patterns of Florida mosquitoes. II. *Culiseta, J. Med. Entomol.,* 9, 429, 1972.

172. **Hayes, R. O. and Doane, O. W., Jr.,** Primary record of *Culiseta melanura* biting man in nature, *Mosq. News,* 18, 216, 1958.

173. **Schober, H.,** Notes on the behavior of *Culiseta melanura* (Coq.) with three instances of its biting man, *Mosq. News,* 24, 67, 1964.

174. **Crans, W. J.,** personal communication, 1985.

175. **Sudia, W. D., Chamberlain, R. W., and Coleman, P. H.,** Arbovirus isolations from mosquitoes collected in south Alabama, 1959—1963, and serologic evidence of human infection, *Am. J. Epidemiol.,* 87, 112, 1969.

176. **Howard, J. J.,** personal communication, 1985.

177. **Scott, T. W. and Olson, J. G.,** Detection of eastern equine encephalomyelitis antigen in avian blood by enzyme immunoassay: a laboratory study, *Am. J. Trop. Med. Hyg.,* 35, 611, 1986.

178. **Snoeyenbos, G. H., Weinack, O. M., and Rosenau, B. J.,** Immunization of pheasants for eastern encephalitis, *Avian Dis.,* 22, 386, 1978.

179. **Eisner, R. J. and Nusbaum, S. R.,** Encephalitis vaccination of pheasants: a question of efficacy, *J. Am. Vet. Med. Assoc.,* 183, 280, 1983.

180. **Sussman, O., Cohen, D., Gerende, J. E., and Kissling, R. E.,** Equine encephalitis vaccine studies in pheasants under epizootic and pre-epizootic conditions, *Ann. N.Y. Acad. Sci.,* 70, 328, 1958.

181. **Stamm, D. D.,** Studies on the ecology of equine encephalomyelitis, *Am. J. Public Health,* 48, 328, 1958.

182. **Morris, C. D.,** A structural and operational analysis of diurnal resting shelters for mosquitoes (Diptera:Culicidae), *J. Med. Entomol.,* 18, 419, 1981.

183. **Sudia, W. D. and Chamberlain, R. W.,** Battery-operated light trap, an improved model, *Mosq. News,* 22, 126, 1962.

184. **Odum, E. P.,** The value of wetlands: a hierarchical approach, in *Wetlands Functions and Values: The State of Our Understanding,* Creeson, P. E., Clark, J. R., and Clark, J. E., Eds., American Water Resources Association, Minneapolis, 1979, 16.

185. **Kadlec, R. H.,** Wetlands for tertiary treatment, in *Wetlands Functions and Values: The State of Our Understanding,* Cresson, P. E., Clark, J. R., and Clark, J. E., Eds., American Water Resources Association, Minneapolis, 1979, 490.

186. **Johnson, R. and Sheffield, C. W.,** Treated effluent for Florida wetlands, paper presented at FS/AWWA, FPCA, and FW&PCOA Annu. Tech. Conf., Sarasota, November 8, 1985, 27.

187. **Mortenson, E. W.,** Mosquito occurrence in wastewater marshes: a potential new community problem, in *Proc. Calif. Mosq. Vector Control Assoc.,* Grant, C. D., Washino, R. K., Lusk, E. E., and Coykendall, R. L., Eds., CMCA Press, Sacramento, 1982, 65.
188. **Roberts, F. C.,** The implications of a freshwater simulation model on the Alameda County mosquito abatement district, in *Proc. Calif. Mosq. Vector Control Assoc.,* Grant, C. D., Washino, R. K., Lusk, E. E., and Coykendall, R. L., Eds., CMCA Press, Sacramento, 1982, 83.
189. **Schooley, J., Roberts, F. C., and Conner, G. E.,** A preliminary simulation of mosquito control in Coyote Hills freshwater marsh, Alameda County, California, in *Proc. Calif. Mosq. Vector Control Assoc.,* Grant, C. D., Washino, R. K., Lusk, E. E., and Coykendall, R. L., Eds., CMCA Press, Sacramento, 1982, 76.

Chapter 25

GETAH VIRUS DISEASE

Yuji Kono

TABLE OF CONTENTS

I. HISTORICAL BACKGROUND

Getah virus (GEV), a member of the Semliki Forest (SF) virus complex of alphaviruses,[1] was first isolated in 1955 from a pool of *Culex (Cx.) gelidus* collected near Kuala Lumpur, Malaysia by suckling-mouse-brain inoculation. Another isolate was later obtained from *Cx. tritaeniorynchus* collected at Kuantan, Malaysia.[2] Thereafter, many identical strains were isolated from several species of mosquitoes: *Cx. tritaeniorynchus*[3-5] *Cx. bitaeniorynchus,*[6] *Aedes (Ae.) vexans nipponi,*[5,7] and *Anopheles (An.) amictus,*[6] and swine[8] in Japan, Southeast Asian countries, and Australia. Seroepidemiological surveys also indicated that the virus is widely distributed among mammals, including man and birds, over a vast area.[9-17] Association of the virus with disease, however, remained uncertain for many years. During the autumn of 1978, a disease characterized by fever and occasionally by rash and/or edema of the limbs was seen among race horses, and GEV was identified as the causal agent.[18-20] Since then, the etiological significance of the virus has been established.

Several strains of Sagiyama virus which have antigenic properties indistinguishable from those of GEV were isolated from mosquitoes (*Cx. tritaeniorynchus* and *Ae. vexans nip.*) collected in heronries in Japan.[21] "Sagiyama" means heronry in Japanese. Antibodies against the virus have been detected in man, domestic animals, and birds.[21,22]

Since Sagiyama virus is difficult to distinguish from GEV biologically, serologically, and genetically, as is discussed in this chapter, descriptions of Sagiyama virus are included in this chapter.

II. THE VIRUS

A. Antigenic Relationships

As concerns members of the genus *Alphavirus* in the family Togaviridae, greater antigenic specificity is revealed by the cross-virus neutralization (VN) test than is evident in the hemagglutination-inhibition (HI) or complement-fixation (CF) tests. Alphaviruses are divided into six complexes, and each complex is divided into species[1] on the basis of the results obtained by cross-VN tests in cell cultures[23-25] and by HI, CF, and cross-protection tests in mice.[26] Getah species belongs to the SF complex and consists of four subtypes: GEV, Sagiyama virus, Bebaru virus, and Ross River virus (RRV). The first two have a very close, if not identical, antigenic relationship with each other and the latter two have a weak antigenic relationship with the first two in cross HI and VN tests.[19,20,24,25,27,28] Sagiyama virus, however, has only weak reactivity with antibody to GEV in the CF test, although GEV antigen reacts with antibody to Sagiyama virus as well as with homologous antibody.[2,19]

B. Host Range

A wide range of vertebrates and invertebrates is known to be susceptible to the virus. The virus has been isolated from horses with illness[18,20,29,30] and pigs without signs of illness.[31] Also, infection with the virus is evident serologically in man,[9-12,14,15,17,21] monkeys,[16] horses,[32-34] cattle,[13,14,35] water buffalo,[16] pigs,[14,16,21,36] goats,[16] dogs,[14,16] domestic rabbits,[14] chickens,[14,16,21] and night herons.[21]

Experimental infection of horses was established by subcutaneous,[18] intramuscular, and intranasal[18,20,37] inoculation with the virus. Concentration of the virus as high as 10^7 and 10^6 suckling mouse LD_{50} was found in blood and nasal swabs, respectively, 1 to 4 days after inoculation into suckling mouse brains. No infectivity was detected in feces or urine.[18] The virus present in various tissues,[37] including lymph nodes, especially axillary and inguinal lymph nodes, contained as much as $10^{3.5}$ to $10^{6.5}$ $TCID_{50}/g$. Specific antibodies detectable by CF, HI, and VN tests are found 6 days after infection. VN antibody titers increased gradually to 512 to 1280, and CF and HI titers to 64. VN and HI antibodies persisted for at least 6 months and CF antibodies for 1 year.[32]

Pigs and cattle have been shown to be infected subclinically with the virus.[35,38,39] Pigs developed viremia consistently 1 to 5 days after subcutaneous inoculation with several strains of GEV and Sagiyama virus, including a strain isolated from an infected horse with signs of illness, but the pigs revealed no signs of illness.[22,38,39] The levels of viremia were between $10^{1.6}$ and $10^{5.2}$ suckling mouse $LD_{50}/m\ell$. HI and VN antibodies were detectable 7 days after subcutaneous and intravenous inoculations. Cattle developed no signs of illness after subcutaneous and intravenous inoculations with an isolate pathogenic for horses, but showed only fleeting and inconsistent viremia.[35] Calves infected subclinically sometimes show histopathological evidence of encephalitis.[40] Serological responses may be observed after experimental inoculation.[35,41]

Among experimental animals, the suckling mouse is highly sensitive to the virus, with a fatal infection resulting from intracerebral, intraperitoneal, or intranasal inoculation, while weanling and adult mice are infected subclinically.[22,42]

It has been reported that weanling guinea pigs and rabbits showed no clinical infection, but develop CF and VN antibodies after intracerebral and intraperitoneal inoculation.[22] Scherer et al.[22] have demonstrated that Sagiyama virus grows in baby chicks, causing viremia, and can then be propagated in embryonated eggs by inoculation into the yolk sac. No regular deaths or signs of illness occurred in the animals or embryos after serial passage. In contrast, Chung[43,44] reported failure of virus propagation in embryonated eggs and in 1-day-old chicks.

Among invertebrates, mosquitoes are the only known arthropod to support the multiplication of GEV. GEV and Sagiyama virus have been isolated from *Cx. gelidus*,[2,45] *Cx. tritaeniorynchus*,[3-5,16,21,45] *Cx. bitaeniorynchus*,[6] *Ae. vexans nip.*,[5,7,21] and *An. amictus*[6] in Australia, Southeast Asian countries, and Japan. Furthermore, propagation of the virus in mosquitoes after feeding of virus-containing blood was shown in *Cx. tritaeniorynchus*,[46,47] *Ae. vexans nip.*,[46-48] *Armigeres subalbatus*, *Ae. japonicus*, *Ae. aegypti*, *Ae. albopictus*, *Cx. pipens pallens*, *Tripteroides bambusa*,[47] *Ae. funereus*,[49] *Ae. vigilax*,[50] and *Cx. annulirostris*.[50] The last eight of these species had not been regarded as potential vectors of the virus since the virus was not isolated from them in the field.

Infection of mosquitoes with the virus develops upon the route of infection. Carley et al.[51] reported that virus multiplication in *Cx. fatigans* did not occur after feeding infected blood, but it was observed after intrathoracic inoculation of the virus. Other factors which influence the infection of mosquitoes with the virus and growth of the virus in mosquitoes is described in the following section.

GEV propagates well and has an apparent cytopathic effect (CPE) in various types of immortalized mammalian, amphibian, and mosquito cell lines. In particular, high sensitivity of hamster and simian kidney cell lines such as BHK-21, VERO, and LLC-MK2 to alphaviruses and plaque-forming ability of the viruses in the cultures have been documented by Karabatsos and Buckley[52] and Stim.[53] Ueba and Kimura[54] demonstrated induction of polykaryocytosis by the virus in BHK-21 cells. Other cell lines, MA104 from simian kidney, HmLu from hamster lung, RK13 from rabbit kidney, and a primary human embryonic lung and various horse cells support the virus propagation well.[19,20,28] The virus matures on cell membrane by the budding process (Figure 1).

GEV multiplication has been seen in the XTC$_2$ cell line[55] derived from the South African clawed toad (*Xenopus laevis*). Plaque formation in the cells occurred inconsistently.[27]

Mosquito cells, including the *Cx. molestus* cell line[56] and *Ae. albopictus* clone C6/36[57] have been used to study plaque formation by GEV.[4,58] In the former cells, plaques were produced only by the virus passaged once in the *Culex* cells. Plaque counts in those cells were almost the same as in BHK-21 cells. The *Ae. albopictus* clone was used for virus isolation from mosquitoes and was better than inoculation into suckling mice.[4] Another line of mosquito cells, *Ae. pseudoscutellaris* M61 cell line,[59] also supports the growth of GEV without plaque formation.[27]

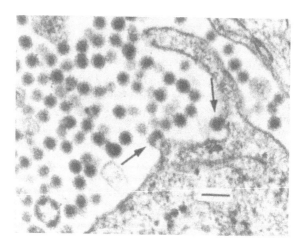

FIGURE 1.   An electron micrograph of equine dermal cells infected with the MI-110 strain of Getah virus. Buddings (arrows) on the marginal membrane are seen. Bar indicates 100 nm. (Micrograph courtesy of Dr. M. Kamada.)

C. Strain Variation

Many virus strains of Getah virus consist of virus populations with a wide range of plaque sizes.[27,60] Each variant has a different biological activity such as antigenicity, hemagglutination, and virulence. Large and small plaque variants can be easily isolated in the presence of antiserum against a heterologous variant, suggesting a difference in antigenicity between the two.[27] The plaque properties are relatively stable in mammalian cell cultures. A small plaque variant showed maximal hemagglutination at pH 6.2 to 6.4, and was neutralized by antiserum in the absence of complement. In contrast, a large plaque variant failed to cause hemagglutination at any pH in the range of 6.0 to 6.8, and its plaquing efficiency increased in the presence of high concentration of antiserum without complement.[60]

A large plaque variant has high virulence for mice and a small plaque variant has low virulence.[27,60] Chanas et al.[27] reported that approximately equal numbers of large and small plaque variants appeared in mice after inoculation with a small plaque variant. Similarly, a population of large plaque variants changed to a mixed population containing small plaque variants after passage once in *Ae. pseudoscutellaris* M61 cells.[27] In contrast, Kimura and Ueba[60] reported that a small plaque variant of low virulence for mice remained stable after several passages in suckling mice. Differences in degree of variability of strains used for the different experiments might be one reason for inconsistency in the stability of plaque properties.

Data are accumulating that indicate the presence of various variants or mutants in mosquitoes. Igarashi et al.[61,62] isolated host-dependent temperature-sensitive (TS) mutants from two strains of GEV by inoculation of mosquito C6/36 cells. These mutants do not grow well in BHK cells at 37°C, but they grow well in C6/36 cells at 28°C. In addition, Igarashi and colleagues successfully isolated many host-dependent TS mutants from a strain of GEV which was isolated from a mosquito homogenate by inoculation into suckling mouse brain.[61]

GEV and Sagiyama virus show different behavior in hemagglutinin production.[22] GEV produces high titers of hemagglutinating antigen in mouse brain, which needs no protamine treatment for its expression, while Sagiyama virus produces low titers of the antigen which become detectable only after protamine treatment. Among the strains of Sagiyama virus, there were marked differences in the capacity to yield potent hemagglutinins.

Thus, strains of Getah virus are highly variable in nature, and differences in host susceptibility to the variant or mutant may play an important role in determining the composition of a virus strain and the evolution of a new strain.

Recently, Morita and Igarashi[63] examined various strains of GEV by oligonucleotide fingerprint analysis, which can detect differences among strains at the RNA level. Sagiyama virus, which was isolated in Japan in 1956, had a relatively high ratio of similarity (0.72) with the prototype strain of GEV, which was isolated at Malaysia in 1955. On the other hand, though 12 recent Japanese isolates (from 1978 to 1981) of GEV had similarity ratios as high as 0.82:0.96, they had ratios of 0.68:0.72 with the prototype strain, and 0.71:0.78 with Sagiyama virus. These results indicate that the GEV genome had undergone independent changes in Japan over the past 20 years. Moreover, among the recent Japanese isolates, there were no identical fingerprint patterns even among strains isolated in the same locality in the same year. These results suggest that the GEV genome undergoes mutation rather frequently. This idea is also supported by the fact that TS mutants of GEV which can be promptly isolated in mosquito cells have different fingerprint patterns from that of the parental strain.[63]

D. Methods of Assay

Inoculation of test materials into suckling mouse brains has commonly been used for the isolation of arboviruses.[64] This method is sensitive enough to isolate GEV. Mammalian cell cultures are also used for virus isolation, and are as sensitive as suckling mouse brain.[20,31,65] Mosquito cell cultures are sometimes more sensitive than mouse brain. Igarashi et al.[4,61] found a higher rate of isolation of GEV from mosquitoes in C6/36 cells than in suckling mouse brain.

To isolate the virus from mammals and birds, heparinized blood, plasma, or serum is used as an inoculum. A higher virus isolation rate may be expected when plasma with platelets and leukocytes is employed as an inoculum.[20,31,37] Igarashi et al.[3] found that among eight strains of GEV obtained from swine blood in C6/36 cells, four were isolated from plasma with platelets, two from either lymphocyte fractions or plasma with platelets, two from lymphocyte fractions, and none from whole blood. They thought that frequent association of the virus with lymphocytes may be one reason for higher efficiency in detecting the virus, and that the toxic effect of swine red blood cells on C6/36 cells may be a critical reason for lower efficiency of isolation.

To isolate the virus from mosquitoes, a pool of approximately 100 mosquitoes for each species is emulsified with various volumes of phosphate-buffer saline, pH 7.2, containing 0.2% bovine serum albumin and antibiotics, and centrifuged at 3000 to 10,000 RPM for 30 min. Volumes of the diluent vary with each species: *Cx. tritaeniorynchus*, 2 mℓ; *Cx. pipens pallens*, 4 mℓ; *Cx. bitaeniorynchus*, 2 mℓ; *Mansonia uniformis*, 4 mℓ; *Ae. vexans nip.*, 6 to 8 mℓ; *Armigeres*, 8 mℓ; and *Anopheles*, 8 mℓ. With these concentrations of mosquitoes, the toxic effects on suckling mouse brain are avoided.[5]

The emulsions are inoculated intracerebrally, in amounts of 0.02 mℓ, into a litter of 1- to 4-day-old suckling mice. If a mouse dies within 2 days after inoculation without signs of central nervous system illness, the death is considered accidental and the mouse is discarded. Brains of mice at crisis with signs of the illness are harvested, tested for antigenicity, and serially passaged in suckling mouse brain or in cell cultures. The mosquito emulsions are also inoculated into cell cultures of VERO,[20,31,65] MA104,[35] HmLu,[31] RK13,[20] and mosquito C6/36[4] cells. Before the inoculation, mammalian cells are washed several times with minimal essential medium (MEM) to remove antibodies which may be present in the serum used for the medium, and then inoculated with the samples. After adsorption, maintenance medium consisting of MEM containing 0.01% bovine albumin or 2% antibody-free calf serum is added, and the cultures are incubated for 7 days. The initial CPE is observed 3 to 5 days after inoculation. The CPE becomes apparent and complete after

Table 1
CLINICAL SIGNS OF HORSES INFECTED WITH
GETAH VIRUS[a]

Group	Pyrexia alone	Rash	Limb edema	Rash and limb edema	Total
Febrile	230[b]	106	103	151	590
	(31.9)	(14.9)	(14.3)	(20.9)	(81.7)
Afebrile	0	78	20	34	132
		(10.8)	(2.8)	(4.7)	(18.3)
Total	230	184	123	185	722
	(31.9)	(25.5)	(17.0)	(25.6)	(100)

[a] Revised from a result observed in a horse training center by Fukunaga et al.[37]

[b] Number of horses. Number in parentheses shows percent.

passage. Plaques and CPE are observed in mosquito cells after the primary inoculation. Details of the culture methods are described by Igarashi et al.[4]

The isolate is identified by HI, CF, and VN tests.

III. ASSOCIATION WITH DISEASE

A. Humans and Other Mammals

GEV infection in man is indicated seroepidemiologically in Japan,[14,21] Southeast Asian countries,[11,17] Australia,[9,10,12] and in the Pacific Islands,[9,11] although some of the positive reactions have been regarded as cross reactions with antibody to Ross River virus. There is, however, no evidence of clinical infection of man. Even during enzootics of GEV infection in race horses in Japan, there was no case suggesting clinical infection with the virus among grooms and their families who live at the same site as the horses, though serological tests were not performed.[96]

Consistent clinical signs of infection with GEV are seen only in horses.[18,20,37,65] The main clinical signs are fever (temperature of 38.5 to 41°C), accompanied in some animals by rash and/or edema of the limbs. Some horses manifest rash or edema without fever (Table 1). The fever persists for 1 to 4 days and is sometimes biphasic. Rash forming a papule without a crust appears principally in the neck, around the brachial region, and on the thigh, with a size ranging from 3 to 5 mm in diameter (Figure 2). Edema in the limbs, mainly in the fetlock, is usually noninflammatory. Rash and edema tend to appear symmetrically a few days after the onset of fever, and disappear within 1 week after the onset.[37]

The rate of illness among the horses that showed seroconversion was reported as 22% by Sentsui and Kono,[18] and as 70% (calculated from the data presented) by Imagawa et al.[34] The difference in the rate could be due to the difference in observations for signs of illness.

Infected horses have been studied histopathologically only by Wada et al.[68] They examined five horses experimentally exposed to the MI-110 strain of GEV. The main gross lesions were moderate enlargement of the lymph nodes of the whole body in all cases, and scattered maculae in the dermis and edema in the subcutaneous tissue in two horses with a rash. Histopathologically the most outstanding lesion was lymphoid hyperplasia in the lymph nodes of the whole body and in the spleen. Atrophy of the follicles of the spleen with degenerated lymphocytes was observed in two horses killed during the convalescent stage of the disease. These histological findings in the lymphatic tissue appeared to be reactive changes attributable to immune responses. The cutaneous lesions, which were limited to the dermis, showed grossly the scattered maculae and were characterized by perivascular and/or

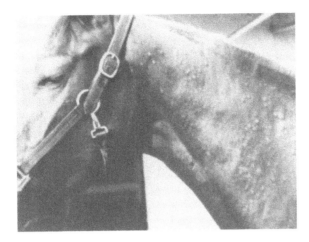

FIGURE 2. Rash on the neck of a horse experimentally infected with the MI-110 strain of Getah virus by the intranasal route. Photograph was taken 8 days after infection. (Photograph courtesy of Dr. M. Kamada.)

diffuse infiltration of lymphocytes, histiocytes, and eosinophils, edematous alteration in the wall of blood vessels, and hemorrhagic foci. Perivascular cuffs of mononuclear cells were observed in the cerebrum in two cases, and small hemorrhagic foci in the spinal cord in another case.

There had been no report on pigs showing signs of illness in natural and experimental infection with GEV until recently. Fujisaki et al.[66] investigated seroepidemiologically 1700 primary pregnant pigs and suspected some correlation between abortion and stillbirth and the virus infection, but failed to isolate the virus from aborted fetuses.[67] However, recently a case was found by Yago and associates.[97] A litter of 12 piglets were depressed and lost appetite 2 days after birth, and showed systemic trembling on the 3rd day. By the 5th day, eight died. The remaining four survived, but their growth was markedly retarded. The dam showed no signs of illness during that time. GEV was isolated from every tested tissue of the dead animals. The surviving animals, including the dam, developed antibodies to GEV. Kawamura et al.[98] succeeded in reproducing the disease in 5- and 18-day-old germ-free piglets by intramuscular inoculation. Signs of illness appeared within 24 hr after inoculation. The authors thought natural infection might be caused by mosquito bites in the neonatal stage, but not by transplacental transmission. Experimental inoculation with the virus caused transplacental infection and fetal death in sows at 26 and 28 days of pregnancy.[39]

Mice are highly susceptible to GEV and Sagiyama virus, and manifest various signs of illness.[22,42] The pathogenicity of the viruses depends on the age of the host and the route of inoculation. Intracerebral inoculation causes death in mice less than 9 days old. It induces retarded growth, insufficient hair growth, paralysis, and sporadic death in 11-day-old mice, but no signs in 13-day-old mice. The virus multiplies rapidly and extensively in every tissue tested, but particularly in the muscular tissues.[42] Many virus particles are seen in the tissue (Figure 3). Histopathologically, focal myositis, neuronal degeneration in the spinal cord and brain, and brown fat necrosis are commonly observed.[22]

B. Applicable Diagnostic Procedures

Since horses and pigs show viremia for relatively long periods as mentioned above, virus isolation in suckling mice or in cell cultures from an animal with signs of disease seems to be an ideal way to diagnose the infection. Methods of assay were described in a previous

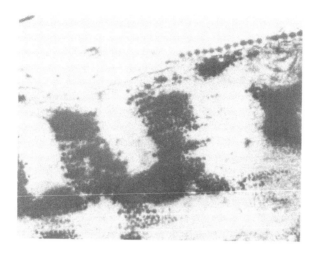

FIGURE 3.   Viral particles on the sarcoplasmic membrane and vi-
roplasma in cytoplasms of striated muscle of mouse limb. (Micrograph
courtesy of Drs. R. Wada and M. Kamada.)

section. For serological diagnosis, VN,[24,25] HI,[18] and CF[18] tests using sera collected in the
acute and convalescent stages of illness are most commonly used. The VN test is the one
recommended most since it has higher specificity than HI and CF tests and shows only
limited cross reaction with other alphaviruses.[24,25,27]

VN tests are performed by a routine procedure, the constant virus-variable serum method,
in VERO, BHK-21, or MA104 cell cultures and judged by either CPE inhibition or plaque
reduction.[18,20,24,25] HI tests are performed with sucrose-acetone extracted mouse brain antigen[70]
or infected cell culture fluid as hemagglutinin,[19] acetone-ether extracted serum,[71] or kaolin
and goose-erythrocyte treated serum,[19] and goose erythrocytes. Since the HA activity is
temperature dependent, pH dependent, and dependent on the molarity of NaCl,[19] it is
recommended that HI tests be performed at pH 6.2 with 0.35 M NaCl solution and at room
temperature. CF tests are performed by Kolmer's method with microtitration procedures.
The antigens are the same as for the HA test.[19]

IV. EPIDEMIOLOGY

A. Incidence among Horses

Epizootics of GEV infection among horses have been recorded only in Japan. After the
first appearance of the disease during the autumn of 1978,[18,20,34,37] it was seen again from
July to November of 1979 and 1983 in many racetracks and horse training centers.[29,30,33]
According to the investigation of the Bureau of Animal Industry, Japan, the number of
horses affected in two horse training centers and stock farms was 727 out of 2478 (29.3%)
in 1978.

Signs of illness as well as infection were observed in both sexes and every age group
without significant differences. The infection, which was confirmed by serological tests,
spread relatively slowly and involved 56% of the horses in one horse training center[18] and
61% in another.[34] It took about 40 days to infect 722 of 1908 horses kept in the latter
center.[34]

Two seroepidemiological investigations[32,34] disclosed that the virus infection among horses
has spread all over Japan with positive rates in individual regions ranging from 3.3 to 50%
(average 29%, 421/1465) and from 3.6 to 55% (average 26%, 126/496).

Antibody to GEV was found at a very high rate (46%) even in the Soya district (45°N), the northernmost part of Japan, while antibodies to Japanese encephalitis (JE) virus were only rarely detected in that district.[72] This is probably due to the distribution of vectors in the district. *Ae. vexans nip.*, which transmits GEV efficiently but rarely transmits JE, is abundant in that area, but *Cx. tritaeniorynchus*, which transmits both GEV and JE, is rare.[73]

Chronological examination revealed that epizootics or endemics of GEV infection among the horse population have occurred successively among horses in Japan since the discovery of Sagiyama virus in 1956.[32]

B. Seroepidemiology in Man and Domestic Animals

Evaluation of seroepidemiological tests of GEV infection has been difficult since there are strong cross reactions among related alphaviruses, particularly with RRV, which is highly prevalent in western New Guinea, Papua New Guinea, and Australia.[9,11] In those countries, most GEV-positive human and animal sera also reacted at higher titers with RRV, suggesting that the GEV positivity is due to a cross reaction with RRV. On the other hand, it is considered in Japan that GEV positivity is a reflection of GEV infection since no antigenically related virus has been found, except for Sagiyama virus, which is virtually similar or identical to GEV.

Among the human population, individuals seropositive to GEV have been reported mainly in Japan and Malaysia. The positive rate was 1.2 to 9% (23/1851, 4/47, 18/192) in Japan,[7,14,21] and 0.82 to 37% (36/4384, 99/269) in Malaysia.[2,15,17] In other investigations, only four sera from Vietnam, Malaysia, and the Philippines gave monospecific reactions with GEV,[11] and one serum from a child in Australia was suspected as being specifically positive.[9] Although many positive reactors to GEV were found in the indigenous human population in Australia,[9,10,12] western New Guinea,[11] Papua New Guinea,[11] and some Pacific islands,[11] all of these reactions were considered to be cross reactions with RRV.

Pigs have antibodies to GEV at the high rate of 20 to 80% in Japan and Malaysia.[14,16,36,74] The calculated monthly infection rate is 13.3% in Sarawak, Malaysia.[36] Seroconversion to GEV occurs gradually and does not reach 100% in a group of pigs, while that to JE occurs at one time and reaches to 100% in most cases.[3,14,31] This phenomenon suggests that the vector mosquito has lower sensitivity to GEV than to JE and/or that levels and duration of viremia in pigs are not sufficient to allow a number of mosquitoes to be infected.

Serological studies on cattle are quite limited so far as I know. In Japan, seropositive cattle could be detected everywhere at an average ratio of 38%,[35] although the numbers tested were limited. The rate is 10% in Malaysia[16] and 2.3% in Australia.[13]

Information on GEV infection in domestic poultry is scarce. Nakamura et al.[14] found HI antibodies, which were distinguished from nonspecific inhibitors, in 18/27 sera in Japan. Seropositive cases were also found in Malaysia at a low rate,[16] but not in Australia.[9]

V. TRANSMISSION CYCLES

A. Vectors

Many strains of GEV have been isolated from nine species of mosquitoes. More than 30 strains were isolated from *Cx. tritaeniorynchus*,[2-4,21,61,75-80] more than 20 strains from *Ae. vexans nip.*,[3,5,7,21,76] two strains from *Cx. gelidus*,[2,45] and one each from *Cx. vishnui*,[81] *Cx. bitaeniorynchus*,[6] *Anopheles amictus*,[6] *Anopheles* spp.[45] *Armigeres subalbatus*,[66] and *Mansonia bonneae dives*.[4] These species are presumed to serve as vectors of the virus to a certain extent depending upon the local ecology. Other species, such as *Ae. japonicus, Ae. aegypti, Ae. albopictus, Cx. pipens pallens, Triteroides bambusa*,[47] *Cx. annulirostris, Ae. vigilax*,[50] and *Ae. funereus*,[49] are also suspected as possible vectors because they can be infected experimentally with GEV. However, the high frequency of virus isolations from *Cx. tritaeniorynchus* and *Ae. vexans nip.* suggests that the two species may be the main vectors of GEV.

These mosquitoes are persistently infected with the virus by feeding on an animal with viremia. The infection rate in mosquitoes is dependent on the species and the concentration of virus in the blood.[46-48] *Ae. vexans nip.* has higher susceptibility to the virus than *Cx. tritaeniorynchus* and had infection rates of 100, 60, and 13% when they were engorged with blood having infectivities of $10^{6.3}$, $10^{4.3}$, and $10^{3.3}$ plaque-forming units (PFU)/mℓ, respectively, while infection rates of *Cx. tritaeniorynchus* were 86, 11, and 0%, when they fed on blood containing virus concentrations of $10^{7.8}$, $10^{5.1}$, and $10^{4.1}$ PFU/mℓ, respectively.[47] Growth of the virus in the mosquitoes and ability to act as a transmitter are influenced by the ambient temperature,[47] as previously shown with other arboviruses.[82] GEV propagates as rapidly at 20°C as at 28°C in *Ae. vexans nip.*, and the mosquito is capable of transmitting the virus 7 days after infection, while the propagation was inhibited at 20°C in *Cx. tritaeniorynchus* and 14 days' incubation was needed to acquire the transmitting capacity. These findings indicate that *Ae. vexans nip.* is a more suitable vector, particularly in high latitudes, than *Cx. tritaeniorynchus*.

Rates of isolation of the virus from mosquitoes collected in the field have reflected the difference in susceptibility between the two species.[5,21] Kumanomido et al.[76] collected 5875 (127 pools) *Cx. tritaeniorynchus* and 8238 (172 pools) *Ae. vexans nip.* in a pigpen. They isolated 3 strains of GEV and 32 strains of JE from the former and 13 strains of GEV and 1 strain of JE from the latter. The rates of infection with GEV in the two species were 1:2746 and 1:451, respectively. Thus, *Ae. vexans nip.* has higher susceptibility to GEV than *Cx. tritaeniorynchus* in nature as well as experimentally.

B. Vertebrate Hosts

Horses show high levels of virus in the blood and excretion of the virus onto the nasal mucosa for several days,[18] and can be infected intranasally.[20] Therefore, direct transmission of the virus among the horse population is conceivable, besides vector-mediated transmission. Nevertheless, during an outbreak of the disease in a horse training center where about 2000 race horses were kept, it was observed that an affected animal would appear randomly in any stable without distinction between contaminated and noncontaminated stables, suggesting that vector-mediated transmission was the main mode.[37]

In mice the virus transmits horizontally among the litter mates and vertically from the dam to her offspring via milk. A suckling mouse excretes the virus into the oral cavity and the dam into the milk.[42] Further, the virus is transmitted vertically through the placenta to the fetus, but does not cause fetal death. Mice infected *in utero* survive to grow to adulthood.[69] These modes of transmission might occur sometimes in the field.

Though horses qualify as amplifiers because of their high susceptibility,[18] they may not play a role as major amplifiers country-wide since their distribution is limited to particular places such as racetracks and breeding regions in GEV-prevalent areas. The horse as a work animal is decreasing in number in every country. Therefore, the animal will have importance only in local ecology.

In contrast, pigs have been considered as the most suitable amplifying animal because of their high susceptibility to the virus, distribution and density, speed of alternation of generations, and mosquito preference.[21,36] In fact, it was found in a pigpen that GEV was isolated from mosquitoes before virus isolation from sentinel pigs and their seroconversion. Then, following the increase in infected pigs, the number of virus-infected mosquitoes increased.[31,76]

Cattle seem to play a minor or unimportant role as amplifiers since they have low sensitivity to the virus[40] and show fleeting and inconsistent viremia.[35] GEV was not isolated from mosquitoes collected in cow sheds, though it was found in those collected in pigpens and horse-baited stables in the same place.[76]

Fowls possess antibodies[14,16] and are infected experimentally with the virus, showing viremia and signs of disease.[22] Despite a contradictory report describing failure of propagation of the virus in the chicken,[44] the animal seems to qualify as an amplifier because of its wide distribution, population density, and a rapid turnover rate. The ecological role of fowls in GEV maintenance in nature needs to be elucidated.

Wild animals in Australia and Malaysia seem not to be involved in natural cycles of infection.[9,16] Wild monkeys are the only wild animals that have a high frequency of antibodies to GEV, although the ecological significance of these antibodies has not been clarified.[16] In Japan, juvenile black-crowned night herons captured in heronries where mosquitoes containing virus were collected had detectable VN antibody.[21] However, no further information for evaluating the role of the birds in the ecological cycles of GEV has been found.

C. Maintenance and Overwintering Mechanisms

GEV is maintained throughout the year by the pig-mosquito cycle in the tropical zone where breeding of mosquitoes occurs continuously throughout the year. However, the overwintering mechanism of GEV is not known at all in the temperate zone where most, if not all, of the mosquitoes pass their lives during the winter. Transovarial passage of GEV as demonstrated in JE infection in mosquitoes by Rosen et al.[83,84] or of overwintering of some infected mosquitoes as imagos, must be considered as possible mechanisms which could rationally account for the consistent reappearance of the virus year after year in the same localities.

Another important possibility, which has been suspected as the main mechanism in JE, is reintroduction of GEV from the south by migrating birds, though there is no evidence to support the idea.

VI. ECOLOGICAL DYNAMICS

GEV is distributed widely in Far Eastern, Eastern, and southeastern Asian countries and in Australia. The virus was isolated from various species of mosquitoes in the Amour district of Siberia, U.S.S.R., situated at a latitude between 50 and 60°N,[85] in Japan between 30 and 45°N,[4,5,21,61,76,77,79,87] in Taiwan between 22 and 25°N,[86] in Cambodia between 8 and 15°N,[75] in the Philippines between 6 and 19°N,[81] in Malaysia between 0 and 6°N,[2,16,45] and in Australia between 10 and 40°S.[6]

The virus is found from the end of June to the end of October in Japan, in April in Australia, and throughout the year in tropical areas.[45] In Japan the virus appears every year coinciding with the appearance of JE in mosquitoes[4,5,21,76,77] and in pigs.[8,31] The mode of appearance of mosquitoes (*Cx. tritaeniorynchus*) infected with GEV is quite different from that of JE-infected mosquitoes. GEV has a tendency to be isolated sporadically for long periods, while JE appears at one time at a high frequency.[3,4,61,76,78] This is probably due to the low susceptibility of the mosquito to the virus and/or the low degree of viremia in pigs.

Cx. tritaeniorynchus, one of the suspected main vectors, is present as a common species in a wide area from Japan to Indonesia and Malaysia.[2,12,16,45,73,81,87-89] This mosquito has broad feeding habits, with its blood sources including pigs, other domestic and wild animals, man, and birds.[87,88,90] Therefore, this species must have a very important role in the prevalence of the virus among various animal species in the area from the temperate to the tropical zone.

The other main vector, *Ae. vexans nip.*, lives in Siberia, Sakhalin, Japan, Korea, China, and Taiwan, but not in tropical countries.[12] The mosquito feeds on pigs, other mammals, and fowls, and also bites man quite frequently. Therefore, the species is regarded as another important viral transmitter, particularly in the northern part of the temperate zone to the subfrigid zone where the average temperature during the summer season is around 20°C or does not exceed 20°C.

Distribution of mosquitoes has strong local characteristics. Therefore, most of the mosquito species which are known to be virus carriers may assume the major vector role according to the local ecology. *Cx. gelidus*, a species which is predominantly found only where there are pigs or, less commonly, cattle[87,88,90] seems to be important for maintenance of the virus by the pig-mosquito cycle. In Australia, GEV was isolated from *Ae. amictus* and *Cx. bitaeniorynchus* which were found in only a limited region.[6] It is hard to evaluate the role of these species without further information. *Cx. annulirostris*, which is believed to be the main vector of many Australian arboviruses and to be highly prevalent in Queensland,[6] may be assumed to be a vector of GEV since infection of the mosquito has been shown experimentally.[50]

VII. PREVENTION AND CONTROL

The relationship of GEV to human disease has been given attention since the virus is often isolated from mosquitoes and pigs with JE and man has antibodies to it. Also, studies on the relationship to abortion and stillbirths in pigs have been conducted because of the high percentage of positive reactors. Most of them, however, were not published because of their negative results. Since then, interest in the virus has declined with the decrease in pathogenic significance to man and economically important animals. Therefore, no study on surveillance of prevalence of the virus has been conducted.

GEV causes a transient but distinct disease in horses. Nowadays, the importance of horses for work is declining in every country. Therefore, the disease does not cause any economic impact except in race horses. In racetracks it has been shown that immunization caused prevalence of the disease.[29] Vaccination against the virus with inactivated vaccine has been carried out in the main racetracks in Japan.[92] Since then, wide prevalence such as that which occurred in 1978 has markedly diminished.

VIII. FUTURE RESEARCH

Interesting questions remain on the pathogenicity of GEV. Before an outbreak of a disease characterized by fever followed by rash and/or edema among race horses in 1978, there was no record of such disease, although presence of the virus had been demonstrated serologically for more than 20 years.[29] This observation poses the possibility that passage through a large group of horses may result in variation in virulence for horses.[19] Moreover, the recent appearance of a strain pathogenic for piglets, as described in this chapter, also suggests the same possibility. Therefore, studies on the mechanisms of variation and of acquiring virulence for the host are very important, not only in GEV but also in other alphaviruses. It has been proposed that o'nyong nyong (ONN) virus, an alphavirus, which caused a pandemic in East Africa, developed from chikungunya virus by changes in antigenic structure and vector potential which gave it new pathogenicity.[93]

Plaque size has been considered to be one of the markers of pathogenicity. It appears, however, that virulence and plaque morphology are not necessarily correlated in alphaviruses.[94,95] Detection of a virulence marker at the molecular and biochemical levels may be one of the main approaches for analysis of virulence.

A common important question among mosquito-borne Togaviridae is the mechanism of overwintering of the virus in temperate and subfrigid zones. As the mechanisms for regular seasonal reappearance of the virus, transovarial passage of the virus in mosquitoes, and overwintering of infected mosquitoes as imagos should be considered.

Finally, Sagiyama virus, classified as a subtype of the Getah species along with Bebaru and Ross River viruses,[1] should be considered a "variety" of GEV, since the two viruses, GEV and Sagiyama virus, are indistinguishable serologically[19-21,25,28] and genetically.[63] More precise antigenic, biochemical, and molecular biological investigations focused on this point are necessary.

REFERENCES

1. **Calisher, C. H., Shape, R. E., Brandt, W., Casals, J., Karabatsos, N., Murphy, F. A., Tesh, R. B., and Wiebe, M. E.,** Proposed antigenic classification of resistered arboviruses. I. Togaviridae, Alphavirus, *Intervirology,* 14, 229, 1980.
2. **Berge, T. O., Ed.,** *International Catalogue of Arboviruses Including Certain Other Viruses of Vertebrates,* Publ. No. (CDC) 75-8301, U.S. Department of Health, Education and Welfare, Washington, D.C., 1975.
3. **Igarashi, A., Morita, K., Bundo, K., Matsuo, S., Hayashi, K., Matsuo, R., Harada, T., Tamoto, H., and Kuwatsuka, M.,** Isolation of Japanese encephalitis and Getah virus from *Culex tritaeniorynchus* and slaughtered swine blood using *Aedes albopictus* clone C6/36 cells in Nagasaki, 1981, *Trop. Med.,* 23, 177, 1981.
4. **Igarashi, A., Buei, K., Ueba, N., Yoshida, M., Ito, S., Nakamura, H., Sasao, F., and Fukai, K.,** Isolation of viruses from female *Culex tritaeniorynchus* in *Aedes albopictus* cell cultures, *Am. J. Trop. Med. Hyg.,* 30, 449, 1981.
5. **Matsuyama, T., Oya, A., Ogata, T., Kobayashi, I., Nakamura, T., Takahashi, M., and Kitaoka, M.,** Isolation of arboviruses from mosquitoes collected at live-stock pen in Gumma Prefecture in 1959, *Jpn. J. Med. Sci. Biol.,* 13, 191, 1960.
6. **Doherty, R. L., Carley, J. G., Mackerras, M. J., and Marks, E. N.,** Studies of arthropod-borne virus infection in Queensland. III. Isolation and characterization of virus strains from wild-caught mosquitoes in North Queensland, *Aust. J. Exp. Biol. Med. Sci.,* 47, 17, 1963.
7. **Shichijo, A., Mifune, K., Chinn, C. C., Hayashi, K., Wada, Y., Ito, S., Oda, T., Omori, N., Suenaga, O., and Miyagi, I.,** Isolation of Japanese encephalitis virus from *Aedes vexans nipponii* caught in Nagasaki area, *Jpn. Trop. Med.,* 12, 91, 1970.
8. **Matsuyama, T., Nakamura, T., Isahai, K., Oya, A., and Kobayashi, M.,** Haruna virus, a group A arbovirus isolated from swine in Japan, *Gunma J. Med. Sci.,* 16, 131, 1967.
9. **Doherty, R. L., Garman, B. M., Whitehead, R. H., and Carley, J. G.,** Studies of arthropod-borne virus infections in Queensland. V. Survey of antibodies to group A arboviruses in man and other animals, *Aust. J. Exp. Med. Sci.,* 44, 356, 1966.
10. **Stanley, N. F. and Choo, S. B.,** Studies of arboviruses in Western Australia, serological epidemiology, *Bull. WHO,* 30, 221, 1964.
11. **Tesh, R. B., Gajdusek, C., Garruto, R. M., Cross, J. H., and Rosen, L.,** The distribution and prevalence of group A arbovirus neutralizing antibodies among human populations in Southeast Asia and the Pacific islands, *Am. J. Trop. Med. Hyg.,* 24, 664, 1975.
12. **Kanamitsu, M., Taniguchi, K., Urasawa, S., Ogata, T., Wada, Y., and Saroso, J. S.,** Geographic distribution of arbovirus antibodies in indigenous human populations in the Indo-Australian archipelago, *Am. J. Trop. Med. Hyg.,* 28, 351, 1975.
13. **Sanderson, C. J.,** A serologic survey of Queensland cattle for evidence of arbovirus infection, *Am. J. Trop. Med. Hyg.,* 18, 433, 1969.
14. **Nakamura, T., Isahai, K., Matsumoto, M., Matsuyama, T., Oya, A., and Okuno, T.,** Antibody responses following natural infection of swine to group A arbovirus Getah and following survey for its antibody among domestic animals, *Nihon Koshueisei Zasshi,* 14, 569, 1967.
15. **Bowen, E. T. W., Simpson, D. I. H., Platt, G. S., Hilary, W., Bright, W. F., Day, J., and Lim, T. W.,** Arbovirus infections in Sarawak, October 1968—February 1970: human serological studies in Land Dyak village, *Trans. R. Soc. Trop. Med. Hyg.,* 69, 182, 1975.
16. **Marchette, N. J., Rudnick, A., Garcia, R., and MacVean, D. W.,** Alphavirus in peninsular Malaysia. I. Virus isolations and animal serology, *Southeast Asian J. Trop. Med. Public Health,* 9, 317, 1978.
17. **Marchette, N. T., Rudnick, A., and Garcia, R.,** Alphaviruses in peninsular Malaysia: serological evidence of human infection, *J. Trop. Med. Public Health,* 11, 14, 1980.
18. **Sentsui, H., and Kono, Y.,** An epidemic of Getah virus infection among racehorses: isolation of the virus, *Res. Vet. Sci.,* 29, 157, 1980.

19. **Kono, Y., Sentsui, H., and Ito, Y.,** An epidemic of Getah virus infection among racehorses: properties of the virus, *Res. Vet. Sci.,* 29, 162, 1980.
20. **Kamada, M., Ando, Y., Fukunga, Y., Kumanomido, T., Imagawa, H., Wada, R., and Akiyama, Y.,** Equine Getah virus infection: isolation of the virus from racehorses during an enzootic in Japan, *Am. J. Trop. Med. Hyg.,* 29, 984, 1980.
21. **Scherer, W. F., Funkenbusch, M., Buescher, E. L., and Izumi, T.,** Sagiyama virus, a new group A arthropod-borne virus from Japan. I. Isolation, immunological classification, and ecologic observations, *Am. J. Trop. Med. Hyg.,* 11, 255, 1962.
22. **Scherer, W. F., Izumi, T., McCown, J., and Hardy, J. L.,** Sagiyama virus. II. Some biologic, physical and immunological properties, *Am. J. Trop. Med. Hyg.,* 11, 269, 1962.
23. **Porterfield, J. S.,** Cross-neutralization studies with group A arthropod-borne viruses, *Bull. WHO,* 24, 735, 1961.
24. **Karabatsos, N.,** Antigenic relationships of group A arboviruses by plaque reduction neutralization testing, *Am. J. Trop. Med. Hyg.,* 24, 527, 1975.
25. **Chanas, A. C., Johnson, B. K., and Simpson, D. I. H.,** Antigenic relationships by a simple micro-culture cross-neutralization method, *J. Gen. Virol.,* 32, 295, 1976.
26. **Casals, J.,** Relationships among arthropod-borne animal viruses determined by cross-challenge tests, *Am. J. Trop. Med. Hyg.,* 12, 587, 1963.
27. **Chanas, A. C., Johnson, B. K., and Simpson, D. I. H.,** A comparative study of related alphaviruses — a naturally occurring model of antigenic variation in the Getah subgroup, *J. Gen. Virol.,* 35, 455, 1977.
28. **Kamada, M., Kumanomido, T., Ando, Y., Fukunaga, Y., Imagawa, H., Wada, R., and Akiyama, Y.,** Studies on Getah virus: some biological, physicochemical and antigenic properties of the MI-110 strain, *Jpn. J. Vet. Sci.,* 44, 89, 1982.
29. **Sentsui, H. and Kono, Y.,** Reappearance of Getah virus infection among horses in Japan, *Jpn. J. Vet. Sci.,* 47, 333, 1985.
30. **Sugiura, T., Fukunaga, Y., and Hirasawa, K.,** Epizootics of Getah virus infection in horses in the Kanto region of Japan in 1983, *Bull. Equine Res. Inst. (Jpn.),* No. 21, 19, 1984.
31. **Kumanomido, T., Fukunaga, Y., Ando, Y., Kamada, M., Imagawa, H., Wada, R., Akiyama, Y., and Tanaka, Y.,** Ecological survey on Getah virus among swine in Japan, *Bull. Equine Res. Inst. (Jpn.),* No. 19, 89, 1982.
32. **Sentsui, H. and Kono, Y.,** Survey on antibody to Getah virus in horses in Japan, *Natl. Inst. Anim. Health Q. (Jpn.),* 20, 39, 1980.
33. **Sugiura, T., Ando, Y., Imagawa, H., Kumanomido, T., Fukunaga, Y., Kamada, M., Wada, R., Hirasawa, K., and Akiyama, Y.,** An epizootiological study of Getah virus among light horses in Japan in 1979, *Bull. Equine Res. Inst. (Jpn.),* 18, 103, 1981.
34. **Imagawa, H., Ando, Y., Kamada, M., Sugiura, T., Kumanomido, T., Fukunaga, Y., Wada, R., Hirasawa, K., and Kumanomido, T., Fukunaga, Y., Wada, R., Hirasawa, K., and Akiyama, Y.,** Sero-epizootiological survey on Getah virus infection in light horses in Japan, *Jpn. J. Vet. Sci.,* 43, 797, 1981.
35. **Morozumi, K., Sentsui, H., and Kono, Y.,** Experimental infection of calves with Getah virus and field survey on antibody against the virus, *Bull. Natl. Inst. Anim. Health,* 80, 1, 1980 (in Japanese).
36. **Simpson, D. I. H., Smith, C. E. G., Marshall, T. F. C., Platt, G. S., Way, H. J., Bowen, E. T. W., Bright, W. F., Day, J., McMahon, D. A., Hill, M. N., Bendell, P. J. E., and Heathocote, O. H. U.,** Arbovirus infection in Sarawak: the role of the domestic pig, *Trans. R. Soc. Trop. Med. Hyg.,* 70, 66, 1976.
37. **Fukunaga, Y., Ando, Y., Kamada, M., Imagawa, H., Wada, R., Kumanomido, T., Akiyama, Y., Watanabe, O., Niwa, K., Takenaga, S., Shibata, M., and Yamamoto, T.,** An outbreak of Getah virus infection in horses — clinical and epizootiological aspects at the Miho Training Center in 1978, *Bull. Equine Res. Inst. (Jpn.),* No. 18, 94, 1981.
38. **Kubota, M., Izumida, A., Takuma, H., Kodama, K., and Sasaki, F.,** Experimental infection of Getah virus in pigs, in *Proc. 87th Annu. Meet. Jpn. Soc. Vet. Sci.,* 1978, 47.
39. **Izumida, A., Inagaki, S., Takuma, H., Kubota, M., Hirahara, T., Kodama, K., and Sasaki, F.,** Experimental infection of swine with Getah virus. *Jpn. J. Vet. Sci.,* 50(3), 1988.
40. **Spradbrow, P. B.,** Arbovirus infection of domestic animals in Australia, *Aust. Vet. J.,* 48, 181, 1972.
41. **Sanderson, C. J.,** The immune response to viruses in calves. II. The response in young calves, *J. Hyg.,* 66, 461, 1968.
42. **Sentsui, H. and Kono, Y.,** Pathogenicity of Getah virus for mice, *Natl. Inst. Anim. Health Q. (Jpn.),* 21, 7, 1981.
43. **Chung, Y. S.,** Propagation of Sindbis virus, Murray Valley encephalitis virus and Getah virus, *J. Comp. Pathol.,* 79, 245, 1969.
44. **Chung, Y. S.,** Susceptibility of chickens, mice, and chick embryos to Sindbis, Murray Valley encephalitis and Getah viruses, *J. Comp. Pathol.,* 79, 335, 1969.

45. **Simpson, D. I. H., Way, H. J., Platt, G. S., Bowen, E. T. W., Hill, M. N., Kamath, S., Bendell, P. J. E., and Heathcote, O. H. U.,** Arbovirus infections in Sarawak, October 1968—February 1970: Getah virus isolations in mosquitoes, *Trans. R. Soc. Trop. Med. Hyg.,* 69, 35, 1975.

46. **Takashima, I., Hashimoto, N., Arikawa, J., and Matsumoto, K.,** Getah virus in *Aedes vexans nipponii* and *Culex tritaeniorynchus:* vector susceptibility and ability to transmit, *Arch. Virol.,* 76, 299, 1983.

47. **Takashima, I. and Hashimoto, N.,** Getah virus in several species of mosquitoes, *Trans. R. Soc. Trop. Med. Hyg.,* 79, 546, 1985.

48. **Kumanomido, T., Fukunaga, Y., Kamada, M., and Wada, R.,** Experimental transmission studies of Getah virus in *Aedes vexans, Bull. Equine Res. Inst. (Jpn.),* No. 19, 93, 1982.

49. **Kay, B. H., Carley, J. G., and Filippich, C.,** The multiplication of Queensland and New Guinean arboviruses in *Aedes funereus* (Theobald) (Diptera:Culicidae), *J. Med. Entomol.,* 13, 451, 1977.

50. **Kay, B. H., Carley, J. G., and Filippich, C.,** The multiplication of Queensland and New Guinean arboviruses in *Culex annulirostris* Skuse and *Aedes vigilax* (Skuse) (Diptera:Culicidae), *J. Med. Entomol.,* 12, 279, 1975.

51. **Carley, J. G., Standfast, H. A., and Kay, B. H.,** Multiplication of viruses isolated from arthropods and vertebrates in Australia in experimentally infected mosquitoes, *J. Med. Entomol.,* 10, 244, 1973.

52. **Karabatsos, N. and Buckley, S. M.,** Susceptibility of the baby-hamster kidney-cell line (BHK-21) to infection with arboviruses, *Am. J. Trop. Med. Hyg.,* 16, 99, 1967.

53. **Stim, T. B.,** Arbovirus plaquing in two simian kidney cell lines, *J. Gen. Virol.,* 5, 329, 1969.

54. **Ueba, N. and Kimura, T.,** Polykaryocytosis induced by certain arboviruses in monolayers of BHK-21-528 cells, *J. Gen. Virol.,* 34, 369, 1977.

55. **Pudney, M., Varma, M. G. R., and Leake, C. J.,** Establishment of a cell line (XTC-2) from the S. African clawed toad, *Xenopus laevis, Experientia,* 29, 466, 1973.

56. **Kitamura, S.,** Establishment of cell line from *Culex* mosquito, *Kobe J. Med. Sci. (Jpn.),* 16, 41, 1970.

57. **Igarashi, A.,** Isolation of a Singh's *Aedes albopictus* cell clone sensitive to dengue and chikungunya viruses, *J. Gen. Virol.,* 40, 531, 1978.

58. **Ando, K. and Kitamura, S.,** Plaque formation by an Alphavirus (Getah) in *Culex* mosquito cells, *Acta Virol. (Engl. Ed.),* 21, 168, 1977.

59. **Varma, M. G. R., Pudney, M., and Leake, C. J.,** Cell lines from larvae of *Aedes (stegomyia) malayensis* (Colless) and *Aedes (S.) psudoscutellaris* (Theobald) and their infection with some arboviruses, *Trans. R. Soc. Trop. Med. Hyg.,* 68, 374, 1974.

60. **Kimura, T. and Ueba, N.,** Some biological and serological properties of large and small plaque variants of Getah virus, *Arch. Virol.,* 57, 221, 1978.

61. **Igarashi, A., Makino, Y., Matsuo, S., Bundo, K., Matsuo, R., Higashi, F., Tamoto, Y., and Kuwatsuka, M.,** Isolation of Japanese encephalitis and Getah viruses by *Aedes albopictus* clone C6/36 and by suckling mouse brain inoculation in Nagasaki, 1980, *Trop. Med.,* 23, 69, 1980.

62. **Igarashi, A., Sasao, F., Fukai, K., Ueba, N., and Yoshida, M.,** Mutants of Getah virus and Japanese encephalitis viruses from field-caught *Culex tritaeniorynchus* using *Aedes albopictus* clone C6/36 cells, *Ann. Virol.,* 132E, 235, 1981.

63. **Morita, K. and Igarashi, A.,** Oligonucleotide fingerprint analysis of strains of Getah virus isolated in Japan and Malaysia, *J. Gen. Virol.,* 65, 1899, 1984.

64. **Work, T. H.,** Isolation and identification of arthropod-borne viruses, in *Diagnostic Procedures for Viral and Rickettsial Diseases,* Lennette, E. H. and Schmidt, N. J., Eds., American Public Health Association, New York, 1964, 312.

65. **Kamada, M., Ando, Y., Fukunaga, Y., Imagawa, H., Wada, R., Kumanomido, T., Tabuchi, E., Hirasawa, K., and Akiyama, Y.,** Studies on Getah virus — pathogenicity of the virus for horses, *Bull. Equine Res. Inst. (Jpn.),* No. 18, 84, 1981.

66. **Fujisaki, Y., Morimoto, T., Terasaki, S., Murakami, M., and Yamori, T.,** Studies on viral abortion and stillbirth of pigs. II. Correlation between abortion or stillbirth and Japanese encephalitis, Getah and porcine parvovirus infections, *Jpn. J. Vet. Sci.,* 34(Suppl.), 23, 1972.

67. **Morimoto, T., Fujisaki, Y., Shiraishi, T., Tada, Y., and Hanyu, A.,** Studies on viral abortion and stillbirth of pigs. III. Isolation of viruses from fetuses at stillbirth, *Jpn. J. Vet. Sci.,* 34(Suppl.), 23, 1972.

68. **Wada, R., Kamada, M., Fukunaga, Y., Ando, Y., Kumanomido, T., Imagawa, H., Akiyama, Y., and Oikawa, M.,** Equine Getah virus infection: pathological study of horses experimentally infected with the MI-110 strain, *Jpn. J. Vet. Sci.,* 44, 411, 1982.

69. **Aaskov, J. G., Davies, C. E., Tucker, M., and Dalglish, D.,** Effect on mice of infection during pregnancy with three Australian arboviruses, *Am. J. Trop. Med. Hyg.,* 30, 198, 1981.

70. **Clarke, D. H. and Casals, J.,** Technique for hemagglutination-inhibition with arthropod-borne viruses, *Am. J. Trop. Med. Hyg.,* 7, 561, 1958.

71. **Porterfield, J. S.,** The hemagglutination-inhibition test in the diagnosis of yellow fever in man, *Trans. R. Soc. Trop. Med. Hyg.,* 48, 261, 1954.

72. **Matsumura, T., Goto, H., Shimizu, K., Sugiura, T., Ando, Y., Kumanomido, T., Hirasawa, K., and Akiyama, Y.**, Prevalence and distribution of antibodies to Getah and Japanese encephalitis viruses in horses raised in Hokkaido, *Jpn. J. Vet. Sci.*, 44, 967, 1982.

73. **Kamimura, K.**, The distribution and habit of medically important mosquitoes of Japan, *Jpn. J. Sanit. Zool.*, 19, 15, 1968 (in Japanese with English summary).

74. **Fujisaki, Y., Morimoto, T., Terasaki, S., Takahashi, S., and Watanabe, K.**, Studies on viral abortion and stillbirth of pigs. I. Prevalence of HI antibodies to Japanese encephalitis, Getah and parvoviruses, *Jpn. J. Vet. Sci.*, 34(Suppl.), 22, 1972 (in Japanese).

75. **Chastel, C. and Rageau, J.**, *Med. Trop. (Marseilles)*, 26, 391, 1966, cited in *Simpson, D. I. H., et al.*, Arbovirus infection in Sarawak, October 1968—February 1969: Getah virus isolations from mosquitoes, *Trans. R. Soc. Trop. Med. Hyg.*, 69, 35, 1975.

76. **Kumanomido, T., Fukunaga, Y., Ando, Y., Kamada, M., Imagawa, H., Wada, R., Akiyama, Y., Tanaka, Y., Kobayashi, M., Ogura, N., and Yamamoto, H.**, Getah virus isolations from mosquitoes in an enzootic area in Japan, *Jpn. J. Vet. Sci.*, 48, 1135, 1986.

77. **Kumanomido, T., Fukunaga, Y., Kamada, M., Imagawa, H., Ando, Y., Wada, R., Nitta, M., and Akiyama, Y.**, Getah virus isolations from mosquitoes collected at two horse habitations in the western areas of Japan, *Jpn. J. Vet. Sci.*, 48, 1191, 1986.

78. **Yamamoto, H.**, Arbovirus infection in the mosquitoes of Fukuoka area, Kyushu, Japan. II. Natural infection of mosquitoes with the viruses of Getah complex in the period from 1963 to 1972, *Jpn. J. Sanit. Zool.*, 31, 23, 1980.

79. **Hurlburt, H. S. and Nibley, C.**, Virus isolations from mosquitoes in Okinawa, *J. Med. Entomol.*, 1, 78, 1964.

80. **Heathcote, O. H. V.**, Japanese encephalitis in Sarawak: studies on juvenile mosquito populations, *Trans. R. Soc. Trop. Med. Hyg.*, 64, 483, 1970.

81. **Ksiazek, T. G., Trosper, J. H., Cross, J. H., and Basaca-Sevilla, V.**, Isolation of Getah virus from Nueva Ecija Province, Republic of Philippines, *Trans. R. Soc. Trop. Med. Hyg.*, 75, 312, 1981.

82. **Chamberlain, R. W. and Sudia, W. D.**, The effect of temperature upon the extrinsic incubation of Eastern equine encephalitis in mosquitoes, *Am. J. Hyg.*, 62, 295, 1955.

83. **Rosen, L., Tesh, R. B., Lien, J. C., and Cross, J. H.**, Transovarial transmission of Japanese encephalitis virus by mosquitoes, *Science*, 199, 909, 1978.

84. **Rosen, L., Donald, A., Shroyer, D. C., and Lien, J. C.**, Transovarial transmission of Japanese encephalitis virus by *Cx. tritaeniorynchus* mosquitoes, *Am. J. Trop. Med. Hyg.*, 29, 711, 1980.

85. **Chumakov, M. P., Moshkin, A. V., and Andreeva, E. B.**, Isolation of five Getah virus strains from mosquitoes in the Southern Amur region, U.S.S.R., *Trans. R. Soc. Trop. Med. Hyg.*, 69, 35, 1974.

86. **Herbert, S. H.**, The pig-mosquito cycle of Japanese encephalitis virus in Taiwan, *J. Med. Entomol.*, 1, 301, 1964.

87. **Macdonald, W. W., Smith, C. E. G., and Webb, H. E.**, Arbovirus infections in Sarawak: observations on the mosquitoes, *J. Med. Entomol.*, 1, 335, 1965.

88. **Macdonald, W. W., Smith, C. E. G., Dawson, P. W., Ganapathipillai, A., and Makadevans, S.**, Arbovirus infections in Sarawak: further observations on mosquitoes, *J. Med. Entomol.*, 4, 146, 1967.

89. **Hill, M. N.**, Japanese encephalitis in Sarawak: studies on adult mosquito populations, *Trans. R. Soc. Trop. Med. Hyg.*, 64, 489, 1970.

90. **Hill, M. N., Varma, M. G. R., Mahadevans, S., and Meers, P. D.**, Arbovirus infection in Sarawak: observations on mosquitoes in the premonsoon period, September to December 1966, *J. Med. Entomol.*, 6, 398, 1969.

91. **Bendell, P. J. E.**, Japanese encephalitis in Sarawak: studies on mosquito behaviour in a land of Dayak village, *Trans. R. Soc. Trop. Med. Hyg.*, 64, 497, 1970.

92. **Akiyama, Y.**, Getah virus infection among horses, *J. Jpn. Vet. Med. Assoc.*, 33, 567, 1980.

93. **Haddow, A. J., Davies, C. W., and Walker, A. J.**, O'nyong-nyong fever: an epidemic virus disease in East Africa. I. Introduction, *Trans. R. Soc. Trop. Med. Hyg.*, 54, 517, 1960.

94. **Schlesinger, R. W.**, Virus-host interactions in natural and experimental investigations with alphaviruses and flaviviruses, in *The Togaviruses, Biology, Structure, Replication*, Schlesinger, R. W., Ed., Academic Press, New York, 1980, 4.

95. **Strauss, E. G. and Strauss, J. H.**, Mutants of alphaviruses: genetics and physiology, in *The Togaviruses, Biology, Structure, Replication*, Schlesinger, R. W., Ed., Academic Press, New York, 1980, 14.

96. **Kono, Y., et al.**, unpublished data.

97. **Yago, S. et al.**, *Jpn. J. Vet. Sci.*, 49, 989, 1987.

98. **Kawamura, A. et al.**, *Jpn. J. Vet. Sci.*, 49, 1003, 1987.

Chapter 26

THE EPIDEMIOLOGY OF DISEASES CAUSED BY VIRUSES IN GROUPS C AND GUAMA (BUNYAVIRIDAE)

Robert E. Shope, John P. Woodall, and Amelia Travassos da Rosa

TABLE OF CONTENTS

I. HISTORICAL BACKGROUND

Viruses of group C and group Guama were among the first agents isolated in 1954 when researchers of the Special Service of Public Health of Brazil and the Rockefeller Foundation[1] set out to find the causes of human fevers at Belém, Brazil, near the mouth of the Amazon River. They noted that the fevers occurred in workers imported from northeastern Brazil to clear forests for planting pepper and rubber, and in Japanese colonists settling in forest-fringe communities. They also noted that lifelong residents of the city of Belém became ill if they entered the forest. Initially, five different group C viruses and one virus in group Guama were isolated from sera of nonimmune *Cebus* monkeys placed in the forest as sentinels, and another, different group Guama virus was isolated from acute-phase serum of a febrile worker clearing forest at the Oriboca Plantation, 20 km east of Belém. The project director, Ottis Causey, had the idea (which he and his colleagues later proved to be correct) that the viruses were being transmitted by forest and forest-fringe mosquitoes, that those persons who entered the forest were at highest risk, and that the viruses were unique to the Amazon forests, and therefore persons from Japan and northeastern Brazil were fully susceptible because they had not previously been exposed.

The viruses were isolated by intracerebral inoculation of sera into baby mice. The isolates were sent for comparison with other arboviruses at the Rockefeller Foundation Virus Laboratories in New York. In the mid-1950s the New York laboratory was making rapid progress in grouping a large variety of newly isolated arboviruses from various parts of the world. Casals and Whitman[2,3] found the new agents to be distinct and to form two serogroups. The five viruses from sentinel monkeys were classified as group C to differentiate them from group A (alphaviruses) and group B (flaviviruses). The other two viruses were initially to constitute group D, but because of the growing concern that there would be more groups than there were letters, it was decided to utilize the name Guama, which was the first virus of the group to be isolated.

In 1957, the Brazilian workers substituted mice for monkeys as sentinels; mice proved highly efficient as a surveillance mechanism. This observation led others to try sentinel mice and hamsters in Trinidad, Peru, Mexico, Panama, Colombia, and Florida, with the result that group C and/or group Guama viruses were isolated in each of these locations.[4] There are now known to be 13 viruses in group C and 12 in group Guama.[4,5] All are found associated with low-lying, usually swampy forests in the New World tropics and subtropics, and all are transmitted by culicine mosquitoes to small rodents, primates, and marsupials.

The primary public health impact of the viruses in the two groups is on persons whose occupation takes them into the forest, for instance, members of geological expeditions, harvesters of nuts, lumber, or rubber, and military personnel. These viruses do not cause epidemics or fatal disease, but they are commonly responsible for severe, self-limited, disabling febrile illnesses which at times have led to significant loss of productivity for those in forest occupations.

II. THE VIRUSES

Group C and Guama viruses belong to the family Bunyaviridae.[6] The particles are 90 to 100 nm diameter spheres with lipid-containing envelopes. The genome is single-stranded, three-segmented, negative-sense RNA which can reassort genetically. This is an important property reflected in the occurrence of strains in nature which have apparently reassorted to yield progeny with properties of each of two commonly encountered parent viruses.[2] The virions contain a nucleocapsid protein and two surface glycoproteins. Replication occurs in the cytoplasm, both in vertebrate and insect cells, in close association with the Golgi apparatus.

Table 1
DISTRIBUTION AND SOURCES OF VIRUSES IN GROUPS C AND GUAMA

Group	Complex	Type	Subtype	Geographic distribution	Source
C	Caraparu	Caraparu		Brazil, Trinidad, Panama, French Guiana, Suriname	Mosquitoes, humans, rodents
			Ossa	Panama	Mosquitoes, humans, rodents
		Apeu		Brazil	Mosquitoes, humans, marsupials
		Vinces		Ecuador	Mosquitoes, sentinel hamsters
		Bruconha		Brazil	Mosquitoes
	Madrid	Madrid		Panama	Mosquitoes, humans, rodents
	Marituba	Marituba		Brazil, Peru	Mosquitoes, humans, marsupials
			Murutucu	Brazil, French Guiana	Mosquitoes, humans, rodents, marsupials
			Restan	Trinidad, Surinam	Mosquitoes, humans
		Nepuyo		Trinidad, Brazil, Honduras, Mexico, Panama, Guatemala	Mosquitoes, humans, rodents, bats
		Gumbo Limbo		Florida (U.S.)	Mosquitoes, rodents
	Oriboca	Oriboca		Brazil, Trinidad, Suriname, French Guiana	Mosquitoes, humans, rodents, marsupials
			Itaqui	Brazil, Venezuela	Mosquitoes, humans, rodents, marsupials
Guama	Guama	Guama		Brazil, Trinidad, French Guiana, Peru, Suriname, Panama	Mosquitoes, phlebotomines, humans, rodents, marsupials, bats
		Moju		Brazil	Mosquitoes, rodents, marsupials
		Ananindeua		Brazil	Mosquitoes, rodents, birds, marsupials
		Mohogany Hammock		Florida (U.S.)	Mosquitoes, rodents
		Bimiti		Trinidad, Brazil, French Guiana, Suriname, Peru	Mosquitoes, rodents
		Timboteua		Brazil	Rodents
		Catu		Brazil, Trinidad, French Guiana	Mosquitoes, humans
	Mirim	Mirim		Brazil	Mosquitoes, sentinel mice
		Bertioga		Brazil	Mosquitoes, sentinel mice
		Cananeia		Brazil	Mosquitoes, sentinel mice
		Guaratuba		Brazil	Mosquitoes, birds, sentinel rodents
		Itimirim		Brazil	Rodents

Antigenic relationships of the viruses within each group have been shown by complement-fixation (CF), neutralization (N), hemagglutination-inhibition (HI), immunofluorescence, and enzyme-linked immunosorbent assay (ELISA). There are four serologic complexes in group C and two in group Guama (Table 1). The N and HI tests measure reactivity of the surface glycoproteins; the major CF antigen is the nucleocapsid protein. Monkeys immunized with a group C virus resisted infection when challenged with a heterologous group C virus,[7] and therefore the antigenic classification has epidemiologic significance.

Some of the viruses in group C form pairs which share CF reactivity, i.e., Caraparu and Itaqui, Murutucu and Oriboca, and Apeu and Marituba. These same viruses form different pairs which are closely related by neutralization and HI, i.e., Itaqui and Oriboca, Caraparu and Apeu, and Murutucu and Marituba. These pairings of viruses which are found in the same Brazilian forest were useful for rapid identification,[8] and may possibly be explained on the basis of genetic reassortment of viruses which coinfect hosts and/or vectors in the forest.

The group Guama viruses are closely related by the CF test, but most serotypes are quite distinct by HI and N. It was originally thought that group C viruses were not related to group Guama viruses,[2,3] but subsequently HI cross-reactivity was noted at a low level,[9] consistent with the remarkably similar biological and ecological properties of viruses in both groups.

Serum of mice and hamsters infected with viruses of both groups agglutinate goose red blood cells. The HI test has been used effectively as a type-specific survey method in ecologic studies of rodents and marsupials, and for serosurvey of monkey and human populations in the American tropics.

III. DISEASES CAUSED BY VIRUSES OF GROUPS C AND GUAMA

Persons infected by group C and group Guama viruses develop fever of sudden onset ranging from 38 to 40°C. Patients with group Guama virus infections usually have lower grade fevers than those with group C infections. Severe headache is the dominant complaint. Vertigo, backache, and/or muscle and joint pains are seen in nearly 2/3 of cases, and nausea and photophobia in about 1/3. The fever lasts from 2 to 5 days and is sometimes biphasic. Rash is absent; however, conjunctival injection is sometimes noted. Leukopenia is recorded in some cases. Patients recover with weakness and anorexia lasting 1 or 2 weeks, but without sequelae. The initial observations were made in 1955 of Brazilian laborers in the Oriboca Forest, near Belém, Brazil.[1] These men earned hourly wages, were not paid when sick, and therefore were assuredly incapacitated when they sought medical assistance. They commonly were out of work for 3 to 7 days when infected with viruses of group C or group Guama.

The incubation period is not known, but is probably less than 10 days based on known periods of forest exposure. Patients gave a history of exposure in the forest or at the forest fringe, and often of exposure during the evening. One Brazilian member of a field epidemiology team had been employed for 5 years during the daytime in the Utinga Forest without serious febrile illness. He was reassigned in 1964 to collect mosquitoes at dusk. Within a 2-week period he developed fever, backache, weakness of his legs, and leukopenia. Caraparu virus (group C) was isolated from his blood. He returned to work in the forest to continue evening mosquito collections, and within another 2 weeks his fever recurred. This time Catu virus (group Guama) was isolated from his blood.

Several field-acquired infections and one laboratory infection have had similar febrile courses. The source of infection in the laboratory case of Apeu fever is not known, but the exposure was assumed to be aerosol.[10] Ossa and Madrid viruses were isolated from entomological field workers in Panama,[11] Catu virus from a febrile patient in Trinidad,[12] and Nepuyo virus from febrile field workers in Guatemala.[13] There have been 10 of the 13 group C serotypes and 2 of the 12 group Guama serotypes isolated from febrile cases (Table 1). The apparent-to-inapparent infection ratio is not known.

IV. EPIDEMIOLOGY

A. Distribution

Group C and group Guama viruses are found only in the American tropics and subtropics, including Florida, Central America, northern South America, and southern Brazil. Their distribution is limited by their requirement for specific mosquitoes and vertebrate hosts. Most of the types and subtypes have focal distribution which coincides with forest or forest-fringe habitats, usually in low-lying, swampy areas. Of the 25 types and subtypes listed in Table 1, 15 are found only in one country and restricted to one known focus. Seven others, Caraparu, Murutucu, Oriboca, Itaqui, Guama, Bimiti, and Catu, are found in northern Brazil plus Trinidad and/or French Guiana, Suriname, and Panama. Itaqui is found in Brazil and Venezuela, and Restan is found in Trinidad and Suriname. Nepuyo virus has the widest distribution, possibly because it infects bats which fly long distances. In contrast, the viruses which infect less mobile animals, i.e., rodents and marsupials, are limited to one focus.

B. Incidence

The incidence of human infection depends directly on exposure. A study of 500 Dutch military personnel stationed for 1 year in Suriname showed that seroconversion occurred 7.4 times per 100 soldier years using three group C antigens.[14] The rate in some units, however, was as high as 22%, indicating that these soldiers probably had been stationed at some time during the year in a focus of group C virus-infected mosquitoes. Bensabath[15] compared antibody rates to group C viruses in persons living in the center of Belém with antibody rates in persons living at the periphery of the city, bordering the forest. People living near the forest had a higher prevalence of antibody.

C. Season

An accurate tabulation of seasonality cannot be made since only small numbers of patients with groups C and Guama virus infections have been recorded. It can be taken as an article of faith, however, that human infections will occur at the time of greatest natural transmission, which is the rainy season in tropical rain forest climates.[16]

D. Other Risk Factors

There is no reason to believe that susceptibility varies with sex or race. The group C and Guama fever cases which have been recorded have been predominantly in young adult males. The major risk factor is occupation, which almost certainly explains the age and sex bias. Persons are at risk who enter the forest to collect nuts, tap rubber trees, or fell trees for lumber. Military forces engaged in jungle operations, especially if exposed at dusk and dawn, are also at higher risk of infection.

There are no epidemics, and the apparent-to-inapparent infection ratio is not recorded.

E. Serological Epidemiology

The HI and N tests have been used to survey for antibody to group C viruses. Results of HI tests (Shope and Travassos da Rosa, unpublished, Table 2) of human sera from the Brazilian Amazon revealed widespread antibody to the Oriboca, Marituba (represented by Murutucu subtype), and Caraparu complexes. Oriboca antibody was localized mostly in Pará State, and Murutucu in Pará and Amazonas States. These were not common infections. Caraparu antibody, on the other hand, was widely distributed and common, especially in Amazonas, where over a third of the indigenous people of some communities reacted.

A survey in the Ribeira Valley of coastal São Paulo State, Brazil revealed that 4.6% of 516 residents, and 14.5% of 83 road workers who camped close to the forest, had Caraparu antibody.[17,18] Antibodies have been found to three group C viruses in Suriname[14] and to Nepuyo virus in Honduras.[19]

Table 2
GROUP C HI ANTIBODY IN HUMANS OF THE BRAZILIAN AMAZON

State or territory	Town	No. sera	% positive with antigen of		
			Oriboca	Murutucu	Caraparu
Pará	Belém	636	3.8	4.4	10.7
	Braganca	23	0	0	0
	Soure	50	0	2.0	2.0
	Abaetetuba	29	6.9	3.4	10.3
	Cameta	55	3.6	0	7.3
Amazonas	Tefe	45	0	4.4	11.1
	Borba	39	0	0	38.5
	Barcelos	18	0	11.1	38.9
	Labrea	59	0	0	15.3
	Eirunepe	84	0	1.2	9.5
	Codajas	26	0	0	19.2
	Benjamin Constant	66	0	0	9.1
	Manaus	36	0	0	0
Acre	Río Branco	15	0	0	0
Guaporé	Porto Velho	35	2.9	0	8.6
Río Branco	Boa Vista	22	0	0	0

Serological survey results of group Guama antibodies have not been reported except for the finding of neutralizing antibody to Bimiti virus in one of six persons sampled in Trinidad.[20]

V. TRANSMISSION CYCLES

A. Evidence from Field Studies

Extensive studies to determine the natural vectors and hosts of groups C and Guama viruses were carried out as part of the Rockefeller Foundation arbovirus program in collaboration with the governments of Brazil and Trinidad. The evidence for transmission of group C viruses as reviewed by Woodall[21] implicates culicine mosquitoes and small mammals. Similar cycles exist for group Guama viruses.

The potential arthropod vectors of viruses in groups C and Guama are listed in Tables 3 and 4. The listing is based on isolation of virus from the arthropod, or in some cases the transmission of the virus from naturally infected mosquitoes to laboratory mice or hamsters. Guama was transmitted three times by *Culex portesi*, and once each by *Cx. taeniopus*, *vomerifer*, and B17; Mirim once by *Cx. taeniopus*; Caraparu once by *Culex* spp.; Ossa five times by *Cx. vomerifer* and once by *Cx. taeniopus*; Marituba once by *Cx. portesi*; Nepuyo once by *Culex* spp.; and Oriboca once by *Cx. portesi*. Transmission by naturally infected mosquitoes is the best evidence so far available that these species are involved. All of these transmissions were by mosquitoes of the *Culex* subgenus *Melanoconion*, which also predominate among the sources of virus isolations. The evidence favors transmission by *Melanoconion* mosquitoes of all group C viruses with the possible exception of Apeu, which has been isolated also from *Aedes*. Isolates from Diptera other than *Melanoconion* may mean that the insect had recently fed on a viremic animal and retained infected gut contents, or that it may be an occasional vector only.

The evidence also points to *Culex (Melanoconion)* mosquitoes as vectors of group Guama viruses, but Timboteua and Itimirim viruses have not yet been isolated from mosquitoes, and others have been isolated too infrequently to draw a conclusion.

The *Melanoconion* are nocturnal forest mosquitoes of the Americas, most active at dusk

Table 3
POTENTIAL ARTHROPOD VECTORS OF GROUP C VIRUSES[4,5,21]

Virus	Culex (Melanoconion)	Other Culex	Aedes	Other Diptera
Caraparu	aikenii complex (2)[a]	(Cx.) coronator (1)		Limatus durhamii (1)
	spissipes (2)	(Cx.) nigripalpus (1)		Wyeomyia medioalbipes (1)
	portes (22)	(Eubonnea) accelerans (3)		Undesignated Sabethini (1)
	vomerifer (14)	(Eu.) amazonensis (1)		Mixed pool (1)
	Undesignated (2)	Undesignated (5)		
Ossa	taeniopus (1)			
	vomerifer (6)			
Apeu	aikenii complex (4)		arborealis (1)	
	Undesignated (1)		septemstriatus (1)	
Vinces	vomerifer (3)			
Bruconha	sacchettae (5)			
	Undesignated (1)			
Madrid	vomerifer (5)			
Marituba	aikenii complex (2)			Mixed species (2)
	portesi (1)			
	Undesignated (2)			
Murutucu	portesi (5)	Undesignated (2)		Coquillettidia venezuelensis (1)
	aikenii complex (4)			Undesignated Sabethini (1)
	vomerifer (1)			
	Undesignated (1)			
Restan	portesi (2)	Undesignated (1)		
Nepuyo	iolambdis (1)	(Eu.) accelerans (1)		
		Undesignated (5)		
Gumbo Limbo	Undesignated (61)		taeniorhynchus (2)	
Oriboca	portesi (46)	Undesignated (6)	taeniorhynchus (4)	Coq. venezuelensis (1)
	aikenii complex (6)			Coq. arribalzagai (1)
	spissipes (1)			Mansonia spp. (1)
	Undesignated (1)			Psorophora ferox (1)
				Sabethini (2)
				Mixed species (1)
Itaqui	spissipes (1)	Undesignated (1)		Mixed species (1)
	aikenii complex (1)			
	portesi (3)			
	vomerifer (10)			

[a] Numbers of isolations in parentheses (includes laboratory transmission from naturally infected vectors).

Table 4
POTENTIAL ARTHROPOD VECTORS OF GROUP GUAMA VIRUSES[4,5,21]

Virus	Culex (Melanoconion)	Other Culex	Aedes	Other Diptera
Guama	portesi (93)[a]	Undesignated (10)	Undesignated (1)	Mansonia (1)
	vomerifer (9)			Limatus (1)
	taeniopus (1)			Psorophora (1)
				Trichoprosopon (1)
				Lutzomyia (sand fly) (1)

Table 4 (continued)
POTENTIAL ARTHROPOD VECTORS OF GROUP GUAMA VIRUSES[4,5,21]

Virus	*Culex* (Melanoconion)	Other *Culex*	*Aedes*	Other *Diptera*
Moju	*portesi* (1)			*Coq. venezuelensis* (2)
	vomerifer (8)			*Mansonia* (2)
	Undesignated (2)			*Sabethini* (1)
Ananindeua	*portesi* (3)	B19 complex (14)		*Coq. venezuelensis* (1)
	taeniopus (2)	B27 (5)		*Culicoides paraensis* (5)
	vomerifer (1)	Undesignated (1)		
	Undesignated (2)			
Mahogany Hammock	Undesignated (44)			
Bimiti	*portesi* (19)			Undesignated (1)
	taeniopus (1)			
	spissipes (1)			
Timboteua	(No isolations from arthropods)			
Catu	*portesi* (80)	Undesignated (12)		*Anopheles numbus* (1)
Mirim	*taeniopus* (1)		*serratus* (1)	*Psorophora ferox* (1)
	Undesignated (1)			
Bertioga	*sacchettae* (1)			
Cananeia	*sacchettae* (1)			
Guaratuba	*taeniopus* (1)		*serratus* (1)	*An. cruzii* (1)
Itimirim	(No isolations from arthropods)			

[a] Numbers of isolations in parentheses (includes laboratory transmission from naturally infected vectors).

and dawn. They feed basically on rodents and marsupials, but some species prefer birds, and all listed in Tables 3 and 4 will also engorge on primates, including humans when available.

The vertebrate hosts of viruses of groups C and Guama are listed in Tables 5 and 6. The most extensive studies of vertebrate hosts have been conducted in Belém, Brazil and Bush Bush Forest in Trinidad. Groups C and Guama viruses have also been isolated from wild vertebrates in Florida, São Paulo State (Brazil), and Panama (Table 1). The indigenous rodent and marsupial species differ in each of these areas, and therefore the viruses use different sets of hosts. For instance, in the forests near Belém, *Oryzomys capito* and *Proechimys guyanensis* made up 92% of the animals entering traps at ground level. In Trinidad, *Oryzomys laticeps* and *Zygodontomys* were the predominant rodents at ground level. These respective sets of animals are hosts in the cycle of Caraparu virus in each site. The Brazilian *Oryzomys* and *Proechimys* are also hosts of Itaqui, Murutucu, Oriboca, Guama, Moju, and Catu; the analogous *Oryzomys* and *Zygodontomys* in Trinidad are hosts of Guama, Bimiti, and Catu viruses in Bush Bush Forest.

Two group C viruses, Apeu and Marituba, and a group Guama virus, Ananindeua, probably use marsupials as their principal hosts. These viruses have been isolated only from marsupials, and marsupials have the highest prevalence of Apeu and Marituba antibody of any animals tested. In contrast, as suggested by Woodall,[21] marsupials may not be susceptible to Caraparu virus. This virus has never been isolated from a marsupial, although the mosquitoes which transmit Caraparu feed on marsupials and marsupials have antibody at low levels.

Some cycles of transmission occur at the level of the forest floor while other cycles take place in the forest canopy. Sentinel mice exposed in the canopy were infected more often with Apeu, Marituba, and Oriboca viruses, while those exposed at ground level were infected more often with Caraparu and Itaqui viruses.[20] This finding correlates with the cycles of Apeu and Marituba viruses in marsupials, which tend to be canopy dwellers, and the cycles of Caraparu and Itaqui in forest-floor-dwelling rodents.

Table 5
VERTEBRATE HOSTS OF GROUP C VIRUSES

Virus	Host	Virus isolations	Serology Brazil[a]	Trinidad[b]
Caraparu	*Oryzomys capito*	7	60	
	O. laticeps	6		77
	Proechimys guyanensis	16	52	
	Nectomys squamipes	2	36	
	Zygodontomys brevicauda	4		43
	Heteromys anomalus	1		13
Ossa	*Proechimys semispinosus*	1		
Apeu	*Caluromys philander*	2	12	
	Marmosa cinerea	1	17	
Madrid	*Proechimys semispinosus*	1		
Marituba	*Didelphis marsupialis*	1		
	Caluromys philander		27	
Murutucu	*Oryzomys capito*	1	5	
	Proechimys guyanensis	9	11	
	Nectomys squamipes	2	8	
	Didelphis marsupialis	2		
	Marmosa cinerea	2	7	
	Bradypus tridactylus	1		
	Proechimys guyanensis	1		
	Nectomys squamipes	1		
	Artibeus literatus	1		
	A. jamaicensis	1		
Gumbo Limbo	*Sigmodon hispidus*	2		
Oriboca	*Oryzomys capito*	2	17	
	Proechimys guyanensis	7	12	
	Marmosa cinerea		36	
	Didelphis marsupialis	2		
Itaqui	*Oryzomys capito*	4	33	
	Proechimys guyanensis	4	34	
	Nectomys squamipes	3	33	
	Marmosa murina	1	17	
	Metachirus nudicaudatus	1		

Note: Vinces, Bruconha, and Restan have not been isolated from wild vertebrates.

[a]   Maximum monthly HI percent antibody prevalence in Utinga Forest, Brazil, 1962—1966.
[b]   Maximum quarterly HI percent antibody prevalence in Bush Bush Forest, Trinidad, 1960—1963.

Table 6
VERTEBRATE HOSTS OF GROUP GUAMA VIRUSES

Virus	Host	Virus isolations	Serology Brazil[a]
Guama	*Oryzomys capito*	31	42
	Proechimys guyanensis	25	46
	Nectomys squamipes	4	28
	Caluromys philander		30
	Marmosa cinerea		15
	Oryzomys laticeps	4	
	Heteromys anomalus	2	
	Zygodontomys brevicauda	7	
	Coendu spp.	1	

Table 6 (continued)
VERTEBRATE HOSTS OF GROUP GUAMA VIRUSES

Virus	Host	Virus isolations	Serology Brazil[a]
Moju	*Oryzomys* spp.	17	28
	Proechimys guyanensis	18	50
	Nectomys squamipes	4	8
	Oecomys	1	
	Didelphis marsupialis	1	
Ananindeua	*Caluromys philander*	7	
	Didelphis marsupialis	1	
Mahogany Hammock	*Sigmadon hispidus*	1	
Bimiti	*Oryzomys laticeps*	7	
	Zygodontomys brevicauda	2	
	Heteromys anomalus	1	
	Proechimys guyanensis	1	
Timboteua	*Proechimys* spp.	1	
Catu	*Nectomys squamipes*	3	39
	Proechimys guyanensis	16	53
	Oryzomys capito	22	67
	Oryzomys laticeps	2	
	Zygodontomys brevicauda	4	
	Didelphis marsupialis	2	
	Molossus obsurus (bat)	1	
Guaratuba	*Xanthomyias virescens*	1	
Itimirim	*Oryzomys* spp.	1	

Note: Mirim, Bertioga, and Cananeia have not been isolated from wild vertebrates.

[a]   Maximum monthly HI percent antibody prevalence in Utinga Forest, Brazil, 1962—1966.

B. Evidence from Experimental Infection Studies

Evidence from natural cycles is preferred to evidence from experimental infection because viruses change on laboratory passage, and vectors and vertebrate hosts may be altered physiologically by laboratory conditions. Nevertheless, the results of laboratory experiments reinforce conclusions when they agree with observations in nature. Jonkers and his colleagues[16,22,23] working in Trinidad showed the susceptibility of rodents to groups C and Guama viruses. Their results correlated closely with field observations. Three of the four group C viruses of Trinidad were inoculated into nonimmune rodents and each circulated virus to high titer. *Oryzomys* and *Zygodontomys* were susceptible generally to group C viruses. *Heteromys* was tested with Caraparu virus and did not develop viremia. *Heteromys* also was uncommonly infected with Caraparu virus in nature.

The three Trinidadian group Guama viruses induced viremia when inoculated into *Oryzomys* and *Zygodontomys* in the laboratory. The viremia titers were lower than those of group C viruses, but again were comparable to titers found in nature. *Heteromys* was not susceptible to Guama infection in the laboratory.

Infection of laboratory animals has also been used to determine if immunity to one virus in groups C or Guama inhibits viremia with another virus of the same group. Rhesus monkeys were susceptible to subcutaneous inoculation of six Brazilian group C viruses, although viremia with Itaqui and Caraparu was not consistently observed.[24] The monkeys were challenged with heterologous group C viruses and were immune to challenge with seven different combinations of primary and challenge viruses.

Cross-immunity was also demonstrated with group Guama viruses in Trinidadian rodents, although the inhibition of viremia varied with the virus combination.[23] Catu virus still circulated, although at a lower level, in animals immunized with Bimiti or Guama, while

Guama viremia was almost completely inhibited in animals immune to either Bimiti or Catu viruses.

These cross-challenge experiments offer a plausible explanation why people and rodents exposed in the forest are not found serially viremic with different viruses of the same serogroup.

VI. ECOLOGIC DYNAMICS: MAMMAL RECAPTURE PROGRAMS IN BRAZILIAN AND TRINIDADIAN FORESTS

A. Experimental Design

The interaction of groups C and Guama viruses with their rodent and marsupial hosts was studied in Brazil and Trinidad. A mammal recapture program was instituted by workers of the Belém Virus Laboratory in the Utinga Forest near Belém. Wild rodents and marsupials were captured, bled, marked, released, and periodically recaptured during the period July 1962 to November 1964.[24,25] All together, 1389 animals were sampled 8750 times. Approximately half the animals were *Proechimys guyanensis*, the spiny rat, and a further third were *Oryzomys capito*, the rice rat. The remainder were water rats and marsupials of five genera and species. Animals were tested for viremia at each capture and, when possible, for HI antibody at 3-week intervals.

Additional biometric data were collected. The animals were weighed at each capture, and an age curve for each species was estimated by constructing a weight curve on the time axis and extrapolating the weight curve from birth weight to the weight at first capture. Animals were observed for pregnancy and lactation status. Rectal temperature readings were recorded for a number of the rodents as they were trapped and retrapped; a battery-operated thermometer with calibrated hypodermic thermistor probe was used.

The animals were captured in box traps baited with corn and banana. The time of capture was recorded by rigging a string from the trap treadle to the flywheel of an alarm clock which stopped when the trap closed. The traps were located in a 25-ha square at 150 to 200 m intervals. During part of the study, additional traps were placed in an attempt to add precision to home-range estimates. Other traps were placed on platforms which were elevated to the tree canopy in order to sample the arboreal marsupials. The home range of an animal was estimated by recording which traps the animal entered and the distance between traps which the animal entered at different recapture events.

B. Virus Isolations and Seroconversions in Rodents and Marsupials

It was unusual at any given capture to find viremia. Only 46 or about 0.5% of the blood samples contained viruses of groups C and Guama. Virus was never isolated in serial samples. The length of viremia is probably less than 5 days since several animals were recaptured at intervals of 3 to 5 days. The highest titer viremia was 4.5 log mouse $LD_{50}/m\ell$, and only seven animals had titers greater than 2.7 log mouse LD_{50}. It is probable that samples were not taken at the peak of viremia, but even so, it can be inferred from the brief and low-titered viremia that the *Culex (Melanoconion)* mosquitoes must have a low threshold of infection for groups C and Guama viruses.

Antibody studies indicated that animals were frequently infected later in life with one or more additional viruses in groups C and Guama. No second episode of viremia was detected in any rodent or marsupial, either of the same virus or of any other virus in the two groups. This is in contrast to human infection in which group C followed by group Guama viremia has been documented.

It should be noted that nearly all the viremias detected in Utinga Forest rodents occurred in young animals. The experience with naturally infected *P. guyanensis* at initial capture is illustrated in Figure 1. By the time rodents of this species were old enough to range in

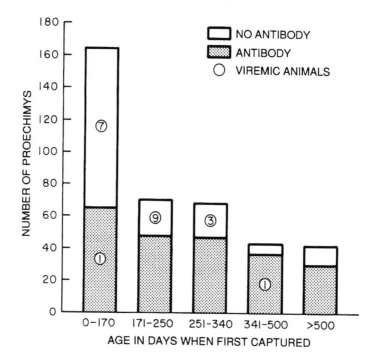

FIGURE 1.   Number of *Proechimys guyanensis* with group Guama antibody by age when first captured in Utinga Forest. The circles show the numbers of viremic animals with and without antibody in each age group.

search of food and thus find baited traps, 70%, as determined by antibody studies, had already been infected with one or more of the Guama group viruses present in the forest. At the time of viremia, 19 of the rats were negative for Guama group HI antibody; only 2 viremic animals were positive for Guama group antibody, although a figure of 27 would have been expected on the assumption that rodents with antibody are exposed in the same degree as those without. These data thus teach the lesson that young, nonimmunized animals were the source of virus. The data also offer an explanation for the apparently very low rate of viremia in animals captured in the study. Most animals were older and already immune when first trapped; if the very young animals entered the traps, many more individuals with viremia would probably have been found.

C. Influence of Host Species, Age, Habitat, Time of Day, and Season on Prevalence of Vertebrate Infections

The study in Utinga Forest clarified the many factors which influenced the prevalence of vertebrate infections, and especially of vertebrate viremias. Only two rodent species were found frequently enough at ground level to account for the majority of group C and group Guama infections. These were the spiny rat (23/46 viremias) and the rice rat (18/46 viremias). Other animals such as *Nectomys squamipes* (the water rat) and marsupials were viremic in small numbers in proportion to the small numbers trapped. In Utinga Forest, the groups C and Guama viruses utilized each of the available rodents and marsupials, but the spiny and rice rats were of greatest importance because of their larger numbers.

To be most effective in maintaining the transmission cycle, the spiny rat and rice rat must be active at the same hours as *Culex (Melanoconion)* mosquitoes. The time of maximum activity of over 400 rodents and marsupials in Utinga Forest, as determined by the clocks attached to the traps, was between 6 and 10 p.m. (Figure 2). Spiny rats had a second lesser peak at 3 a.m. This feeding activity correlated almost exactly with the *Culex (Melanoconion)* biting frequency.[26]

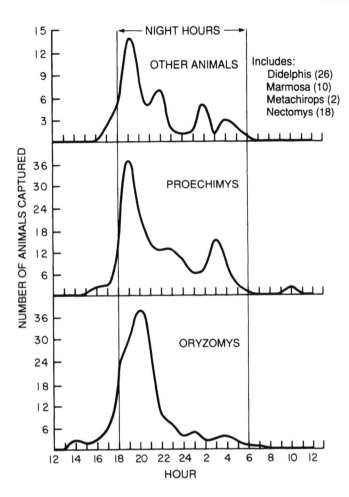

FIGURE 2.   Hour at which animals entered traps in Utinga Forest mammal recapture program.

Animals were almost never caught during the day. Nylon strings were attached to the rats to find their daytime resting sites. The rats were found to inhabit burrows, sometimes 1 m or more underground and thus inaccessible to daytime-biting mosquitoes.

Viruses of groups C and Guama are in general focally distributed, and restricted to tropical rainforest swamp habitat. This is in part explained by the short home ranges of the host rodents. *P. guyanensis* and *O. capito* ranged from 12 to 325 m. The spiny rat ranged farther than the rice rat, older animals ranged farther than younger animals, and males farther than females. Since viremia was mostly in younger animals, the chance that a viremic animal would range very far is small.

Groups C and Guama viruses were transmitted in the Utinga Forest of the Brazilian Amazon during periods in which many young, nonimmune rodents were entering the population. Figure 3 shows the estimated date of birth of rice rats based on weight at first capture. The figure also shows that within 2 or 3 months of birth of each new cohort, the prevalence of group C virus HI antibody rose rapidly. Antibody levels then fell between November and January during each dry season as older animals died, as transmission slowed, and as HI antibody decayed.

D. Influence of Vector Associations on Prevalence of Vertebrate Infections

Infections of vertebrate animals with groups C and Guama viruses correlated with high

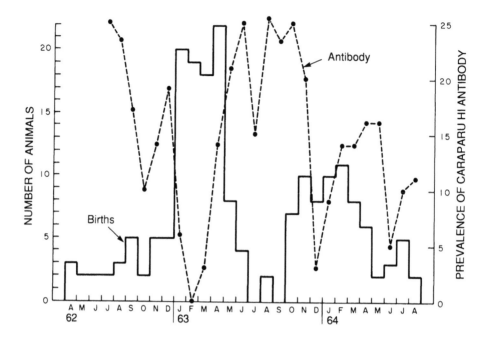

FIGURE 3.   Estimated birth dates of juvenile *Oryzomys capito* captured between June 1962 and November 1964 in Utinga Forest. Prevalence of Caraparu HI antibody in the same population is shown by the broken line.

population levels of *Culex (Melanoconion)* mosquitoes. In Utinga Forest these periods followed the onset of heavy rains and probably were in response to flooding of the swamp. High levels of transmission also coincided with the rapid influx of young, nonimmune *Oryzomys*.

In the Bush Bush Forest of Trinidad, as in Brazil, groups C and Guama viruses were transmitted most frequently when populations of *Culex (Melanoconion) portesi* mosquitoes were abundant.[16] This subgenus of mosquito, in turn, was believed to depend on high levels of water in the swamp for its breeding.[26] A population crash of rodents during 1964 in Bush Bush Forest, however, left the viruses without vertebrate amplifying hosts, and transmission (as measured by infection of sentinel mice) ceased almost completely.[16]

Woodall[21] has reviewed the host preferences, vector susceptibility, and abundance of the *Culex (Melanoconion)* mosquitoes involved in transmission of viruses in group C. Those species of mosquitoes which had the highest infection rates fed preferentially on rodents. Bird-feeding species had low infection rates. It was noted, in addition, that closely related subtypes of Caraparu virus in Brazil and Trinidad were transmitted by different *Melanoconion* species. Woodall postulated that there was either a difference in vector susceptibility or a genetic difference between the Brazilian and Trinidadian virus subtypes.

Transmission rates as measured by infection of sentinel mice generally correlated with the abundance of *Culex (Melanoconion)*, but there were some unexplained discrepancies which may relate to infection rates of mosquitoes, and indirectly to age and parous rates.[27]

VII. SURVEILLANCE, PREVENTION, AND CONTROL

Illnesses caused by viruses in groups C and Guama are not reportable, and surveillance is not maintained. The true impact of these diseases is therefore not known. Nor is there a vaccine or any other useful method of prevention or control other than personal measures. These include mosquito repellents, bed nets, and avoidance of the forest habitat of *Culex (Melanoconion)* mosquitoes, especially from dusk to dawn.

VIII. FUTURE RESEARCH

There are important practical and theoretical questions which need to be answered. First, what is the true incidence of illness caused by viruses of groups C and Guama? Diseases which cause disability and loss of time at work at often neglected unless they also kill or have sequelae. Tropical fevers are an impediment to colonization and exploitation of much of the earth, and as shown in the 1950s, those in the Amazon are closely linked to mosquito-borne viruses.

Second, are there generic approaches to prevention? It is not yet practical to develop vaccines for each of scores of arboviruses, nor is it easy to control mosquitoes or intervene at the level of the small mammal in the forest. It might, however, be feasible to find an antiviral drug for human use which is active prophylactically against viruses in the family Bunyaviridae. Prevention by drug would be ideal because exposure to these viruses is usually over a short term, and the time of risk can be anticipated in those who enter the forest.

Third, the viruses of groups C and Guama are ideal models to study reassortment in a natural forest setting. They are transmitted in a compact ecosystem which may well favor coinfection of mosquitoes and/or small mammals with two viruses of a group. There is already evidence of natural reassortment in the form of group C viruses which share antigens with other viruses in the same group.[2]

ACKNOWLEDGMENTS

Studies reported here were supported by the Rockefeller Foundation and the Servico Especial de Saude Publica of the Government of Brazil. Many scientists participated in collection and analysis of data involving the mammal recapture project in Utinga Forest, including O. R. Causey, C. E. Causey, G. Bensabath, and A. P. Souza.

REFERENCES

1. **Causey, O. R., Causey, C. E., Maroja, O. M., and Macedo, D. G.,** The isolation of arthropod-borne viruses including members of two hitherto undescribed serological groups, in the Amazon region of Brazil, *Am. J. Trop. Med. Hyg.,* 10, 227, 1961.
2. **Casals, J. and Whitman, L.,** Group C, a new serological group of hitherto undescribed arthropod-borne viruses. Immunological studies, *Am. J. Trop. Med. Hyg.,* 10, 250, 1961.
3. **Whitman, L. and Casals, J.,** The Guama group: a new serological group of hitherto undescribed viruses. Immunological studies, *Am. J. Trop. Med. Hyg.,* 10, 159, 1961.
4. **Karabatsos, N., Ed.,** *International Catalogue of Arboviruses 1985,* American Society of Tropical Medicine and Hygiene, San Antonio, Tex., 1985.
5. **Calisher, C. H., Coimbra, T. L., Lopez, O. de S., Muth, D. J., Sacchetta, L. de A., Francy, D. B., Lazuick, J. S., and Cropp, C. B.,** Identification of new Guama and Group C serogroup bunyaviruses and an ungrouped virus from southern Brazil, *Am. J. Trop. Med. Hyg.,* 32, 424, 1983.
6. **Bishop, D. H. L. and Shope, R. E.,** Bunyaviridae, in *Comprehensive Virology,* Fraenkel-Conrat, H. and Wagner, R. R., Eds., Plenum Press, New York, 1979, 1.
7. **Allen, W. P., Belman, S. G., and Borman, E. R.,** Group C arbovirus infections in rhesus monkeys, *Am. J. Trop. Med. Hyg.,* 16, 196, 1967.
8. **Shope, R. E. and Causey, O. R.,** Further studies on the serological relationships of group C arthropod-borne viruses and the association of these relationships to rapid identification of types, *Am. J. Trop. Med. Hyg.,* 11, 283, 1962.
9. **Shope, R. E., quoted in Casals, J.,** New developments in the classification of arthropod-borne animal viruses, *An. Microbiol.,* 11, 13, 1963.
10. **Gibbs, C. J., Jr., Bruckner, E. A., and Schenker, S.,** A case of Apeu virus (Casals' group C) infection, *Am. J. Trop. Med. Hyg.,* 13, 108, 1964.

11. **De Rodaniche, E., De Andrade, A. P., and Galindo, P.,** Isolation of two antigenically distinct arthropod-borne viruses of group C in Panama, *Am. J. Trop. Med. Hyg.,* 13, 839, 1964.

12. **Tikasingh, E. S., Ardoin, P., and Williams, M. C.,** First isolation of Catu virus from a human in Trinidad, *Trop. Geogr. Med.,* 26, 414, 1974.

13. **Scherer, W. F., Dickerman, R. W., and Ordonez, J. V.,** Enfermedad humana causada por el virus Nepuyo, un bunyavirus de Mesoamerica transmitido por mosquitoes, *Bol. Of. Sanit. Panam.,* 95, 111, 1983.

14. **Jonkers, A. H., Spence, L., and Karbaat, J.,** Arbovirus infections in Dutch military personnel stationed in Surinam. Further studies, *Trop. Geogr. Med.,* 20, 251, 1968.

15. **Bensabath, G.,** personal communication, 1960.

16. **Jonkers, A. H., Spence, L., Downs, W. G., Aitken, T. H. G., and Worth, C. B.,** Arbovirus studies in Bush Bush Forest, Trinidad, W. I., September 1959—December 1964. VI. Rodent-associated viruses (VEE and agents of groups C and Guama): isolations and further studies, *Am. J. Trop. Med. Hyg.,* 17, 285, 1968.

17. **Iversson, L. B., Travassos da Rosa, A. P. A., and Travassos da Rosa, J. F.,** Estudos sorologicos para pesquisa de anticorpos de arbovirus im populacao humana de regiao do Vale do Ribeira. II. Inquerito em pacientes do Hospital Regional de Pariquera-cu, 1980, *Rev. Saude Pub.,* 15, 587, 1981.

18. **Iversson, L. B., Travassos da Rosa, A. P. A., Travassos da Rosa, J. F., Eleuterio, G. C., and Prado, J. A.,** Estudos sorologicos para pesquisa de anticorpos de arbovirus na populacao humana da regiao do Vale do Ribeira. I. Seguimento sorologico de grupo populacional residente em ambiente silvestre, in *Internacional Simposium on Tropical Arboviruses and Haemorrhagic Fevers,* Pinheiro, F. P., Ed., Academia Brasileira de Ciencias, Rio de Janeiro, 1982, 229.

19. **Calisher, C. H., Chappell, W. A., Maness, K. S. C., Lord, R. D., and Sudia, W. D.,** Isolations of Nepuyo virus strains from Honduras, 1967, *Am. J. Trop. Med. Hyg.,* 20, 331, 1971.

20. **Spence, L., Anderson, C. R., Aitken, T. H. G., and Downs, W. G.,** Bimiti virus, a new agent isolated from Trinidadian mosquitoes, *Am. J. Trop. Med. Hyg.,* 11, 414, 1962.

21. **Woodall, J. P.,** Transmission of group C arboviruses (Bunyaviridae), in *Arctic and Tropical Arboviruses,* Kurstak, E., Ed., Academic Press, New York, 1979, 123.

22. **Jonkers, A. H., Spence, L., Downs, W. G., and Worth, C. B.,** Laboratory studies with wild rodents and viruses native to Trinidad. II. Studies with the Trinidad Caraparu-like agent TRVL 34053-1, *Am. J. Trop. Med. Hyg.,* 13, 728, 1964.

23. **Jonkers, A. H., Spence, L., and Olivier, O.,** Laboratory studies with wild rodents and viruses native to Trinidad. III. Studies with three Guama-group viruses, *Am. J. Trop. Med. Hyg.,* 17, 299, 1968.

24. **Causey, C. E.,** The role of small mammals in maintenance of arboviruses in the Brazilian Amazon forests, *An. Microbiol.,* 11(Parte A), 119, 1963.

25. **Shope, R. E., de Andrade, A. H. P., and Bensabath, G.,** The serological response of animals to virus infection in Utinga Forest, Belém, Brazil, *Atas do Simposio sobre a Biota Amazonica* 6 (Pathologia), 225, 1967.

26. **Aitken, T. H. G., Worth, C. B., and Tikasingh, E. S.,** Arbovirus studies in Bush Bush Forest, Trinidad, W. I., September 1959—December 1964. III. Entomologic studies, *Am. J. Trop. Med. Hyg.,* 17, 253, 1968.

27. **Davies, J. B., Corbet, P. S., Gillies, M. T., and McCrae, A. W. R.,** Parous rates in some Amazonian mosquitoes collected by three different methods, *Bull. Entomol. Res.,* 61, 125, 1971.

Chapter 27

ISSYK-KUL FEVER

Dimitri K. Lvov

TABLE OF CONTENTS

I. HISTORICAL BACKGROUND

A. Discovery of the Agent

In 1970—1971, a study of bats and their ectoparasites was carried out in the northern part of the Kirghiz S.S.R. on the shores of Issyk-Kul Lake. From 432 bats studied, 7 antigenically similar virus strains were recovered. The prototype LEIV-315 K strain was isolated from pooled liver, spleen, kidney, and brain of *Nyctalus noctula*, caught in the attic of a house in May 1970. This new virus has been called Issyk-Kul after the site of collection. The other six strains were isolated from *Myotis blythi* and *Vespertilio serotinus* bats and from *Argas (Carios) vespertilionis* ticks collected from bats. Issyk-Kul (IK) virus belongs to the family Bunyaviridae.[1] In 1966, Keterah virus (strain P6-1361), antigenically closely related to IK virus, was isolated by Marchette and Rudnick from *A. pusillus* ticks infesting *Scotophilus temmenckii* in Keterah, Kelantan, Malaysia.[2] At the present time, a total of three Keterah virus strains are available, two from ticks and one from 12 specimens of *S. temmenckii*, but not from other Malaysian bats.

B. Human Infections

Single cases of febrile illness etiologically attributed to the IK virus have been detected in Tajik S.S.R.[3-5] In May to August 1981, about 20 cases of the disease, later called IK fever, were reported in a village in the southwestern part of Tajik S.S.R.[6]

II. THE VIRUS

A. Antigenic Relationships

By complement-fixation (CF) tests, Issyk-Kul and Keterah viruses are virtually indistinguishable.[39] Cross-neutralization tests have not been reported, and it remains to be determined whether the two viruses are antigenically distinct. No serologic relationship was found with over 80 arboviruses isolated in the U.S.S.R. and elsewhere.[1,7]

Issyk-Kul virus has been shown to be a member of the family Bunyaviridae on the basis of electron microscopic morphology.[8]

B. Host Range

The infection is lethal by the 5th to 6th day for suckling mice and 3- to 4-week-old mice after both intracranial and subcutaneous inoculation. The titers reach 9.0 and 8.5 $\log_{10}$ LD$_{50}$, respectively. No clinical manifestations appear after intraperitoneal injection of adult Syrian hamsters, white rats, and guinea pigs nor after allantoic-sac inoculation of 9-day-old chick embryos.[1,2] The virus replicates in cultures of chick embryo fibroblasts (up to 6.0 log/mℓ) and duck embryo fibroblasts (up to 8.0 log/mℓ). Under natural conditions, IK virus has been shown to infect birds and mosquitoes, as well as bats and their obligate ectoparasites *A. vespertilionis* and *Ixodes vespertilionis*.[7,8]

C. Strain Variation

There have been 19 strains isolated from bats (6 strains), *A. vespertilionis* (10 strains), birds (1 strain), and mosquitoes (2 strains) studied by cross-titration in CF tests.[9] No significant antigenic variation was found among the strains isolated in 1970 to 1974 from various sources at different places in Kirghiz S.S.R. A virus strain isolated in 1976 in the Tajik S.S.R. from a redstart (*Phoenicurus phoenicurus*) seems to be the only exception.[10]

D. Methods for Assay

Intracranial inoculation of suckling mice proved to be the most useful method for virus isolation. Both CF and N (neutralization) have been successfully applied in chick fibroblast cultures for virus identification.[11]

FIGURE 1.   Area of Issyk-Kul virus distribution in the Kirghiz and the Tajik S.S.R., according to data from virological and serological studies. (1) Chujsk valley; (2) Issyk-Kul hollow; (3) Inner Tyan-Shan; (4) eastern regions adjacent to Ferghana; (5) southwestern regions adjacent to Ferghana; (6) Zeravashan region, Gissar-Darvaz Province; (7) South-Tajik Province. ● = Laboratory-verified human cases of disease; ■ = isolation of the virus from natural sources (bats, birds, ticks, mosquitoes). Population immunity according to complement-fixation test: light diagonal shading 1%; heavy diagonal shading 3%; light crosshatch 3.7 to 5%; heavy crosshatch 5.1 to 9%.

III. DISEASE ASSOCIATIONS

A. Humans

The clinical syndrome is characterized by abrupt onset followed by fever of 39 to 41°C; 80% of patients complained of headache and 50% of dizziness. Pharynx erythema, cough, and myalgia occurred in 30%, nausea and vomiting in 25%, and rash, abdominal pain retro-orbital pain, lacrimation, and photophobia in ≤6%. The average duration of the acute illness was 8 days, but return to full health took 1 to 1.5 months. No fatal cases have been registered.[6,7]

B. Domestic Animals

No clinical disease has been described in domestic animals. Nevertheless, on the basis of serological surveys, domestic animals are involved in virus infection cycles. In large samples of cattle in Kirghiz S.S.R., CF antibodies were found in only 0.5%, but in Tajik S.S.R., the antibody prevalence was fourfold higher, 2%. Pigs were found to be seronegative.[12]

C. Wild Animals

No clinical manifestations of the disease have been revealed in wild animals.

IV. EPIDEMIOLOGY

A. Geographic Distribution

Figure 1 summarizes data on virus isolations from various natural sources and from human patients. Virus circulation has been demonstrated in both the Tajik and the Kirghiz S.S.R. The virus has also been isolated from a horsefly (*Tabanus agrestis* Wied), caught at the bank of the Irghiz River in the Kazakh S.S.R.,[15] representing the northernmost site of virus isolation. However, human disease has been detected only in Tajik S.S.R., where sporadic cases and one outbreak have been registered.[6] The epidemic focus was located on the banks of the Pyandzh River in the arid part of the subtropical climate zone, having a very warm summer and a mild winter.

In addition to the virus isolations, serological data reveal a wide area of virus distribution in the U.S.S.R., including at least the Kirghiz S.S.R. and Tajik S.S.R., the Turkemenian S.S.R., and the southern part of Kazakh S.S.R. In respect to zoogeography, the virus is distributed within the borders of west Tyan-Shan and Pamyr-Alaj districts, Mountain Asia Province, Central-Asian subregion, as well as adjacent regions of Turan district, Iran-Turan Province, of the Mediterranean subregion of the Holarctic region. If Keterah virus is considered, the Zond Province, Malaya subregion of the Indo-Malaysian zoogeographical region, is also included.

B. Incidence

A total of 21 cases of IK disease have been reported, of which 17 were virologically or serologically diagnosed. The single outbreak occurred in 1983 in one of a number of small villages in southwestern Tajik S.S.R.

C. Seasonal Pattern

The first of 17 laboratory-verified cases of the disease occurred at the end of March, coinciding with the arrival of bats from their hibernation places. Two cases were reported in April and one in May. Mosquito activity increases dramatically during the second half of May. This may have caused the increase in case incidence to four in June, four in July, and five in August.

D. Risk Factors

Human cases were reported only in adults. Disease was associated with presence of bats in houses, mainly *Vespertilio pipistrellus* and *V. murinus*. Several possible modes of infection were proposed, including transmission by the respiratory route in buildings inhabited by bats, the alimentary route (eating food contaminated by feces and urine of these animals), and transmission by bites of ticks or mosquitoes.[6]

E. Serologic Epidemiology

The results of serologic surveys of human population groups by the CF test are shown in Figure 1. In the Kirghiz S.S.R., from 0% (most areas) to 1.5% (in the lowland of the Tchu River basin) of 4800 persons examined had detectable antibodies.[9] In the Tajik S.S.R., 2.5% of 2200 persons were seropositive; the prevalence reached 5.7%[11] and 7.8%[16] in the southern parts of the region. In the southwestern part of Turkmen S.S.R., 9% of the population was immune, the prevalence being five times higher in autumn (November to December) than in spring (April to May),[17,18] indicating a high infection rate of the population in summer. Thus, the antibody prevalence in the eastern part of Middle Asia increases gradually from the Kirghiz S.S.R. to the south of Tajik S.S.R. and the Turkmen S.S.R.

V. TRANSMISSION CYCLES

A. Data from Field Studies
1. Vectors

In total, at least 20 virus strains of IK virus have been isolated from blood-sucking arthropods. IK virus has been repeatedly isolated from ticks collected from bats in the Kirghiz S.S.R. and the Tajik S.S.R. (Table 1, Figure 1). The infection rate in *A. vespertilionis* ticks, which dominates in Middle Asia, reaches 32%.[19] The virus has also been isolated

Table 1
ISOLATION OF ISSYK-KUL VIRUS FROM ARTHROPODS

Species	Place of collection	Month/year of collection	No. strains	Ref.
Ticks collected from bats				
Argas vespertilionis	Kirghiz S.S.R. (Sokuluk region)	5/1970	1	1
	(Sokuluk region)	8/1971	1	1
	(Osh town)	6/1972	1	1
	(Frunzhe town)	8/1971	1	9
	(Lenin region)	5/1973	6	9, 20
	(Kalinin region)	5/1972	2	9, 19
		4/1974	1	19
	Tajik S.S.R.	7/1982	1	10
Ixodes vespertilionis	Kirghiz S.S.R.	4/1972	1	1
Mosquitoes				
Aedes caspius	Kirghiz S.S.R. (Moscow region)	4/1973	1	9
Anopheles hyrcanus	(Moscow region)	8/1974	1	9
Horseflies				
Tabanus agrestis	Kazakh S.S.R. (Aktubinsk region, bank of Irghiz River)	8/1981	1	15

from *Ix. vespertilionis*, also an obligatory parasite of bats. *A. vespertilionis* may the principal vector of the virus in bat colonies of Middle Asia. These ticks have been noted inside human dwellings and have been found to engorge on humans.[20] This mode of transmission may contribute to human infection.

Isolation of the virus from *Aedes caspius* and *Anopheles hyrcanus* (two predominant species in Middle Asia) indicate that mosquitoes may also be involved in circulation of the virus. Host-preference studies by blood-meal precipitin testing revealed that *Ae. caspius*, *Culex pipiens*, and *A. hyrcanus* feed on bats.[21,22] These mosquito species (especially *Cx. pipiens*) also readily attack both humans and birds.

2. Vertebrate Hosts

Bats are believed to be the principal vertebrate host of IK virus (Table 2). The agent has been isolated from seven bat species; transmission between bats and their obligate tick parasites provides stable maintenance of the virus. An average of 2.4% of bats studied have been found infected.

Involvement of avian species in virus circulation is probably due to the transmission of the virus by mosquitoes. The virus has been isolated from five species of passeriform birds, including *Hurundo rustica* and *Motacilla alba*, and also from Coraciiformes and Piciformes (one species in each order).[8,9,14,23] The involvement of other species in virus cycles cannot be excluded. The role of birds is seasonally controlled and coincides with mosquito activity.

In the Kirghiz S.S.R., 0.3% of wild birds had CF antibodies and in Tajik S.S.R., 2%.[9,13,14] Antibodies have been found in passeriform, anseriform, charadriiform, and columbiform species.

Antibodies have also been detected in the southern part of the Tajik S.S.R. in 3.8% of *Mus musculus* and 1.9% of *Meriones erythrourus* rodents.[11] In the former species, rodents may have been infected due to contacts with bats, in the latter, they are due to arthropod transmission.

Table 2
ISOLATIONS OF ISSYK-KUL VIRUS FROM VERTEBRATES

Species	Place of collection	Month/year of collection	No. strains	Ref.
Bats				
Nyctalus noctula	Kirghiz S.S.R. (Dzhety-Oghuz region)	5/1970	1	1
Myotis blythi	(Sokuluk region)	9/1971	1	1
	(Sokuluk region)	5/1973	1	
Eptesicum serotinus	(Sokuluk region)	6/1971	2	1,26
	(Lenin region)	5/1973	1	1,35
Vespertilio pipistrellus	Tajik S.S.R. (Kumsanchir region)	7/1974	2	23,27
	(Kumsanchir region)	6—8/1982	3	10
V. murinus	(Kumsanchir region)	7/1982	1	10
Rhinolopus ferrumequnum	(Komsomolabad region)	7/1982	1	10
Birds				
Passer hispaniolensis	Kirghiz S.S.R. Moscow region)	4/1973	1	8,9
Motacilla alba	(Moscow region)	4/1973	1	8,14
M. cinerea	Tajik S.S.R. (Komsomolabad region)	4/1976	1	10,13
Phoenicurus phoenicurus	(Komsomolabad region)	4/1976	1	10,13
Hirundo rustica	Kirghiz S.S.R. (Moscow region)	4/1973	1	8,14
Jynx torquilla	Tajik S.S.R. (Dzhirghital region)	4/1976	1	10,13
Alcedo atthis	Kirghiz S.S.R. (Moscow region)	4/1973	1	8,14

B. Experimental Data

1. Vectors

No experimental studies using ticks have been reported. *Ae. caspius* mosquitoes collected in the Tajik S.S.R. have been studied experimentally.[24] Mosquitoes were infected by feeding on suckling mice (titer of the virus in blood, 5.5 to 6.3 $\log_{10}$ $LD_{50}/0.02$ mℓ). The rate of infection reached 72%. The virus was detected in mosquitoes for up to 20 days (duration of the experiment); suspensions of infected mosquitoes contained between 2.8 and 4.6 $\log_{10}$ $LD_{50}/0.02$ mℓ. The virus has been biologically transmitted to suckling mice on the 7th day after infection; transmission rates were 50% on day 9, 80% on day 13, and 78% on day 20. These data support the results of field studies implicating *Ae. caspius* in transmission.

2. Vertebrate Hosts

After experimental subcutaneous inoculation of *Vespertilio pipistrellus* bats with a large virus dose, they were incubated at 24°C. The virus was isolated on the 3rd day from blood, liver, spleen, and brown fat; on days 5, 8, and 12, it was recovered from brown fat only.[9] Under conditions of artificial hibernation, the virus persisted in bat tissues for longer periods up to day 46.[4,25]

Experimental infection of chickens, doves, and myna birds have been performed. One-day-old chickens infected subcutaneously and intracerebrally, had viremia (up to 2.4 $\log_{10}$ $LD_{50}/0.02$ mℓ) after the first day. Chickens showed clinical symptoms of encephalitis, and virus was isolated on days 7 and 15 after inoculation from the brains of dead birds.

No clinical symptoms have been described after subcutaneous infection of *Columba livia*. Nevertheless, the virus was isolated from the brain on day 25 after infection.[9,14] After infection of *Streptopelia senegalensis*, the virus was occasionally isolated on day 8 from blood or from viscera on day 12. Viremia titers in *Acridotheres tristis* were also low (titers 1.0 to 1.5 $\log_{10}$ $LD_{50}/0.02$ mℓ).[10,13] The titers of virus detected in blood of the birds studied are not sufficient for the infection of mosquitoes. The species studied experimentally do not play a significant role in natural transmission cycles; however, these studies do not exclude a role for other birds.

VI. PROPHYLAXIS AND CONTROL

At present, preventive measures include elimination of bats and their ticks from houses and food stores. Residents of the endemic area should also avoid visits to caves and other natural resting places for bats. Personal measures against mosquito bite should be used in the endemic regions. No vaccine exists, and vaccine application is unlikely to be a practical public health measure.

VII. SYNTHESIS AND FUTURE INVESTIGATIONS

Argas vespertilionis and its bat hosts are the main reservoirs of IK virus in nature. An agent (Keterah virus), antigenically closely related to IK virus, has been isolated from *A. pusillus* in Malaysia; this tick species is close to *A. vespertilionis* in its biology and ecology. Other representatives of the complex subgenus *Carios*, both species closely related to *A. vespertilionis* (*A. pusillus* in Southeast Asia and *A. dewae* in southwestern Australia) and those less closely related (*A. australiensis* in northeastern Australia and *A. daviesi* in southeastern Australia), may be involved in transmission of IK virus (Figure 2). If so, the area of virus activity may cover vast areas of southern Europe, Africa, southern Asia, and Australia. To verify this hypothesis, collaborative studies by arbovirologists in these areas are encouraged.

Data from field studies on bats and their ticks suggest a stability of natural foci of IK virus infection. Therefore, data on oral infection and transstadial and transovarian transmission of the virus by *A. vespertilionis* would be of great value. Field and experimental studies are needed on mechanisms of virus overwintering in ticks and bats. The possibility of dissemination of the virus by migrating bats and birds also should be investigated.

FIGURE 2.   Area of the main reservoir of the IK virus — ticks of the *Argas (Carios) vespertilionis* complex.[12,28-38] Zoogeographical regions: (I) Holarctic region — (1) Arctic subregion, (2) Circumboreal subregion, (3) Mediterranean subregion, (4) Central Asian subregion, (5) China-Himalayan subregion. (II) Indo-Malaysian region — (6) Indian-Indochinese subregion, (7) Malaysian subregion. (III) Australian region — (8) Australian subregion. (IV) Ethiopean region — (9) West African subregion, (10) East African subregion, (11) Madagascan subregion. (Dots) Areas of ticks, *Argas vespertilionis* group; (pinwheel) regions of IK virus isolation.

REFERENCES

1. **Lvov, D. K., Karas, F. R., Timofeev, E. M., Tsyrkin, Yu. M., Vargina, S. G., Veselovskaya, O. V., Osipova, N. Z., Grebenyuk, Yu. I., Gromashevski, V. L., Steblyanko, O. N., and Fomina, K. B.,** Issyk-Kul virus, a new arbovirus isolated from bats and *Argas (Carios) vespertilionis* (Latr. 1802) in the Kirghiz S.S.R., *Arch. Ges. Virusforsch.*, 42, 207, 1973.
2. **Berge, T. O., Ed.,** *International Catalogue of Arboviruses*, Publ. No. (CDC) 75-8304, U.S. Department of Health, Education and Welfare, Washington, D.C., 1975.
3. **Lvov, D. K., Kostyukov, M. A., Pak, T. P., and Gromashevski, V. L.,** Isolation of an arbovirus antigenically related to Issyk-Kul virus from the blood of a sick person, *Vopr. Virusol.*, 1, 61, 1980.
4. **Pak, T. P., Kostyukov, M. A., Danijarov, O. A., Kuima, A. U., Bulychev, V. P., and Gordeeva, Z. E.,** Arbovirus infections in the Tajikistan, in *Arboviruses*, Ivanovsky Institute of Virology, Moscow, 1978, 35.
5. **Pak, T. P.,** Clinico-epidemiological characteristics of arboviral infections in Tadjikistan, in *Arboviruses*, Ivanovsky Institute of Virology, Moscow, 1981, 101.
6. **Lvov, D. K., Kostyukov, M. A., Danijarov, O. A., Tukhtaev, T. M., Sherikov, B. K., Bunietbekov, A. A., Bulychev, V. P., and Gordeeva, Z. E.,** An outbreak of arbovirus infection in the Tajik S.S.R. caused by Issyk-Kul virus (Issyk-Kul fever), *Vopr. Virusol.*, 1, 89, 1984.

7. **Lvov, D. K. and Iljichev, V. D.**, *Migrations of Birds and Transmission of Infectious Agents*, Nauka, Moscow, 1979.

8. **Lvov, D. K., Kostjukov, M. A., Berezina, L. K., Gordeeva, Z. E., Gushchina, E. A., and Danijarov, O. A.**, Issyk-Kul fever — a new Bunyavirus infection in Middle Asia, in *Abstr. 6th Int. Conf. Virology*, Sendai, Japan, 1984, 37.

9. **Karas, F. R., Vargina, S. G., Osipova, M. Z., Grebenyuk, Yu. I., Steblyanko, S. N., Usmanov, R. K., Tsyrkin, Yu. M., Timofeev, E. M., Gromashevsky, V. L., and Lvov, D. K.**, Study of foci of arbovirus infections in the territory of the Kirghiz S.S.R., in *Ecology of Viruses*, Ivanovsky Institute of Virology, Moscow, 1973, 74.

10. **Gordeeva, Z. E., Lapina, T. F., and Kostyukov, M. A.**, Comparative study of biological and antigenic properties of Issyk-Kul virus (strain 620 Taj) and strain 218 Taj isolated from birds in the Tajikistan, in *Arboviruses*, Ivanovsky Institute of Virology, Tallinn, 1984, 69.

11. **Gordeeva, Z. E.**, Population immunity of man, domestic and wild mammals and birds to Issyk-Kul, West Nile and Sindbis viruses in Tadjikistan, in *Arboviruses*, Ivanovsky Institute of Virology, Tallinn, 1982, 132.

12. **Kaiser, M. N. and Hoogstraal, H.**, Bat ticks of the genus *Argas* (Ixodoidea:Argasidae). X. *A. (Carios) dewae*, new species, from southeastern Australia and Tasmania, *Ann. Entomol. Soc. Am.*, 67, 213, 1974.

13. **Gordeeva, Z. E.**, Ecological relationships of arboviruses with wild birds in Tajikistan, in *Ecology of Viruses*, Ivanovsky Institute of Virology, Moscow, 1980, 126.

14. **Steblyanko, S. P.**, Role of birds in dissemination of arboviruses, in Proc. Conf. Reg. Pathol. Hyg., Frunze, 1977, 72.

15. **Drobishchenko, N. I., Kiryushchenko, T. V., and Karimov, S. K.**, Isolation of Issyk-Kul virus in the Kazakhstan, in Conf. Natural Foci, Proc. Infections (Tyhmen, 1984), Moscow, 1984, 57.

16. **Kostyukov, M. A., Gordeeva, Z. E., Rafiev, H. K., Danijarov, O. A., Lvov, D. K., and Pak, T. P.**, Immunological structure of a population to Issyk-Kul and Wad Mendani viruses in the Tadjik S.S.R., in *Ecology of Viruses*, Ivanovsky Institute of Virology, Moscow, 1976, 108.

17. **Andreev, V. P., Lvov, D. K., and Sokolova, N. N.**, Results of serological investigation on arboviruses and influenza viruses of wild and domestic animals in the West Turkmenia (1972—1974), in Proc. 10th Symp. Ecology of Viruses, Min. Public Health Association, S.S.R., Baku, 1976, 75.

18. **Kurbanov, M. M., Berezina, L. K., Zakaryan, B. A., Kiseleva, N. S., and Vatolin, V. P.**, Results of serological examination of blood sera of humans and domestic animals with 13 arboviruses in the zone of Karakum canal and southeastern part of T.S.S.R., in *Ecology of Viruses*, Ivanovsky Institute of Virology, Moscow, 1974, 120.

19. **Vargina, S. T., Kuchuk, L. A., Gershtein, V. I., and Karas, F. R.**, The transmission of Issyk-Kul virus by ticks *Argas vespertilionis* in experiment, in *Ecology of Viruses*, Ivanovsky Institute of Virology, Moscow, 1982, 123.

20. **Galuso, I. G.**, Argasid ticks (Argasidae) and their epizootiological significance, Academy of Sciences Kazach S.S.R., Alma-Ata, 1957.

21. **Bulychev, V. P., Alekseev, A. N., Kostyukov, M. A., Gordeeva, Z. E., Danijarov, O. A., and Pak, T. P.**, Hosts of mass mosquito (Diptera, Culicidae) species in Tajikistan, *Med. Parasitol. Parasitol. Dis.*, 6, 65, 1978.

22. **Bulychev, V. P., Alekseev, A. N., Kostyukov, M. A., Gordeeva, Z. E., and Lvov, D. K.**, Transmission on Issyk-Kul virus by *Aedes caspius* Pall. mosquitoes by bite experimentally, *Med. Parasitol. Parasitol Dis.*, 6, 53, 1979.

23. **Kostyukov, M. A.**, A study of the ecology of Issyk-Kul virus, in *Arboviruses*, Ivanovsky Institute of Virology, Moscow, 1981, 78.

24. **Kostyukov, M. A.**, Biocenotic relationships of Tahyna and Issyk-Kul viruses with bloodsucking mosquitoes *Aedes caspius caspius*, *Ecology of Viruses in Kazakhstan and Middle Asia*, Alma-Ata, 1980, 135.

25. **Pak, T. P.**, Problems of maintenance mechanism of arboviruses in interepidemic period, in *Ecology of Viruses*, Ivanovsky Institute of Virology, Moscow, 1980, 118.

26. **Vargina, S. G., Kuchuk, L. A., and Karas, F. R.**, Chiroptera as an ecological niche of arboviruses, in *Viruses and Virus Infections of Man*, Ivanovsky Institute of Virology, Moscow, 1981, 94.

27. **Kostyukov, M. A., Bulychev, V. P., Danijarov, O. A., and Pak, T. P.**, Of 2 strains of Issyk-Kul virus from bats *Vespertilio pipistrellus* Schreber in the Tadjikistan, in Proc. 9th Symp. Ecology of Viruses, Dushanbe, 1975, 31.

28. **Hoogstraal, H. and Kohls, G. M.**, Bat ticks of the genus *Argas* (Ixodoidae, Argasidae). V. Description of larvae from Australian and New Guinea Carios-group populations, Proc. Linn. Soc. N.S.W., 1962, 87, 275.

29. **Hoogstraal, H., Santana, F. J., and Dirkvan Peenen, P. F.**, Ticks (Ixododea) of Mt. Sontra, Danang, Republic of Vietnam, *Ann. Entomol. Soc. Am.*, 61, 722, 1968.

30. **Kaiser, M. N., and Hoogstraal, H.**, Bat ticks of the genus *Argas* (Ixodoidea, Argasidae). IX. *A. (Carios) daviesi*, new species, from western Australia, *Ann. Entomol. Soc. Am.*, 66, 423, 1973.

31. **Kohls, G. M.,** Ticks (Ixodoidea) of the Philippines, *Natl. Inst. Health Bull.,* 192, 28, 1950.
32. **Kohls, G. M. and Hoogstraal, H.,** Bat ticks of the genus *Argas* (Oxodoidea, Argasidae). IV. *A. (Carios) australiensis* n. sp. from Australia, *Ann. Entomol. Soc. Am.,* 55, 555, 1962.
33. **Philippova, N. A.,** *Argas* ticks (Argasidae), in *Phauna U.S.S.R.,* Vol. 4, *Arachnoidae,* Nauka, Moscow-Leningrad, 1966.
34. **Roberts, F. H. S.,** Australian ticks, *Commonw. Sci. Ind. Res. Organ.,* 11, 267, 1970.
35. **Vargina, S. G., Karas, F. R., Kuchuk, L. A., and Shukurov, E. D.,** Study of Chiroptera in natural foci of arboviruses in the Chuisk valley of the Kirghizia, in *Ecology of Viruses of Kazakhstan and Middle Asia,* Alma-Ata, 1980, 78.
36. **Yamaguti, N., Tipton, V. J., Kugan, H. L., and Toshioka, S.,** Ticks of Japan, Korea, and the Ryukyu Islands, *Brigham Young Univ. Sci. Bull. Biol. Ser.,* 15, 1, 1971.
37. **Hoogstraal, H.,** Bat ticks of the genus *Argas* (Ixodoidea, Argasidae). III. The subgenus *Carios,* a redescription of *A. (Carios) vespertilionis* (Latreille, 1802) and variation within an Egyptian population, *Ann. Entomol. Soc. Am.,* 51, 159, 1958.
38. **Hoogstraal, H.,** African Ixodoidea. I. Ticks of the Sudan, U.S. Navy, Washington, D.C., 1965.
39. **Main, A.,** personal communication.

Chapter 28

JAPANESE ENCEPHALITIS

Donald S. Burke and Colin J. Leake

TABLE OF CONTENTS

I. HISTORICAL BACKGROUND

A. Discovery of Agent and Vector(s)

Japanese encephalitis (JE) virus is a mosquito-borne flavivirus which was first isolated in Japan in 1935 from the brain tissue of a fatal encephalitis case.[1] The seasonal occurrence of the disease in Japan suggested a vector relationship, and in 1938 the virus was isolated from *Culex tritaeniorhynchus* mosquitoes.[2] Subsequently, this mosquito has been shown to be a principal vector over much of the geographic distribution of JE.

B. History of Epidemics

Although minor epidemics of "summer encephalitis" were recorded in Japan as far back as 1870, the disease attracted little attention until the great epidemic of 1924 resulted in 6125 cases and 3797 deaths.[3] Based on clinical and epidemiologic features, the disease was termed "Type B" epidemic encephalitis to differentiate it from epidemic encephalitis lethargica, "Type A".[4] The qualifying suffix "B" of Japanese B encephalitis has subsequently been dropped. Severe epidemics of JE occurred in Japan in 1935 and 1948. Epidemics have not been recorded in Japan since 1968, although sporadic cases (less than 100/year) continue to occur.[5-9]

JE was probably recognized in Korea as early as 1926, but the first major epidemic was not documented until 1949, when 5548 cases were recorded.[10] For the next 2 decades the incidence of disease fluctuated, then dropped dramatically between 1965 and 1970, from an average of 2000 to less than 200 cases per year. However, JE epidemic activity resurged in 1982,[8,11] when nearly 3000 cases were reported from the southwest provinces of South Korea. JE activity has also been detected in the maritime provinces of the U.S.S.R.[12]

JE was first recognized in China in 1940;[13] in 1966, disease incidence peaked sharply with over 40,000 cases reported throughout the country.[14] In recent years, more than 10,000 cases are reported annually. The disease is widespread; all provinces except Sinkiang, Chinghai, and Tibet are involved.[14-16]

Farther south, in tropical Southeast and south Asia, JE is endemic. Serologic evidence, coupled with sporadic isolates from mosquito pools, indicates substantial virus activity in Malaysia, the Philippines, Indonesia, Sri Lanka, and southern Thailand.[16-25] However, cases of clinical encephalitis are relatively infrequent in these regions, and epidemics have not been recorded.

In contrast to the endemic pattern of southern Thailand, a predictable pattern of recurrent annual epidemics has been observed since the late 1960s in northern Thailand.[26-28] Attack rates in severely affected provinces along the Chiang Mai Valley may reach 2 to 3/1000 children per year. In recent years 2000 to 5000 cases were reported annually.[8] Northern Vietnam is similarly affected.[29,30] Minor epidemic activity has been recorded in the Shan states of northern Burma and in Bangladesh.[31-33] Little information is available on JE in Kampuchea and Laos; it is nonetheless likely that clinical cases of JE do occur in these countries.

JE was recognized in India in 1954; sporadic clinical cases were confined to the southern regions of the country until 1978, when severe epidemics erupted in the northeastern states of Bihar, Uttar Pradesh, and West Bengal.[34-36] Over 6000 cases were reported in these three states. Simultaneously, epidemic JE first appeared in the Terrai region of Nepal.[37,38] Minor summer epidemics continue to recur in these regions.[39]

Autochthonous transmission of JE has not been detected in Irian Jaya, Australia, or Pakistan.

A valuable chronological reference list was published in 1974 which itemized over 1000 publications on JE epidemiology and ecology.[40] Statistics on the occurrence of JE are regularly reviewed by the World Health Organization.[6,8]

C. Social and Economic Impact

Clinical cases of JE in humans have been documented in at least 16 countries in eastern, southern, and Southeast Asia with a combined population in excess of 2 billion persons. Despite a lack of exact figures, the authors estimate (see Table 1) that approximately 50,000 cases occur annually in this region; of these, approximately 25% are rapidly fatal and 50% may result in permanent neurologic or psychiatric sequelae. Although these aggregate attack rate figures for JE are seemingly low (2/100,000/year), epidemics are focal and intense. The impact of JE is best appreciated by observing the level of public concern during the midst of an epidemic: 1 of every 300 children is stricken within a 1-month period; hospital wards are promptly filled with comatose patients; alarming headlines appear in the local news; and public health officials are pressed for action.

Cases among tourists and expatriates have generated political and economic consequences for affected countries.[41,42] Epidemic JE has been a problem among foreign military in Asia.[30,43,44]

Although the impact of JE is primarily measured in terms of human illness, JE can also cause agricultural losses. Fatal encephalitis occurs in domesticated equines.[45,46] Morbidity rates in horses may be as high as 2% during an epidemic.[14] JE may cause abortion and fetal wastage by infected sows.[47]

Table 1
ESTIMATED NUMBER OF JE CASES AND
FATALITIES AMONG HUMANS IN ASIA

Countries	Population	Incidence/ 100,000/ year[a]	Cases/ year[b]	Deaths[c]
Southeast Asia				
Burma	38,900,000	1.0	400	100
Cambodia	6,100,000	1.0	60	15
Indonesia	161,600,000	0.1	160	40
Laos	3,700,000	5.0	200	50
Malaysia	15,300,000	0.1	15	4
Philippines	54,500,000	1.0	500	120
Thailand	51,700,000	5.0	2,500	620
Vietnam	58,300,000	5.0	3,000	750
East Asia				
China	1,034,500,000	3.0	30,000	7,500
Japan	110,900,000	0.2	220	60
N. Korea	19,600,000	5.0	1,000	250
S. Korea	42,000,000	5.0	2,100	520
Taiwan	19,200,000	1.0	200	50
Mid-South Asia				
India	746,400,000	0.5	3,700	900
Nepal	16,600,000	3.0	500	120
Sri Lanka	16,100,000	3.0	480	120
Totals		2.0	45,000	11,000

[a] Approximate average over last decade; estimated where exact figures not available.
[b] Calculated as (population) × (incidence rate); rounded off to two significant places (10s, 100s, etc.).
[c] Calculated as 25% of cases per year; rounded as above.

II. THE VIRUS

A. Antigenic Relationships

Casals and Brown[48] grouped JE with yellow fever, dengue, and other viruses on the basis of serologic cross-reactions detected by hemagglutination inhibition (HI). Originally termed the "Group B" arboviruses, these viruses are now classified in the family Flaviviradae, all of which share the following features: virion size approximately 50 nm; a lipid envelope; three structural proteins (C, M, and E); a single-stranded RNA, positive (message) sense genome of 10 to 12 kb in length; and antigenic cross-reactivity.[49]

Group-reactive epitopes are present on both the E and M proteins of most, if not all, flaviviruses.[50,51] Preliminary genome sequence data show considerable predicted amino acid homology between yellow fever, dengue, and JE, so it is likely that flavivirus group-reactive epitopes may be identified on all the JE virus-coded proteins.[52-54] The specific molecular configurations that confer encephalitogenic properties to JE and other related flaviviruses are currently unknown.

Flavivirus serotypes are classically defined on the basis of virus neutralization (N) tests. Hyperimmune JE sera show some limited cross neutralization of other encephalitogenic flaviviruses: St. Louis encephalitis (SLE), West Nile (WN), and Murray Valley encephalitis (MVE). These viruses can collectively be referred to as belonging to the encephalitis virus complex.[55,56] Since neutralization of flaviviruses is associated with antibodies to the envelope

(E) protein, it is probable that the viruses of the encephalitis virus complex share one or more neutralizing epitopes, as well as group (non-neutralizing) epitopes, on their E proteins.[57,58] JE may also share cross-reactivity with a nuclear antigen in mammalian cells.[59]

B. Host Range

Most strains of JE grow quite well in a wide variety of vertebrate (human, monkey, pig, hamster, mouse, cat, chicken) cells.[60] Compared to most other flaviviruses, growth of JE can be both quite rapid (1 to 2 days) and lead to high yields ($\geq 10^8$ plaque forming units [PFU] per milliliter). Experimental inoculation studies indicate that the in vivo host range is equally broad. Although cytolytic CPE is typical of in vitro growth in cell culture, JE rarely produces illness in animals unless inoculated directly into the central nervous system. Fatal JE is regularly produced by intracerebral (i.c.) inoculation in vertebrate species (monkeys, horses, pigs, hamsters, mice).

C. Strain Variation

Strain variation among isolates of JE has most clearly been demonstrated by cross-neutralization tests, by reactivity with mouse monoclonal antibodies, and by two-dimensional electrophoresis of T1 RNase digests of virion RNA.

Although reproducible variations in titer have been observed using human or animal immune serum in plaque-reduction N tests with various strains of JE, the degree of variations is minor.[61,62] Somewhat greater differences have been detected using mouse hyperimmune ascitic fluids in mouse passive protection tests,[63] but the significance of these variations with regard to epidemiology or vaccine development is unknown. In the opinion of the authors, these variations are probably not of practical importance; immunization of children with a vaccine derived from the prototype Nakayama strain (isolated in 1935 in Japan) provided full protection against JE in Thailand in 1985.[64]

Examination of a large number of JE isolates from Japan with mouse monoclonal antibodies reveals substantial strain variation.[65,66] Strains could be roughly grouped according to the decade of isolation.[66] This analysis suggests that new strains evolved in or were introduced into Japan. However, there is as yet no evidence that the mouse monoclonal antibody patterns define groups of strains that differ in any fundamental properties of epidemiological importance.

JE strains from widely separated geographic locations (Japan, China, northern Thailand, southern peninsular Thailand) differ substantially in their RNA fingerprint patterns.[67] Again, the significance of these variations, other than as phenotypic strain markers, is uncertain. Strain variation according to the host from which the isolate is obtained is less well studied. In a study of JE strains isolated in one northern Thai village in 1982—1983, RNA fingerprint patterns of strains from humans, pigs, and mosquitoes were compared.[67] Eight of nine strains obtained from human post-mortem brains were virtually identical, while fingerprints of isolates from pigs in the same province differed from the human brain isolates and from each other by two to three spots. Isolates from mosquitoes showed even greater variation (four to six spots). These observations suggest that some JE strains, as differentiated by RNA fingerprint analysis, are more virulent than others. Huang[68] has suggested that naturally attenuated strains may evolve by continuous passage of JE through mosquitoes by transovarial transmission.

Strain variations in other antigenic and chemico-physical properties of JE proteins have also been reported.[69-72] As reviewed by Huang,[68] investigations in China have revealed JE strain variations in peripheral pathogenicity in mice, in stability to heat, and in hemagglutinin and hemolytic activities. The virulence of strains, as measured by subcutaneous (s.c.) challenge in mice, could be substantially altered by serially passing strains in mice by i.c. or s.c. inoculation.

D. Methods for Assay

JE grows rapidly in many laboratory animals. Intracerebral inoculation of 2- to 3-day-old suckling mice is an extremely sensitive assay method for JE; titers are conventionally expressed as suckling mouse i.c. lethal doses (SMICLD$_{50}$). Many conventional vertebrate cell lines such as VERO, PS, and LLC-MK2 also support JE growth. Mosquito cell lines, particularly the *Aedes albopictus* and *Ae. pseudoscutellaris* cell lines, are probably the most sensitive cell lines available.[73] Intrathoracic inoculation of live mosquitoes combined with fluorescent-labeled antibody (FA) detection of JE antigens is a very sensitive assay system,[74] and localization of JE antigens within cells of sectioned mosquitoes has been achieved using FA and immunoperoxidase.[75,76]

III. DISEASE ASSOCIATIONS

A. Humans

Although JE can produce a mild febrile illness, asceptic meningitis, or (rarely) visceral inflammatory changes, by far the most important pathological manifestation of infection in humans is acute meningomyeloencephalitis.[77] After an asymptomatic incubation period of 1 to 2 weeks, patients typically present themselves for medical care with 1 to 3 days of fever, headache, stupor, and especially in children, generalized motor seizures.[78] Physical findings are usually limited to a depressed state of consciousness, but evidence of focal motor impairment of the cranial nerves or limbs is not infrequent. Examination of the cerebrospinal fluid (CSF) shows a normal or moderately increased pressure, normal or slightly increased total protein, and a lymphocyte pleocytosis of 10 to 1000 mononuclear cells per cubic millimeter.[79,80] Stupor progresses rapidly to coma, which in nonfatal cases resolves in 1 to 2 weeks. Examination of brains in fatal cases shows that JE is a pan-encephalitis, with infected neurons scattered throughout the entire central nervous system (CNS).[77,81,82] Necrotic foci, when present, are invariably of microscopic proportions. The thalamus is typically heavily involved. Approximately one fourth of cases are rapidly fatal, half lead to neuropsychiatric sequelae (including persistently abnormal EEGs) and one fourth fully resolve.[83,84] The human fetus can probably be infected *in utero*, resulting in abortion.[85] Treatment with hyperimmune antibodies or interferon inducers has been reported for animal models;[86-88] however, to date no therapy has been proven efficacious in man.

B. Domestic Animals

Although essentially all domestic animals can be infected by JE, adult animals rarely develop signs of illness. Fatal encephalitis (originally designated "Japanese equine encephalitis") due to JE has been recognized in horses.[45,46] Abortion and fetal wastage may occur among infected sows.[47]

C. Wildlife

JE is not known to have a significant disease impact on wildlife.

D. Applicable Diagnostic Procedures

JE virus can rarely, if ever, be isolated from the peripheral blood during the acute illness in humans.[80] Presumably, the viremic phase terminates before the appearance of CNS signs (see Table 2). Virus can be isolated from the CSF early in the course of acute encephalitis; successful isolation is an ominous prognostic sign. In the authors' experience, cultivable virus in CSF is associated with 100% mortality. If an adequate sample of fresh brain tissue is obtained at autopsy, virus can be isolated from virtually every case,[89] and intraneuronal viral antigens are readily detected.[82]

A presumptive diagnosis of JE can be made in endemic regions by demonstration of a

Table 2
RECOMMENDED METHODS FOR
ESTABLISHING A LABORATORY DIAGNOSIS OF
JE

Patient samples
1A Detection of virus-specific IgM in CSF
1B Detection of virus-specific IgM in serum
2 Detection of a fourfold antibody titer rise in serum
3 Isolation of virus from CSF

Post-mortem samples
4 Isolation of virus from post-mortem brain
5 Detection of intraneuronal JE antigen by immunohistochemical
staining of post-mortem brain

fourfold or greater antibody titer rise to JE by any of a number of conventional techniques, such as HI, immunofluorescent antibody (IFA), complement fixation (CF), or IgG ELISA.[60] However, this approach must be used with caution, because (1) dengue or other flaviviruses can produce a cross-reactive serologic response to JE antigens, and (2) many JE-infected patients (perhaps 25%) fail to show a fourfold rise in titer because the antibody response has already peaked by the time the patient presents himself to medical care.[90,91]

An alternative approach to the serologic diagnosis of acute JE is to demonstrate JE specific IgM in serum or CSF.[92,93] JE IgM can be detected in the CSF only during acute encephalitis, and not during an asymptomatic infection.[90] For this reason, detection of JE IgM is the preferred method for routine clinical diagnosis. Over 75% of patients have CSF JE IgM detectable at the time of hospital admission; the remainder develop CSF JE IgM within 3 days.

The same techniques used for humans are also generally applicable to diagnosis of JE in animals. A JE IgM ELISA has been developed for diagnosis of acute JE infection in swine.[94]

E. Adverse Effects of Virus on Vector

No data have been obtained to suggest that JE has any adverse effects on mosquito vectors. However, studies with *Culex tritaeniorhynchus* have shown a marked growth of the virus within the nervous system,[75,76] and it seems likely that this may affect the physiology or behavior of the mosquito.

IV. EPIDEMIOLOGY

JE is essentially a rural problem, a zoonotic cycle between rice-field breeding mosquitoes and domestic pigs and/or water birds. Human infections are a consequence of increased vector densities associated with increased rainfall or with irrigation. JE has a broad geographic distribution; the fine details of its epidemiology vary in different areas.

A. Geographic Distribution

Countries with proven epidemics of JE, or that harbor the virus are India, Nepal, Sri Lanka, Bangladesh, Burma, Laos, Thailand, Kampuchea, Vietnam, Malaysia, Singapore, Philippines, Indonesia, China, maritime Siberia, Korea, and Japan (see Figure 1). Although low-level, year-round transmission probably occurs in many areas, sharply defined seasonal epidemics occur with regularity mainly in northern Southeast Asia, China, and Korea. Epidemic transmission occurs broadly but unpredictably within the area of 15 to 45° north latitude; the northern boundary of the geographic range extends to Japan and maritime Siberia. In the more equatorial areas transmission is endemic without marked epidemic peaks.

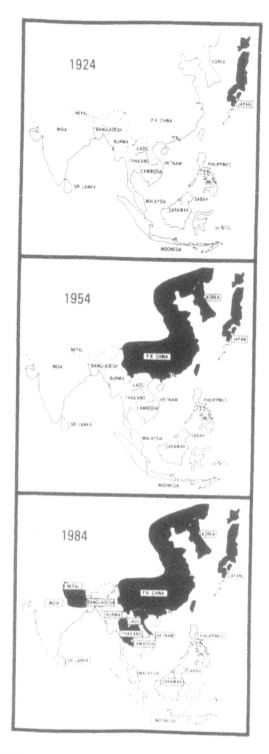

FIGURE 1.   Map of JE distribution in Asia. Regions where epidemics of JE are known to have occurred are shown as solid black. Regions where sporadic clinical cases or virus isolates are known to have occurred in the absence of epidemics of encephalitis are shown as cross-hatched. Distributions as known in 1924, 1954, and 1984 are shown.

In tropical Asia, the eastern limit of JE activity roughly corresponds to the ecological boundary defined by Wallace's Line; farther east, Murray Valley encephalitis virus is the prevalent encephalitogenic flavivirus.[95] Similarly, in India, JE is limited to the eastern and southern regions of the subcontinent; farther west one finds WN virus.[36] This interface in southern India is one of the few regions in the world where two major flaviviruses of the "encephalitis complex" (JE and WN viruses) have been shown to coexist.

B. Incidence

1. Incidence in Humans

The incidence of JE infection in humans varies from nil in disease-free regions to as high as 10 to 20% in a single season during the midst of a severe epidemic.[96] In the chronically involved Chiang Mai Valley in Thailand essentially all children can be shown to have antibodies to JE before reaching adulthood; the lifetime risk of JE infection is 100%.[64,97] It follows that the average incidence of infection in this region is at least 5 to 10%/year.

Limited data on the ratio of apparent to inapparent infections suggest that between 1:25 and 1:400 people bitten by a JE-infected mosquito will develop clinical encephalitis.[98-100] One study in India suggests a ratio as low as 1:20.[36]

2. Incidence in Vectors

The incidence of JE in vectors can be a difficult parameter to measure. Laboratory isolations reflect the number of mosquitoes infected with the virus, not necessarily the number able to transmit the virus. In addition, the method used for collecting mosquitoes may markedly influence results (e.g., animal vs. man-biting vs. light-trap collections). Incidence rates, although variable, have ranged from quite high in Japan (1 JE strain in 233 mosquitoes at Shinama and 0.56 to 2.64% infected mosquitoes in Kyoto) to very low in India (2 strains of JE from 884,235 mosquitoes, and 7 JE strains from 284,844 mosquitoes).[103-104] Variations at a single study site from year to year have been recorded (rates of 1:233 in Japan in 1952 but only 1:11,563 in 1953).[101] Incidence rates may also vary throughout the year. In Thailand, in a year-round study using light traps in the Chiang Mai Valley in 1970, the average JEV isolation rate from *Cx. tritaeniorhynchus* was 1:22,868, but virus strains were isolated only in April (1:8375), May (1:3522), and July (1:38,760).[105] In the same study, overall isolation rates from *Cx. fuscocephala* were 1:71,188 (isolated only in May, 1:10,290) and from *Cx. gelidus* were 1:3832 (isolated only in September, 1:550). On Okinawa, 42 JE isolations were made from 46,798 *Cx. tritaeniorhynchus* (rate 1:1114) but these were concentrated in August (5 isolates, rate 1:822), September (32 isolates, rate 1:909) and October (5 isolates, rate 1:1032), coinciding with the population density peak of *Cx. tritaeniorhynchus*.[106] In Sarawak, human bait catches yielded 38 JE isolates with overall infection rates of 1:4400 for *Cx. tritaeniorhynchus*, 1:7000 for *Mansonia uniformis*, 1:7500 for *M. bonneae/dives*, 1:4600 for mixed *Mansonia* spp., 1:11,000 for mixed *Anopheles* spp., 1:23,000 for *Cx. gelidus*, and 1:79,000 for mixed *Culex* spp.[107] JE isolations from *Cx. tritaeniorhynchus* were fairly evenly distributed year-round, with the peaks of isolates and highest infection rates in April and May, coinciding with harvesting of fields and decline and aging of the *Cx. tritaeniorhynchus* populations, and in November, when vector populations were at their peak. When studied in detail, the timing of appearance of JE in mosquitoes may be very narrow. In Japan in 1981, isolates were obtained only between July 27 and September 8, 1981, with a peak rate of 45% positive pools on August 11.[108] Similar studies in northern Thailand in 1982 detected JE isolates only over a 10-day period, with the last isolate obtained 1 week before the peak of human cases at the local hospital.[109]

In a summary of the early Japanese work, it was concluded that the sequence of events is as follows: (1) the virus can be detected in mosquitoes first during late June, reaching a peak of infected mosquitoes 1 month later, (2) infections in pig and bird amplifying hosts increase in frequency during late July and early August, and (3) only then does a peak of human infections occur.[110] This sequence has generally been confirmed in other studies, although occasionally virus is first detected in sentinel animals.

3. Incidence in Animals

High JE seroprevalence rates have been well documented among pig, horse, and bird populations in Japan and several other countries. Other vertebrate species, including cattle, sheep, dogs, and monkeys have also been found to have appreciable JE seroprevalence rates.[111-115]

In China, serological surveys of potential amplifying hosts show high levels of neutralizing antibodies in pigs, horses, donkeys, cattle, and dogs. As there is a high population turnover, pigs are considered to be the amplifying host of primary importance, compared to the longer-lived mammals. Of birds studied, only 22% of ducks and 3% of chickens were positive, with sparrows, pigeons, magpies, crows, warblers, house mice, and rats all being negative.[68] In Sarawak,[116] comparatively few pigs were found in villages, and with an infection rate of 20%/month or less, pig infection could not solely explain the continued maintenance of JE. Studies on birds indicated that 18% of the wild birds tested had neutralizing antibodies, with the larger rice-paddy species, such as water hens and bitterns, accounting for almost half of those with antibody. Ducks (19%) and dogs (84%) had neutralizing antibodies to JE, but fowl, geese, bats, and rodents appeared to be unimportant in the maintenance of JE in the area. In India, pigs and ardeid birds (cattle egret and pond heron) are considered to be the principal vertebrate hosts. Because bovines are attractive to vector mosquitoes, but develop little or no viremia, Carey et al.[34] have suggested that the presence of bovines may serve to modulate JE activity.

Johnson et al.,[117] in a comprehensive survey in Thailand, found significant JE antibody prevalences in pigs, dogs, cattle, water buffalo, chickens, and ducks. Less than 1% of 726 tree sparrows and none of 54 bats were clearly JE-seropositive. In this study, JE antibodies in ducks and chickens were attributed to cross-reactions due to infection with Tembusu virus, a related flavivirus.

C. Seasonal Distribution

In most regions, the timing of epidemics is difficult to predict with accuracy. Variations in the month of peak activity occur from year to year. In many areas, reliable epidemiological data are scanty. In Nepal and northern India, epidemics usually occur between September and December. In contrast, in the Indian state of Karnataka, early epidemics (between April and June) were observed in 1981 and 1983. In Thailand, epidemic peaks predictably occur between late June and early August or September. In China, Japan, and Korea, epidemic peaks tend to be in August.

D. Risk Factors

There are two independent elements of risk for development of acute JE: (1) the risk for infection and (2) the risk of encephalitis once infected. The risk of infection clearly relates primarily to geography and time, i.e., presence during an epidemic. Within an epidemic, urban dwellers are typically spared, presumably due to their distance from vector breeding sites (rice fields) and amplifying hosts (pigs). Little data exists on age- and sex-specific infection rates.

The risk of encephalitis once infected is probably age related. In Japan, most cases have occurred in the very young and the very old, possibly reflecting susceptibility to encephalitis in these age groups.[8] In northern Thailand, almost all cases occur in children under age 18; this is probably more a reflection of universal exposure to JE during childhood rather than an expression of age-specific susceptibility.[80] In other epidemics, such as in northern India, all ages are affected.[36] A broad age-specific attack curve probably reflects new introduction of JE into an immunologically naive human population.

The male to female ratio among clinical cases is almost always greater than unity.

The probability of virus traversing the blood brain barrier may be affected by local abnormalities within the CNS. A high frequency of underlying cerebral cysticercosis has been reported in some JE autopsy series.[118,119]

A history of previous exposure to other flaviviruses appears to be important in humans. Acute encephalitis is less likely in a dengue-immune subject than in those who are flavivirus naive, and an anamnestic antibody response (presumably a manifestation of dengue-conferred immunity to flavivirus-group antigens) is associated with a favorable outcome in those patients who do develop encephalitis.[80,84,120-122]

Another important determinant of the outcome of a JE infection may be the inherent human neurovirulence of the infecting strain. Data derived from virus RNA fingerprint analyses of JE isolates suggests a strong correlation of certain genome patterns and fatal encephalitis.[67]

E. Serologic Epidemiology

Once infected, antibody is thought to persist in the blood for life.[123-125] Although this appears to be true in patients with encephalitis, the possibility exists that some very mild, unapparent infections may give rise to transient or subdetection levels of antibodies.[124,126,127] Experiments in monkeys have demonstrated that an anamnestic seroresponse may be evoked by JE antigens delivered by a mosquito bite, even in the absence of viral replication.[128]

Nonetheless, cross-sectional serosurveys are useful tools to estimate seroprevalence rates, and hence the past JE exposure histories of defined populations. Such studies must be carefully designed to minimize confounding serologic cross-reactions due to dengue and other flaviviruses. In dengue hyperimmune populations, this is exceptionally difficult because sequential dengue infections may give rise to a broad, low-level flavivirus neutralizing antibodies along with the well-recognized, broad, anamnestic HI and CF group-reactive responses.[129]

Several large population-based studies have been done in human populations using the HI or IgG ELISA techniques. Although the specificity of these techniques is suspect, the data do indicate extensive exposure to JE in adults throughout the region of known distribution of the virus.[130,131] A few studies in which neutralizing serology was performed have tended to confirm the data based solely on HI.[95,96,132,133] Clearly, a new method, perhaps one based on synthetic JE-specific epitopes, will be necessary to accurately measure JE seroprevalence.

V. TRANSMISSION CYCLES

A. Evidence from Field Studies

1. Vectors

JE has been isolated from a large number of mosquito species in numerous field studies (Figure 2, Table 3). This is potentially quite confusing, as is well illustrated by results of studies conducted in Sarawak.[107] Although *Cx. tritaeniorhynchus* yielded by far the most JE isolates, several JE isolates were also obtained from *Cx. gelidus, Mansonia uniformis, M. bonneae/dives*, mixed *Mansonia* spp., *Aedes curtipes*, mixed *Anopheles* spp., and mixed *Culex* spp. Many unimportant species can be excluded either on epidemiological grounds or by laboratory studies on their vector competence. A few examples will serve to illustrate this point. First, in southern India all JE isolates have come from *Cx. tritaeniorhynchus* and other probable vectors including *Cx. pseudovishnui, Cx. vishnui*, and possibly *Cx. whitmorei*. However, the vectors of JE in northeastern India appear to be different. In West Bengal in 1973, JE was isolated from the *Anopheles hyrcanus* group, and *An. barbirostris*, followed

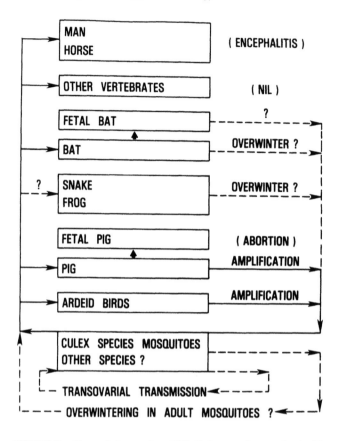

FIGURE 2.   Transmission cycles of JE. Pathways of uncertain significance are shown as dotted lines.

Table 3
SOME MOSQUITO SPECIES FROM
WHICH JE VIRUS HAS BEEN
ISOLATED

Culex annulus	*Mansonia annulifera*
Cx. bitaeniorhynchus	*M. bonneae/dives*
Cx. epidesmus	*M. uniformis*
Cx. fuscocephala	*Aedes aegypti*
Cx. gelidus	*Ae. albopictus*
Cx. pipiens fatigans	*Ae. curtipes*
Cx. p. pallens	*Ae. togoi*
Cx. p. quinquefasciatus	*Ae. vexans*
Cx. pseudovishnui	*Anopheles barbirostris*
Cx. tritaeniorhynchus	*An. hyrcanus*
Cx. vishnui	*An. hyrcanus* group
Cx. vishnui group	*An. sinensis*
Cx. whitmorei	*An. subpictus*
Armigeres obturans	*An. tessalatus*

Table 4
SOME LABORATORY VECTORS OF JE VIRUS

High efficiency	Moderate efficiency	Low efficiency
Culex tritaeniorhynchus	*Cx. pipiens pallens*	*Aedes albopictus*
Cx. gelidus	*Cx. p. quinquefasciatus*	*Ae. aegypti*
Cx. fuscocephalas	*Cx. p. molestus*	*Ae. triseriatus*
	Cx. tarsalis	
	Cx. pseudovishnui	
	Anopheles tessalatus	

in 1974—1975 by two more isolates from the *An. hyrcanus* group and one each from *Cx. bitaeniorhynchus* and *Cx. epidesmus*. Although large numbers of the known vector *Cx. vishnui* were processed, only one isolate was obtained. It is therefore not clear what the principal vector is in West Bengal.[36]

Second, laboratory studies in China indicated that *Ae. chemulpoensis* could vector JE, although no field isolations were obtained. This mosquito was eradicated from Beijing, and although epidemics still occurred, they diminished substantially in scale, suggesting that *Ae. chemulpoensis* may have been a field vector. To further complicate the picture, virus isolations have also been made in China from *Cx. pipiens var. pallens, An. hyrcanus, Ae. albopictus,* and *Armigeres obturans.*[68]

Overall, a consensus that has emerged is that *Cx. tritaeniorhynchus* is the principal vector throughout most of the region, but other species may be locally important, such as *Cx. gelidus* in Southeast Asia, and members of the *Cx. vishnui* group in Taiwan and India.

2. Vertebrate Hosts

As noted above, serum JE antibodies can be detected in most domestic vertebrates and in a wide variety of wild vertebrates. There is no evidence to suggest that JE can be transmitted horizontally from animal to animal by any route other than mosquito vectors. Although laboratory animals can be infected by inhalation of virus aerosols or by intranasal installation of virus suspensions, the doses required are unnaturally great.[87,134] Detection of JE in the secreta or excreta of infected animals has not been reported.

B. Evidence from Experimental Infection Studies
1. Vectors

JE does not appear to be highly host specific; the virus will replicate in a variety of insects and acarines when inoculated intrathoracically and in vitro in a wide variety of *Aedes* and *Culex* cell lines and *Anopheles gambiae* and *Drosophila melanogaster* cells, but not in *Rhipicephalus appendiculatus* tick cells.[73,135] It is therefore hardly surprising that a large number of mosquito species have been shown to be susceptible to JE and capable of transmission in the laboratory. Table 4 summarizes some experimental results of JE vector competence. *Cx. tritaeniorhynchus* has been studied in greatest depth. The first detailed experiments depended upon the establishment of a laboratory colony.[136,137] Infected pigs and baby chicks were used to feed mosquitoes. It was observed that when the incubation period was in the range of 5 to 34 days after infection, the percentage of infected mosquitoes was related to the original concentration of the virus in the blood of the host animal. Subsequent transmission of the virus could be demonstrated to pigs, black-crowned night herons, egrets, and chicks. Interestingly, no significant differences could be found in the virus titer between groups of mosquitoes that transmitted or failed to transmit JE.

Variations in the susceptibility of colonies of *Cx. tritaeniorhynchus* from diverse geographic areas have been demonstrated and may explain the geographic distribution of JE.

Table 5
SOME NATURAL
VERTEBRATE HOSTS
OF JE VIRUS

Man
Domestic animals
   Swine
   Horses
   Donkey
   Cattle
   Water buffalo
   Sheep
   Dog
   Chicken
   Duck
Wild animals
   Bird
   Bat
   Snake
   Frog

In one study, six mosquito strains originating from mainland Japan and Okinawa showed moderate susceptibility, whereas mosquitoes originating from Amami Island were slightly more susceptible than the Japanese mosquito strains, and sixfold more susceptible than Taiwanese mosquitoes.[138] In a subsequent study, two Japanese strains of *Cx. tritaeniorhynchus* were shown to be highly efficient transmitters, one Japanese and one Taiwanese were intermediate transmitters, and three Pakistani strains were susceptible to JE but their transmission efficiency was low.[139] This last result is of particular interest as strains of *Cx. triaeniorhynchus* from Pakistan have been shown to be excellent vectors of the related West Nile virus.[140]

Although there are very little comparative data on virus strain differences, it is likely that this may also be an important source of variations in the rate of virus growth in mosquitoes. Two studies in particular have shown marked differences in the behavior of attenuated and parental JE strains in mosquitoes. In a study in 1969,[141] mosquitoes fed on an attenuated strain of JE failed to become infected. In a more recent study, it was demonstrated in colonized *Cx. tritaeniorhynchus* that 100% infection rates were obtained by oral feeding and 100% transmission rates after intrathoracic inoculation of the parental Chinese SA-14 strain of JE, whereas only 11% of mosquitoes ingesting the attenuated 2-8 strain became infected, and only 3% of inoculated mosquitoes transmitted this attenuated virus.[142] The attenuated strain did not revert to virulence after passage through mosquitoes.

2. Vertebrate Hosts (Summarized in Table 5)

Pigs — Pigs are important in the epidemiology of JE. Virtually all domestic swine becoming infected develop viremia capable of infecting mosquitoes.[143] In controlled experiments, viremia in colostrum-deprived pigs lasted 4 days and ranged up to 10^6 SMIC $LD_{50}/m\ell$ of blood.[144] An interesting study in Kyoto used sentinel pigs to attract *Cx. tritaeniorhynchus* mosquitoes.[145] Pigs became infected 1 week or more after placement. When pig viremia was detected, more than 30% of the engorged mosquitoes captured in a nearby trap were infected, reflecting virus in the blood meal. Given the rapid population turnover of pigs, this domestic animal is clearly a significant source of mosquito infection.

Mice — Rodents probably are not important in natural cycles of JE transmission.[146] Experimental studies using mice have demonstrated that congenital transmission of JE can occur in mammalian species.[147-149]

Horses — Experimental transmission of JE from chick to horse, horse to horse, and horse to chick by *Cx. tritaeniorhynchus* has been demonstrated.[150] Viremia levels in horses were only up to $10^{1.2}$ SMIC LD_{50}; viremia appeared 1 to 4 days after infection and lasted for 2 to 6 days. One animal showed encephalitis.

Bats — Several species of bats are susceptible to JE.[151] Viremia lasts for 6 days or more and is sufficient to infect mosquitoes. Artificial hibernation studies have shown that extended incubation periods are possible. Bats inoculated and held at 10°C up to 107 days all developed viremia after being warmed to room temperature, with blood virus levels sufficient to infect mosquitoes.[152] Transplacental transmission has been documented in experimentally infected bats.[153]

Reptiles and amphibians — Poikilothermic vertebrates were considered to be unimportant in the epidemiology of JE in Japan after studies failed to isolate the virus from 305 wild snakes, and only 5 of 64 artificially infected snakes became infected.[154] In Korea, however, 2 isolates were obtained from 747 wild snakes collected during 1966—1967. HI antibodies were demonstrated in 40% of snakes collected from 1965 to 1970, but more specific plaque N tests showed only 2.8% positives.[155] Experimental overwintering of infected snakes, frogs, and bats (with successful reisolation of the virus from two out of three surviving frogs 5 to 6 months later) has also been reported in Korea.[156] Although these observations are of biological interest, their epidemiological significance remains unclear. The principal JE vector, *Cx. tritaeniorhynchus*, feeds on snakes and frogs only reluctantly, implying inadequate contact for successful field transmission. Nevertheless, this line of inquiry remains relatively unexplored.

Birds — Studies in Japan showed that viremia could be readily produced in the large water birds (the black-crowned night heron, the plumed egret, and the lesser egret), as well as in starlings and chicks, with levels up to $10^{3.5}$ SMIC $LD_{50}/0.03$ mℓ of blood.[157,158] Similar studies in India showed that viremia, ranging up to $10^{3.0}$ SMIC LD_{50}, could be detected in infected pond herons and cattle egrets, but not in little egrets or little cormorants.[159] *Cx. tritaeniorhynchus* mosquitoes were successfully infected by feeding on the viremic birds, and subsequently transmitted the virus to susceptible birds. Viremia was also demonstrated in 2- to 28-day-old domestic ducklings. Viremia levels of 10^1 gave 9% infection, 10^3 produced 50% infection, and 10^4 SMIC LD_{50} or higher gave 100% infection with JE.[160] Low levels of viremia (up to $10^{2.4}$ SMIC $LD_{50}/0.02$ mℓ of blood) were demonstrated in four out of five artificially infected pigeons and all developed antibody.[161] Only 1 to 2% of colony *Cx. tritaeniorhynchus* were infected by feeding on pigeons when the viremia level was less than 10^3 SMIC $LD_{50}/$mℓ of blood.

C. Maintenance / Overwintering Mechanisms

The question of how JE is maintained in areas such as Japan, China, and Korea, where there are very cold winters, is still one of the principal unanswered questions in the epidemiology of JE. Several mechanisms have been proposed,[162,163] and on balance it would seem most likely that JE survives adverse conditions in the vector, either by overwintering in the adult, or by being transovarially passed to the next generation.

There is some limited evidence for maintenance of JE in overwintering mosquitoes. In Japan, JE strains have been isolated from overwintering *Cx. tritaeniorhynchus* mosquitoes,[164,165] and in Korea from overwintering *Cx. pipiens* mosquitoes.[166] Infected *Cx. tritaeniorhynchus* mosquitoes have been shown to transmit JE after experimental hibernation.[167]

Rather more evidence has accumulated than transovarial transmission occurs.[168] As early as 1939, JE was isolated from adult mosquitoes reared from field-collected larvae.[169] Chinese workers have obtained several field isolations of JE from *Cx. pipiens pallens* mosquitoes and seven JE strains were isolated from *Ae. albopictus* reared from larvae collected in Fukien in 1955.[64] Laboratory studies on transovarial transmission have been reviewed,[170] with transovarial transmission of JE in *Cx. tritaeniorhynchus, Ae. albopictus, Ae. togoi, Ae. aegypti,* and *Cx. bitaeniorhynchus* mosquitoes.[170-173] Two points of particular interest emerged. Infection was observed in *Ae. albopictus* and *Ae. togoi* mosquitoes reared from eggs that had been dried and kept at room temperature for 2 months.[172] In studies with *Cx. bitaeniorhynchus,* a very high transmission rate was observed, with 7 out of 32 orally infected females transmitting the virus to at least some of their F1 progeny.[173] Excluding this last example, transovarial transmission rates of flaviviruses are usually low. Now that transovarial transmission has been conclusively demonstrated under experimental conditions, field studies are needed to assess its epidemiological significance.

Alternative explanations for recurrent JE epidemics in cool climates include reintroduction of the virus in migratory birds or bats, or chronic infections in reptiles, amphibians, or hibernating mammals. However, there is no information from direct field studies to prove or disprove these possibilities.

VI. ECOLOGICAL DYNAMICS

A. Macro- and Microenvironment

Environment plays a major role in governing both the population dynamics of vector mosquitoes and disease incidence. Climatic influences are discussed below. The authors are not aware of studies on the role microenvironment plays in interactions between viruses, vectors, and hosts. This is a potentially important area for further study.

B. Climate / Weather

The "summer epidemic" nature of JE epidemiology has long suggested a correlation between human cases and weather conditions. Statistical analysis of data from the Nagasaki prefecture in Japan between 1950 and 1969 demonstrated an inverse correlation between precipitation and the number of cases and a positive correlation between case numbers and temperature. Large epidemics followed dry hot summers, whereas wet summers with low temperatures resulted in fewer cases.[174] Temperature and rainfall were strongly correlated. A multivariate analysis indicated that precipitation alone was sufficient to explain the combined effect of the two factors. Large amounts of rain may effectively flush out mosquito breeding sites, whereas too little rain may restrict population size. Of course, in areas with well-developed irrigation schemes, the role of rainfall becomes less significant. Temperature may have multiple effects:

1. The vector population may be increased through an increased survivorship resulting from a shortened larval breeding period. It was calculated that a reduction of development time from 11 to 7 days could result in quadrupling the number of emerging adults.[175]
2. The extrinsic incubation period of JE in the vector is significantly shortened with increased temperatures (see Section VI.G).
3. High temperature may reduce the survivorship of adult mosquitoes, and serve to modulate an epidemic.

The importance of daylength has also been considered. *Cx. tritaeniorhynchus* feed more readily during longer days,[176] and higher mosquito infection rates and more rapidly rising virus titers in salivary gland secretions are present in mosquitoes maintained on long-day regimes.[175]

In the only study of its type, *Cx. tritaeniorhynchus* mosquitoes were reared, infected, and maintained in a biotron, which simulated temperature, humidity, and daylength not only on a daily cycle but also on a seasonal cycle.[177] Low concentrations of virus, restricted to the posterior midgut, were seen in mosquitoes maintained under autumn or winter conditions, whereas higher titers and disseminated infections were seen in those maintained under summer conditions.

C. Vector Oviposition

The principal culicine vectors lay their rafts of eggs on the surface of water 2 to 3 days after blood feeding. Egg laying occurs in rice fields before the rice becomes too high, or in temporary pools of water. Oviposition is therefore heavily dependent on availability of water and on agricultural practices.

D. Vector Density, Fecundity, and Longevity

The influences of climate and agricultural practices on the population dynamics of *Cx. tritaeniorhynchus* have been reviewed.[178] In Pakistan, Japan, and Korea, with drier and cooler climates, *Cx. tritaeniorhynchus* apparently overwinters as hibernating adults, while breeding continues throughout the year in the warmer maritime climates of Okinawa and Taiwan. In cooler areas, population curves closely follow the annual temperature curves, whereas in tropical climates population patterns more closely follow water availability, either from rainfall or irrigation.

The intimate relationship between *Cx. tritaeniorhynchus* density and agricultural practices has been well demonstrated in Sarawak.[174] In January, the paddy plots are flooded, and little breeding takes place in the clean water. In February and into March, the plots are dried prior to flowering of the rice; numerous temporary rainwater pools have only about 10% *Cx. tritaeniorhynchus* breeding. Harvesting takes place in March and April, leaving behind rice stubble which, while drying, supports heavy *Cx. tritaeniorhynchus* breeding. The fields are then allowed to be overgrown by grass which results in minimal *Cx. tritaeniorhynchus* breeding. Before planting in October, the fields are scythed, leaving a rich infusion; this is followed rapidly by *Cx. tritaeniorhynchus* oviposition. One large synchronous generation of larvae results, reaching the pupal stage rapidly and simultaneously in 4 to 5 days. Although the cut grass is burnt 2 to 3 days after cutting, this does not affect larval numbers; 5 to 6 days after cutting, the plot is cultivated and planted, after which there is little breeding. It was calculated that an average-sized paddy plot (320 m²) produces up to 30,000 adults daily for 3 to 5 days. As cultivation and planting take place over a 6-week period, a large increase in the mosquito population ensues.[179]

Although *Cx. tritaeniorhynchus* may survive for extended periods under laboratory conditions, field survivorship is undoubtedly much shorter. This was emphasized by an analysis of age-composition and survivorship of field populations of *Cx. tritaeniorhynchus* in Osaka prefecture, Japan from 1968 to 1973.[180] Survival kinetics were calculated assuming that an infected mosquito must survive for 10 days before becoming infectious. The original number of *infected* mosquitoes required for *one* to survive to transmit was 629 in July (daily survival 0.525) and 361 in August (daily survival 0.555).[180]

E. Biting Activity, Host Preference

The principal *Culex* vectors bite during the night. Data from Malaysia showed a fairly steady biting rate throughout the night.[181] In Thailand, two biting peaks were observed, feeding beginning shortly after sunset, declining between 10 and 11 p.m., increasing again after midnight, declining again sharply after 5 a.m., and ceasing completely by 6 a.m.[105] In Japan, a similar overall pattern was seen, but the pattern varied with the bait in the trap.[182] Resting *Cx. tritaeniorhynchus* collected at pigsties showed a gradual decline in numbers throughout the night after a peak at dusk. At cowsheds the catch rate fell more noticeably after 9 p.m., but then remained at a fairly steady level before declining toward dawn. In dry-ice traps (comprising mainly unfed mosquitoes) there was a marked fall in catch numbers after 9 p.m.

The principal vectors prefer animals to humans. In Japan, of 130,997 mosquitoes collected in Magoon traps, only 305 were taken in traps baited by humans, whereas 5944 were taken in traps baited with black-crowned night herons and 124,748 in pig-baited traps.[183] The three vector species *Cx. tritaeniorhynchus*, *Cx. gelidus*, and *Cx. fuscocephala* together comprised less than 1%.

F. Vertebrate Host Density, Immunological Background

Pigs are clearly a major amplifying host of JE, and the density of pigs in a given region is a major determinant of epidemic activity. In many countries in Asia, swine are penned in immediate proximity to (often directly underneath) human dwellings. Swine are typically slaughtered near 1 year of age. Up to 100% of 1-year-old pigs taken to abattoirs may show serologic evidence of JE infection.[108,143,184] New litters are delivered all year. Young piglets are probably protected from JE infection by colostrum,[144] but upon weaning enter the pool of potential amplifying hosts.

Perplexingly, this pattern of intense infection of young pigs may be observed not only in regions with severe epidemics of JE, but also in regions where epidemic JE among humans does not occur. In peninsular southern Thailand, where epidemic JE is unknown, pig antibody prevalence rates and seroconversion rates are comparable to, or greater than, rates measured in some severely afflicted provinces in northern Thailand.[25] The relatively low incidence of human cases currently observed in southern Japan (despite a persistence of very high antibody prevalence rates in swine,[108] approaching 100%) may be attributed to near-universal vaccination of children against JE. Alternatively, introduction of a new, perhaps relatively avirulent strain of JE into Japan in the mid-1960s may also account for the low human attack rates.[66]

G. Vector Competence

For a mosquito to be an efficient vector of JE it must (1) be able to acquire an infection from a host which may be circulating relatively low amounts of virus in its blood, (2) it must be able to support rapid replication of the virus and become infective with a short extrinsic incubation period, and (3) it must be able to transmit sufficient virus in its saliva to infect a new host. *Cx. tritaeniorhynchus* fulfills these criteria.

Dose response curves show that 5 and 9% of colonized *Cx. tritaeniorhynchus* can be infected with very low doses of virus around $10^{1.0}$.[159,181,185] With titers $\geq 10^{4.0}$, 100% infection rates can be achieved. The range of host viremias obtained have been discussed above; only pigs and birds develop viremia levels sufficient to infect significant numbers of mosquitoes. JE grows rapidly in infected *Cx. tritaeniorhynchus*, and the rate of growth is dependent on temperature. Maximum titers of virus in orally infected mosquitoes may be reached 5 days after infection at 28°C, whereas at 10°C there is a delay of 6 days in reaching a maximum, which is about ten times lower than that achieved at the higher temperatures.[185] This pattern of replication was confirmed in transmission trials. Held at 28°C, 60% of mosquitoes transmitted virus 9 days after infection, and 100% by 14 days. In contrast, no transmission was detected from mosquitoes held at 20°C for up to 20 days, and the proportion infected never rose above 60% even up to 35 days.[185]

Direct transmission to animals has been widely used to demonstrate JE transmission by infected *Cx. tritaeniorhynchus*. However, the technique is limited by the fact that mosquitoes may refuse to feed, and that no direct measurement of the infecting dose is possible. To overcome these problems, artificial transmission systems have been developed. In one system, infected *Cx. tritaeniorhynchus* mosquitoes were stimulated to salivate directly into capillary reservoirs containing serum agar.[185,186] Virus titers in the capillary tube reservoirs ranged from $\leq 10^1$ to 10^4 (average $10^{3.1}$) SMIC LD$_{50}$. Large amounts of virus can be transmitted by this mosquito.

Infection with a nonpathogenic orbivirus-like agent has been recently shown to interfere with JE transmission in dually infected mosquitoes.[187]

H. Movements and Migrations of Vectors and Hosts

The flight performance of vector mosquitoes is an important factor in the epidemiology of JE. Vector mosquitoes must travel between rice-field breeding sites and feeding areas within villages. Flight range and dispersal patterns have been examined using both unfed and freshly engorged mosquitoes in mark-release-recapture experiments. In Japan, unfed mosquitoes showed a mean dispersal of 1.0 km over a 7-day period, with a maximum dispersal 1 day after release of 5.1 km, and a maximum recorded dispersal of 8.4 km 3 days after release.[188] Dispersal was random, but studies of the topography of the release site indicated that *Cx. tritaeniorhynchus* avoided flying over hills, and at one point, successfully flew over a 2 km stretch of water. In Thailand, most freshly engorged mosquitoes were recaptured 48 hr after release.[189] *Cx. tritaeniorhynchus* was generally recaptured at greater distances than *Cx. geldius* or *Cx. fuscocephala*. *Cx. tritaeniorhynchus* was taken at all capture sites up to 1800 m from the release point. Dispersal appeared to be random. Movements of insects over long distances and potential seasonal introduction of infected insects via major weather factors such as monsoons and prevailing winds has been reviewed.[190]

I. Human Element in Disease Ecology

Compared to pigs or birds, humans are presumed to be a poor amplifying host of JE, directly entering the transmission cycle only as a "dead-end" host. However, the human element is a fundamental indirect component in the ecology of JE. Both the major mosquito vector and the major amplifying host owe their existence to humans: *Cx. tritaeniorhynchus* breeds largely in man-made rice fields, and pigs are raised in close proximity to human habitats.

VII. SURVEILLANCE

A. Clinical Hosts

In most countries in Asia, very few cases of acute encephalitis are evaluated with laboratory tests for JE; most are simply reported as acute encephalitis, etiology unspecified. In Thailand, a national program was established in which blood-saturated, dried filter paper strips were mailed to a central laboratory for HI testing.[191] This probably reflected insensitivity of the test system rather than the existence of other major causes of encephalitis; when the HI method was replaced by a more sensitive JE IgM ELISA system for testing of the filter paper eluates, over 80% of cases could be confirmed as due to JE.[192] Another promising approach to wider laboratory confirmation of JE has been the development of a simple, field-adaptable ELISA kit for detection of JE IgM in patient serum or CSF.[193]

B. Wild Vertebrates

Surveys of wild vertebrates have not been reported to be efficacious for predicting JE epidemics.

C. Vectors

As noted above, JE can usually be detected first in mosquitoes, but routine surveillance systems based on isolation of virus from mosquito pools are too time consuming to be of practical use. In a pilot study in Thailand, the authors found that a rapid immunoassay could detect JE antigens in about 50% of JE-positive mosquito pools. Although the method was less sensitive than virus culture, the logistics of pool processing was improved considerably. Developments of this type may be helpful in forecasting epidemics.

D. Sentinels

Sentinels pigs have been used successfully to monitor JE activity using standard serology or virus isolation techniques.[25,184] In a study in a Thai village in 1984, pig seroconversions were monitored at weekly intervals.[194] From May 16 to June 13 no seroconversions were noted. The first pig seroconversion took place on June 14 and over the next 5 weeks all sentinel animals rapidly seroconverted. The first mosquito isolates were obtained between June 9 and 12 with the first human cases on June 19. Implementation of mosquito control measures could be based on pig seroconversions; this might help to limit the number of human cases.

VIII. INVESTIGATION OF EPIDEMICS

Once an epidemic of acute encephalomyelitis has been defined, the following steps should be undertaken:

1. As a first priority, the etiology of the disease must be established. Samples of CSF and post-mortem brain tissues should be collected from patients with disease and promptly inoculated onto an appropriate cell culture system or into laboratory animals or mosquitoes. Mosquito cell cultures are surprisingly hardy and can be transported to remote epidemic areas, inoculated, and maintained at room temperature for several weeks, with very satisfactory results.[89] CSF and serum can be assayed for detection of specific IgM antibodies; a kit for JE IgM detection has recently been reported to be useful in provincial hospital settings.[193] If the required materials or facilities are not available for immediate isolation attempts or JE IgM detection, then CSF, serum, and brain tissues should be promptly frozen and shipped to a competent laboratory for analysis.
2. Mosquito populations should be sampled, preferably at the homes of patients. Light-trap collections are the most convenient method. Minimal infection rates, by species, coupled with trap yields of each species, should provide sufficient information to deduce the principal vector species.
3. Surveys should be instituted to determine the density of known amplifying hosts, primarily pigs and ardeid birds.
4. More detailed studies of mosquito populations, of additional potential amplifying hosts, and of human risk factors should be undertaken if these standard steps do not clarify the critical components of the epidemic.

IX. PREVENTION AND CONTROL

A. Vector Control

Epidemics of JE are typically associated with population peaks of vector mosquitoes. Source control, or measures designed to reduce adult mosquito populations, may be useful adjuncts to vaccination programs in prevention of an outbreak. Source reduction depends heavily on community participation. Alterations of agricultural practices that promote vector control may be useful, such as short-term draining of rice fields to reduce larval numbers, or introduction of new strains of drought-resistant rice. However, it is difficult to enlist full community participation in JE vector control. Permanent control campaigns by insecticide application are expensive, and because the principal vectors have a very strong flight performance, large areas must be treated repeatedly. In the case of a clearly JE-defined epidemic, emergency control aimed at reducing the adult female vector population is the only viable option. Efficacy of this approach was demonstrated in Korea in 1971 and 1972 where aerial application of ultra-low-volume insecticides reduced numbers of *Cx. tritaeniorhynchus* by

80% for the first 96 hr after application of fenitrothion.[195] In these same studies, malathion showed little effect.[195] Insecticide resistance is an increasing problem, aggravated by the widespread use of agricultural pesticides. *Cx. tritaeniorhynchus* now shows widespread resistance to organophosphate compounds and resistance to carbamates has recently been detected in Japan and Sri Lanka.[196]

B. Control of Vertebrate Hosts

Control of vertebrate hosts has not proven to be efficacious in the prevention and control of JE. Pigs are a major source of family income in many parts of Asia, and no serious effort has ever been mounted to induce farmers to reduce the numbers of pigs raised. Although it is possible to select strains of mice that are relatively resistant or susceptible to flavivirus growth,[197] no efforts to produce swine strains that are genetically resistant to JE have been reported. Extermination of the potential bird amplifying host population is considered impractical.

C. Environmental Modification

The principal vector of JE, *Cx. tritaeniorhynchus*, is predominantly a nocturnal feeder, and any measure taken to avoid mosquito bites at night may decrease the risk of JE. Simple measures include the liberal use of topically applied mosquito repellants (such as *N,N*-diethyl-*m*-toluamide, commonly abbreviated "DEET"), and sleeping within screened rooms.[198] Unfortunately, these methods are beyond the resources of most persons in affected regions. Although *Cx. tritaeniorhynchus* is not highly endophilic, use of mosquito netting during sleep is standard and is probably somewhat helpful.

Another environmental modification that may be quite important in decreasing the risk of JE is alteration of the agricultural practices of rice cultivation. Although this has never been undertaken with the specific intent to decrease JE transmission, the precipitous decline of JE cases in Japan between 1965 and 1970 has been attributed to the widespread changes in rice cultivation and rural environment.[174] In Kyoto, central Japan, the population levels of *Cx. tritaeniorhynchus* fell 100-fold between the mid-1960s and the mid-1970s.[102]

D. Use of Vaccines

In the last 20 years approximately 500 million doses of JE vaccine have been produced and administered to humans, primarily in China, Taiwan, Korea, and Japan.[199] Different vaccine formulations are used in different countries: a formalin-inactivated, baby-hamster-kidney-tissue-culture-derived vaccine in China,[14] and formalin-inactivated, mouse-brain-derived vaccines in Japan and Korea.[11,200] National JE vaccine production programs are currently being developed in India and Thailand.[201,202]

Surprisingly, the efficacy of JE vaccination was not evaluated in a rigorously controlled field trial until 1985 in Thailand, when two doses of an inactivated whole virus, mouse-brain-derived vaccine were shown to provide greater than 95% protection.[64] The complication rate of the vaccine used in this study was nil. Thus, it is now clear that JE is entirely preventable. In the near future it should be possible to produce JE virus envelope protein in mass quantities through genetic engineering techniques. This technology holds the promise of universal availability of safe, inexpensive vaccines. It must be noted that since person-to-person spread of JE does not occur, there is no "herd immunity" effect in immunized human populations; effectiveness of a program is directly proportional to the fraction of the population reached by immunization.

Vaccines have been used to immunize pigs in Japan[144] and China, but the economics and logistics of large-scale pig immunization programs are impractical. Vaccination can be used to protect horses.[203] A live, attenuated JE vaccine has been administered to over a half million horses in China, with apparently good efficacy.[204]

X. FUTURE RESEARCH

Much remains to be learned about the epidemiology of JE. The authors consider the following broad questions to be of paramount importance. Answers to each of these questions will be possible only with a concerted effort in which field and laboratory studies are tightly integrated.

1. Why is the range of epidemic JE expanding? Available evidence seems quite clear that epidemic JE first emerged (with dramatic effect) into Southeast Asia in the late 1960s, and into India and Nepal in the late 1970s. What region(s) will be next? Can this be predicted, or preferably, prevented?
2. Why are there two distinct patterns of JEV transmission, endemic in the tropics, but epidemic in more temperate regions? Why doesn't epidemic JE occur in the tropics?
3. How does JE overwinter in temperate climates? Control measures targeted at over-wintering hosts — adult mosquitoes, mosquito eggs, snakes, frogs, or bats — have a theoretical yet logical appeal.
4. Why does only 1 of 100 persons infected develop encephalitis? Is this primarily a function of a variable relating to the host, the vector, or the virus? Are some forms of virus-human interactions invariably and predictably benign (or even protective), while others invariably and predictably cause encephalitis?
5. Can improved methods be developed, perhaps using "genetically engineered" products, to further dissect and define virus strain variations? Human diagnosis and sero-epidemiology? Vector competence?
6. Can efficacious yet inexpensive JE vaccines be constructed and delivered to all known at-risk human populations?

REFERENCES

1. **Mitamura, T., Kitaoka, M., Watanabe, M., Okuba, K., Tenjin, S., Yamada, S., Mori, K., and Asada, J.**, Study on Japanese encephalitis virus. Animal experiments and mosquito transmission experiments, *Kansai Iji*, 1, 260, 1936.
2. **Mitamura, T., Kitaoka, M., Mori, K., and Okuba, K.**, Isolation of the virus of Japanese epidemic encephalitis from mosquitoes caught in nature, *Tokyo Iji Shinshi*, 62, 820, 1938.
3. **Rappleye, W. C.**, Epidemiology of Japanese B encephalitis, in *Epidemic Encephalitis: Third Report by the Matheson Commission*, Columbia University Press, New York, 1939, 157.
4. **von Economo, C.**, Encephalitis lethargica, *Br. Med. J.*, 2, 896, 1931.
5. **Kono, R. and Kim, K. H.**, Comparative epidemiological features of Japanese encephalitis in the Republic of Korea, China (Taiwan) and Japan, *Bull. WHO*, 40, 263, 1969.
6. **Okuno, T.**, An epidemiological review of Japanese encephalitis, *WHO Stat. Rep.*, 31, 120, 1979.
7. **Berge, T. O., Blender, J. X., Burns, K. F., Casals, J., Cousins, R. F., Keegan, H. G., MaClaren, J. P., Meiklejohn, G., Reeves, W. C., Rees, D. M., and Sather, G.**, Japanese B encephalitis: a complete review of experience on Okinawa 1945—1949, *Am. J. Trop. Med.*, 30, 689, 1950.
8. **Umenai, T., Krzysko, R., Bektimirov, T. A., and Assaad, F. A.**, Japanese encephalitis: current world-wide status, *Bull. WHO*, 63, 625, 1985.
9. **Oya, A.**, Japanese encephalitis vaccine, in *Vaccination, Theory and Practice*, Fukumi, H., Ed., International Medical Foundation of Japan, Tokyo, 1974, 69.
10. **Pond, W. L. and Smadel, J. E.**, Neurotropic viral diseases in the Far East during the Korean war, *Med. Sci. Publ. Army Med. Serv. Graduate Sch.*, 4, 219, 1954.
11. **Paik, S. B.**, personal communication, 1983.
12. **Dandurov, I. V.**, Epidemiologic material from the study of Japanese encephalitis in the Primorsk area in 1965, *Zh. Mikrobiol.*, 45, 136, 1968.
13. **Chu, F. T., Wu, J. P., and Teng, C. H.**, Acute encephalitis in children; clinical and serologic study of ten epidemic cases, *Chin. Med. J.*, 58, 68, 1940.

14. **Chen, B. Q.,** personal communication, 1983.
15. **Grayston, J. T., Wang, S. P., and Yen, C. H.,** Encephalitis on Taiwan, *Am. J. Trop. Med. Hyg.,* 11, 126, 1962.
16. **Mackenzie, J. S.,** *Viral Diseases in Southeast Asia and the Western Pacific: Proc. Int. Semin.,* Academic Press, Sydney, 1982.
17. **Olson, J. G., Ksiazek, T. G., Gubler, D. J., Lubis, S. I., Simanjuntak, G., Lee, V. H., Nalim, S., Juslia, K., and Lee, R.,** A survey for arboviral antibodies in sera of humans and animals in Lombok, Republic of Indonesia, *Ann. Trop. Med. Parasitol.,* 77, 131, 1983.
18. **Olson, J. G., Ksiazek, T. G., Lee, V. H., Tan, R., and Shope, R. E.,** Isolation of Japanese encephalitis virus from *Anopheles annularis* and *Anopheles vagus* in Lombok, Indonesia, *Trans. R. Soc. Trop. Med. Hyg.,* 79, 845, 1985.
19. **O'Rourke, T. F., Hayes, C. G., SanLuis, A. M., Manaloto, C. R., Schultz, G. W., Ranoa, C. P., Beroy, G., Yambao, E., Morales, V., and Bakil, L.,** Epidemiology of Japanese encephalitis in the Philippines, in *Proc. 4th Aust. Symp. Arboviruses,* St. George, T. S., Kay, B. H., and Blok, J., Eds., Queensland Institute of Medical Research, Brisbane, 1986.
20. **Peterson, P. Y., Ley, H. L., Jr., Wisseman, C. L., Jr., Pond, W. L., Smadel, J. E., Diercks, F. H., Hetherington, D. G., Sneath, P. H. A., Witherington, D. H., and Lancaster, W. E.,** Japanese encephalitis in Malaya. I. Isolation of virus and serologic evidence of human and equine infections, *Am. J. Hyg.,* 56, 320, 1952.
21. **Trosper, J. H., Ksiazek, T. G., Cross, J. H., and Basaca-Sevilla, V.,** Isolation of Japanese encephalitis virus from the Republic of the Philippines, *Trans. R. Soc. Trop. Med. Hyg.,* 74, 292, 1980.
22. **Van Peenan, P. F. P., Irsiani, R., Sulianti-Saroso, J., Joseph, S. W., Shope, R. E., and Joseph, P. L.,** First isolation of Japanese encephalitis from Java, *Mil. Med.,* 139, 821, 1974.
23. **Smith, C. E. G., Simpson, D. I. H., Peto, S., Bowen, E. T. W., McMahon, D., Platt, G. S., Way, H., Bright, W. F., Maidment, B.,** Arbovirus infections in Sarawak: serologic studies in man, *Trans. R. Soc. Trop. Med. Hyg.,* 68, 96, 1974.
24. **Ksiazek, T. G., Trosper, J. H., Cross, J. H., and Basaca-Sevilla, V.,** Additional isolations of Japanese encephalitis from the Philippines, *Southeast Asian J. Trop. Med. Public Health,* 11, 507, 1980.
25. **Burke, D. S., Tingpalapong, M., Ward, G. S., Andre, R., and Leake, C. J.,** Intense transmission of Japanese encephalitis virus to pigs in a region free of epidemic encephalitis, *Southeast Asian J. Trop. Med. Public Health.,* 16, 199, 1985.
26. **Yamada, T., Rojanasuphot, S., and Takagi, M.,** Studies on the epidemic of Japanese encephalitis in the northern region of Thailand in 1969 and 1970, *Biken J.,* 14, 267, 1971.
27. **Grossman, R. A., Gould, D. J., Smith, T. J., Johnsen, D. O., and Pantuwatana, S.,** Study of Japanese encephalitis virus in Chiangmai Valley, Thailand. I. Introduction and study design, *Am. J. Epidemiol.,* 98, 111, 1973.
28. **Grossman, R. A., Edelman, R., Chiewanich, P., Voodhikul, P., and Siriwan, C.,** Study of Japanese encephalitis virus in Chiangmai Valley, Thailand. II. Human clinical infections, *Am. J. Epidemiol.,* 98, 121, 1973.
29. **Netler, R.,** Enquette serologique sur l' encephalitis Japonaise B au Viet Nam. I. Recherches chez l'homme, *Bull. Soc. Pathol. Exot.,* 49, 883, 1956.
30. **Ketel, W. B. and Ognibene, A. J.,** Japanese B encephalitis in Vietnam, *Am. J. Med. Sci.,* 261, 271, 1971.
31. **Khan, A. Q., Khan, A. M., Dewan, Z. U., Claquin, P., Myat, A., Joshi, G. P., and Dobrzynski, L.,** An outbreak of Japanese encephalitis in Bangladesh, *Bangladesh Med. J.,* 8, 71, 1980.
32. **Ming, C. K., Swe, T., Thaung, U., and Lwin, T. T.,** Recent outbreaks of Japanese encephalitis in Burma, *Southeast Asian J. Trop. Med. Public Health,* 8, 113, 1977.
33. **Swe, T., Thein, S., and Myint, M. S.,** Pilot sero-epidemiological survey on Japanese encephalitis in northwestern Burma, *Biken, J.,* 22, 125, 1979.
34. **Carey, D. E., Myers, R. M., and Reuben, R.,** Japanese encephalitis in South India. A summary of recent knowledge, *J. Indian. Med. Assoc.,* 52, 10, 1969.
35. **Mathur, A., Chaturvedi, U. C., Tandon, H. O., Agarwal, A. K., Mathur, G. P., Naji, D., Prasad, A., and Mittal, U. P.,** Japanese encephalitis epidemic in Uttar Pradesh India during 1978, *Indian J. Med. Res.,* 75, 161, 1982.
36. **Rodrigues, R. M.,** Epidemiology of Japanese encephalitis in India, in *Natl. Conf. Japanese Encephalitis, 1982,* Indian Council of Medical Research Publication, 1984, 1.
37. **Khatri, I. B., Joshi, D. D., Pradhan, T. M. S., and Pradhan, S.,** Status of viral encephalitis (Japanese encephalitis) in Nepal, *J. Nepal Med. Assoc.,* 21, 97, 1983.
38. **Henderson, A., Leake, C. J., and Burke, D. S.,** Japanese encephalitis in Nepal, *Lancet,* 2, 1359, 1983.
39. **Chakraborty, A. K., Chakraborty, M. S., and Singh, N. C.,** Outbreak of Japanese encephalitis in Manipur (India) during 1982: some epidemiological features, *J. Commun. Dis.,* 16, 227, 1984.

40. **Ryu, E.,** *Chronological Reference of Zoonoses: Japanese B Encephalitis,* International Laboratory for Zoonoses, Taipei, 1974.
41. **Finkelstein, D.,** The lessons of an American's death: in health and hygiene, China is still a Third World country, *Washington Post,* January 24, 1982.
42. **Rose, M. R., Hughes, S. M., and Gatus, B. J.,** A case of Japanese B encephalitis imported into the United Kingdom, *J. Infect.,* 6, 261, 1983.
43. **Sabin, A. B.,** Epidemic encephalitis in military personnel, *JAMA,* 133, 281, 1947.
44. **Sabin, A. B., Schlesinger, R. W., Ginder, D. R., and Matsumoto, M.,** Japanese B encephalitis in American soldiers in Korea, *Am. J. Hyg.,* 46, 356, 1947.
45. **Emoto, O., Kondo, S., and Watanabe, M.,** On the epidemic of equine encephalitis which occurred in the year 1935 in Japan, *J. Vet. Sci.,* 15, 41, 1936.
46. **Goto, H., Shimizu, K., and Shirahata, T.,** Studies of Japanese encephalitis of animals in Hokkaido. I. Epidemiological observation on horses, *Res. Bull. Obihiro Univ.,* 6, 1, 1969.
47. **Burns, K. F.,** Congenital Japanese B infection of swine, *Proc. Soc. Exp. Biol. Med.,* 75, 621, 1950.
48. **Casals, J. and Brown, L. V.,** Hemagglutination with arthropod-borne viruses, *J. Exp. Med.,* 99, 429, 1954.
49. **Westaway, E. G., Brinton, M. A., Gaidamovich, S. Y., Horzinek, C., Igarashi, A., Laariainen, L., Lvov, D. K., Porterfield, J. S., Russell, P. K., and Trent, D. W.,** Flaviviridae, *Intervirology,* 24, 183, 1985.
50. **Henchal, E. A., McCown, J. M., Burke, D. S., Seguin, M. C., and Brandt, W. E.,** Epitopic analysis of antigenic determinants on the surface of dengue-2 virions using monoclonal antibodies, *Am. J. Trop. Med. Hyg.,* 34, 162, 1985.
51. **Feighny, R. and Burke, D. S.,** unpublished observations, 1986.
52. **Takegami, T., Washizu, M., and Yasui, K.,** Nucleotide sequence at the 3′ end of Japanese encephalitis genomic RNA, *Virology,* 152, 483, 1986.
53. **Fournier, M.,** personal communication, 1986.
54. **McAda, P. C., Mason, T. L., Schmaljohn, C. S., Dalrymple, J. M., and Fournier, M. J.,** Synthesis and cloning of cDNA from the Japanese encephalitis virus genome, in *International Workshop on the Molecular Biology of Flaviviruses,* Dalrymple, J. M., Ed., U.S. Army Medical Research Institute of Infectious Diseases, Fort Detrick, Md., 1984, 8.
55. **Hammon, W. McD. and Sather, G. E.,** Immunity of hamsters to West Nile and Murray Valley viruses following immunization with St. Louis and Japanese B, *Proc. Soc. Exp. Biol. Med.,* 91, 521, 1956.
56. **DeMadrid, A. T. and Porterfield, J. S.,** The flaviviruses (group B arboviruses): a cross neutralization study, *J. Gen. Virol.,* 23, 91, 1974.
57. **Shapiro, D., Kos, K. A., and Russell, P. K.,** Japanese encephalitis virus glycoproteins, *Virology,* 56, 89, 1973.
58. **Takegami, T., Miyamoto, H., Nakamura, H., and Yasui, K.,** Biological activities of the structural proteins of Japanese encephalitis virus, *Acta Virol. (Engl. Ed.),* 26, 312, 1982.
59. **Gould, E. A., Alexander, C. C., Buckley, A., and Clegg, C. S.,** Monoclonal immunoglobulin M antibody to Japanese encephalitis virus that can react with a nuclear antigen in mammalian cells, *Infect. Immun.,* 41, 774, 1983.
60. **Shope, R. E. and Sather, G. E.,** Arboviruses, in *Diagnostic Procedures for Viral Rickettsial, and Chlamydial Infections,* 5th ed., Lennette, E. H. and Schmidt, N. H., Eds., American Public Health Association, Washington, D.C., 1979, 26.
61. **Susilowati, S., Okuno, Y., Fukunaga, T., Tadano, M., Juang, R. F., and Fukai, K.,** Neutralization antibody responses induced by Japanese encephalitis virus vaccine, *Biken J.,* 24, 137, 1981.
62. **Banerjee, K.,** Certain characteristics of Japanese encephalitis virus strains by neutralization test, *Indian J. Med. Res.,* 83, 243, 1986.
63. **Fukai, K.,** personal communication, 1983.
64. **Hoke, C. H.,** personal communication, 1986.
65. **Kimura-Kuroda, J. and Yasui, K.,** Topographical analysis of antigenic determinants on envelope glycoprotein V3 (E) of Japanese encephalitis virus, using monoclonal antibodies, *J. Virol.,* 45, 124, 1983.
66. **Kobayashi, Y., Hasegawa, H., Oyama, T., Tamai, T., and Kusaba, T.,** Antigenic analysis of Japanese encephalitis virus using monoclonal antibodies, *Infect. Immun.,* 44, 117, 1984.
67. **Burke, D. S., Schmaljohn, C. S., and Dalrymple, J.,** Strains of Japanese encephalitis virus isolated from human brains have a highly conserved genotype compared to strains isolated from other natural hosts, presented at the Annu. Meet. American Society for Virology, Albuquerque, July 21—25, 1985.
68. **Huang, C. H.,** Studies of Japanese encephalitis in China, *Adv. Virus Res.,* 27, 71, 1982.
69. **Okuno, T., Suzuki, M., Kondo, A., and Ito, T.,** Variability of strains of Japanese encephalitis virus in regard to the pH-dependency in hemagglutination, *Jpn. J. Med. Sci. Biol.,* 18, 227, 1965.

70. **Okuno, T., Okada, T., Kondo, A., Suzuki, M., Kobayashi, M., and Oya, A.,** Immunotyping of different strains of Japanese encephalitis virus by antibody-absorption, haemagglutination-inhibition and complement-fixation tests, *Bull. WHO,* 38, 547, 1968.

71. **Takegami, T., Miyamoto, H., Nakamura, H., and Yasui, K.,** Differences in biological activity of the V3 envelope protein of two Japanese encephalitis virus strains, *Acta Virol. (Engl. Ed.),* 26, 321, 1982.

72. **Fujie, N., Kurata, K., and Sauada, M.,** Studies of immunological differences between the same strains of Japanese B encephalitis virus, *Jpn. Vet. Sci.,* 24, 349, 1962.

73. **Pudney, M., Leake, C. J., and Buckley, S. M.,** Replication of arboviruses in arthropod in vitro systems: an overview, in *Invertebrate Cell Culture Applications,* Academic Press, New York, 1982, 159.

74. **Rosen, L., Roseboom, L. E., Gubler, D. J., Lien, J. C., and Chantiotis, B. N.,** Comparative susceptibility of mosquito species and strains to oral infection and parenteral infection with dengue and Japanese encephalitis viruses, *Am. J. Trop. Med. Hyg.,* 34, 603, 1985.

75. **Doi, R., Shirasaka, A., and Sasa, M.,** The mode of development of Japanese encephalitis virus in the mosquito *Culex tritaeniorhynchus* as observed by the fluorescent antibody technique, *Jpn. J. Exp. Med.,* 37, 227, 1967.

76. **Leake, C. J. and Johnson, R. T.,** unpublished data, 1984.

77. **Miyake, M.,** The pathology of Japanese encephalitis, *Bull. WHO,* 30, 153, 1964.

78. **Thisyakorn, V. and Nimmannitya, S.,** Japanese encephalitis in Thai children, *Southeast Asian J. Trop. Med. Public Health,* 16, 93, 1985.

79. **Dickerson, R. B., Newton, J. R., and Hansen, J. E.,** Diagnosis and immediate prognosis of Japanese encephalitis, *Am. J. Med.,* 12, 277, 1952.

80. **Burke, D. S., Lorsomrudee, W., Leake, C. J., Hoke, C. H., Nisalak, A., Chongswasdi, V., and Laorakpongse, T.,** Fatal outcome in Japanese encephalitis, *Am. J. Trop. Med. Hyg.,* 34, 1203, 1985.

81. **Ishii, T., Matshushita, M., and Hamada, A.,** Characteristic residual neuropathological features of Japanese B encephalitis, *Acta Neuropathol.,* 38, 181, 1977.

82. **Johnson, R. T., Burke, D. S., Elwell, M., Leake, C. J., Nisalak, A., Hoke, C. H., and Lorsomrudee, W.,** Japanese encephalitis: Immunocytochemical studies of viral antigen and inflammatory cells in fatal cases, *Ann. Neurol.,* 18, 567, 1985.

83. **Schneider, R. J., Firestone, M. H., Edelman, R., Chieowanich, P., and Pornpibul, R. V.,** Clinical sequelae after Japanese encephalitis: a one year follow-up study in Thailand, *Southeast Asian J. Trop. Med. Public Health,* 5, 560, 1974.

84. **Edelman, R., Schneider, R. J., and Chieowanich, P.,** The effect of dengue virus infection on the clinical sequelae of Japanese encephalitis: a one year follow-up study in Thailand, *Southeast Asian J. Trop. Med. Public Health,* 6, 308, 1975.

85. **Mathur, A., Tandon, H. O., Mathur, K. R., Sarkari, N. B. S., and Chaturvedi, U. C.,** Japanese encephalitis virus infection during pregnancy, *Indian J. Med. Res.,* 81, 9, 1985.

86. **Hammon, W. McD. and Sather, G. E.,** Passive immunity for arbovirus infection. I. Artificially induced prophylaxis in man and mouse for Japanese encephalitis, *Am. J. Trop. Med. Hyg.,* 22, 524, 1973.

87. **Harrington, D. G., Hilmas, D. E., Elwell, M. R., Whitmire, R. E., and Stephen, E. L.,** Intranasal infection of monkeys with Japanese encephalitis virus. Clinical response and treatment with a nuclease-resistant derivative of poly (I)-poly(C), *Am. J. Trop. Med. Hyg.,* 26, 1191, 1977.

88. **Taylor, J. L., Schoenherr, C., and Grossberg, S. E.,** Protection against Japanese encephalitis virus in mice and hamsters by treatment with carboxymethyl-acridanone, a potent interferon inducer, *J. Infect. Dis.,* 142, 394, 1980.

89. **Leake, C. J., Burke, D. S., Nisalak, A., and Hoke, C. H.,** Isolation of Japanese encephalitis virus from clinical specimens using a continuous mosquito cell line, *Am. J. Trop. Med. Hyg.,* in press.

90. **Burke, D. S., Nisalak, A., Ussery, M. A., Loorakpongse, T., and Chantabivul, S.,** Kinetics of Japanese encephalitis virus immunoglobulin M and G antibodies in human serum and cerebrospinal fluid, *J. Infect. Dis.,* 151, 1093, 1985.

91. **Burke, D. S., Nisalak, A., Lorsomrudee, W., Ussery, M., Laorpongse, W., Ussery, M., and Laorpongse, T.,** Virus-specific antibody-producing cells in blood and cerebrospinal fluid in acute Japanese encephalitis, *J. Med. Virol.,* 17, 283, 1985.

92. **Burke, D. L., Nisalak, A., and Ussery, M. A.,** Antibody capture immunoassay detection of Japanese encephalitis virus immunoglobulin M and G antibodies in cerebrospinal fluid, *J. Clin. Microbiol.,* 16, 1034, 1982.

93. **Gadkari, D. A. and Shaikh, B. H.,** Immunoglobulin M antibody capture ELISA (enzyme-linked immunosorbent assay) in the diagnosis of Japanese encephalitis, West Nile and dengue virus infection, *Indian J. Med. Res.,* 80, 613, 1984.

94. **Burke, D. S., Tingpalapong, M., Elwell, M. R., Paul, P. S., and VanDeusen, R. A.,** Japanese encephalitis virus immunoglobulin M antibodies in porcine sera, *Am. J. Vet. Res.,* 46, 2054, 1985.

95. **Kanamitsu, M., Taniguchi, K., Urasawa, S., Ogta, T., Wada, Y., Wada, Y., and Saroso, S.,** Geographic distribution of arbovirus antibodies in indigenous human populations in the Indo-Australian archipelago, *Am. J. Trop. Med. Hyg.*, 28, 351, 1979.

96. **Scherer, W. F., Kitaoka, M., Okuno, T., and Ogata, T.,** Ecologic studies of Japanese encephalitis virus in Japan. VII. Human infection, *Am. J. Trop. Med. Hyg.*, 8, 707, 1959.

97. **Grossman, R. A., Edelman, R., Willhight, M., Pantuwatana, S., and Udomsakdi, S.,** Study of Japanese encephalitis virus in Chiangmai Valley, Thailand. III. Human seroepidemiology and inapparent infections, *Am. J. Epidemiol.*, 98, 133, 1973.

98. **Benenson, M. W., Top, F. H., Gresso, W., Ames, C. W., and Alstatt, L. B.,** The virulence to man of Japanese encephalitis virus in Thailand, *Am. J. Trop. Med. Hyg.*, 24, 974, 1975.

99. **Halstead, S. B. and Gross, C. R.,** Subclinical Japanese encephalitis. I. Infection of Americans with limited residence in Korea, *Am. J. Trop. Med. Hyg.*, 75, 190, 1962.

100. **Grossman, R. A., Edelman, R., and Gould, D. J.,** Study of Japanese encephalitis virus in Chiangmai Valley, Thailand. VI. Summary and conclusions, *Am. J. Epidemiol.*, 100, 69, 1974.

101. **Buescher, E. L., Scherer, W. F., Rosenberg, M. Z., Gresser, I., Hardy, J. L., and Bullock, H. R.,** Ecologic studies of Japanese encephalitis in Japan. II. Mosquito infection, *Am. J. Trop. Med. Hyg.*, 8, 651, 1959.

102. **Maeda, O., Takenokma, K., Karoji, Y., and Matsuyama, Y.,** Epidemiological studies on Japanese encephalitis in Kyoto City area, Japan. I. Evidence for decrease in vector mosquitoes, *Jpn. J. Med. Sci. Biol.*, 31, 27, 1978.

103. **Dandawate, C. N., Rajagopalan, P. K., Pavri, K. M., and Work, T. H.,** Virus isolations from mosquitoes collected in North Arcot District, Madras State and Chittoor District Andhra Pradesh between November 1955 and October 1957, *Indian J. Med. Res.*, 57, 1420, 1969.

104. **Carey, D. E., Rueben, R., and Myers, R. M.,** Japanese encephalitis studies in Vellore South India. I. Virus isolations from mosquitoes, *Indian J. Med. Res.*, 56, 1309, 1968.

105. **Gould, D., Edelman, R., Grossman, R. A., Nisalak, A., and Sullivan, M. F.,** Study of Japanese encephalitis virus in Chiangmai Valley, Thailand, *Am. J. Epidemiol.*, 100, 49, 1974.

106. **Hurlbut, H. S. and Nibley, C., Jr.,** Virus isolations from mosquitoes in Okinawa, *J. Med. Entomol.*, 1, 78, 1964.

107. **Simpson, D. I. H., Bowen, E. T. W., Way, H. J., Platt, G. S., Hill, M. N., Kamath, S., Wah, L. T., Bendell, P. J. E., and Heathcote, O. H. U.,** Arbovirus infections in Sarawak, October 1968—February 1970: Japanese encephalitis virus isolations from mosquitoes, *Ann. Trop. Med. Parasitol.*, 68, 393, 1974.

108. **Igarashi, A., Morita, K., Bundo, K., Matsuo, S., Hayashi, K., Matsuo, R., Harada, T., Tamoto, H., and Kuwatsuka, M.,** Isolation of Japanese encephalitis and Getah viruses from *Culex tritaeniorhynchus* and slaughtered-swine blood using *Aedes alopictus* clone C6/36 cells in Nagasaki, 1981, *Trop. Med.*, 23, 177, 1981.

109. **Leake, C. J., Ussery, M. A., Nisalak, A., Hoke, C. H., Andre, R. G., and Burke, D. S.,** unpublished data, 1982.

110. **Buescher, E. L. and Scherer, W. F.,** Ecologic studies of Japanese encephalitis virus in Japan. IX. Epidemiologic correlations and conclusions, *Am. J. Trop. Med. Hyg.*, 8, 719, 1959.

111. **Bundo, K., Morita, K., Igarashi, A., Yamashita, K., Aizawa, M., and Miura, N.,** Antibodies against Japanese encephalitis virus in bovine sera in Nagasaki, 1981, *Trop. Med.*, 25, 73, 1983.

112. **Sabai, T., Horimoto, M., and Goto, H.,** Status of Japanese encephalitis infection in cattle: survey of antibodies in various geographical locations in Japan, *Jpn. J. Vet. Sci.*, 47, 957, 1985.

113. **Ogata, M., Ishida, K., and Ochi, Y.,** Natural infection with Japanese B encephalitis virus in monkeys, *Jpn. J. Vet. Sci.*, 17, 7, 1955.

114. **Watanabe, M.,** Japanese encephalitis in domestic animals, *Jpn. J. Trop. Med.*, 10, 190, 1969.

115. **Chen, Y. T.,** Identification of natural infection with Japanese B encephalitis of swine and sheep in Hungchow, *Chin. Med. J.*, 80, 189, 1960.

116. **Simpson, D. I. H., Bowen, E. T. W., Platt, G. S., Way, H. J., Smith, C. E. G., Kamath, S., Liat, L. B., and Wah, L. T.,** Japanese encephalitis in Sarawak: virus isolation and serology in a land dyak village, *Trans. R. Soc. Trop. Med. Hyg.*, 64, 503, 1970.

117. **Johnson, D. O., Edelman, R., Grossman, R. A., Muangman, J., Pomsdhit, J., and Gould, D. J.,** Study of Japanese encephalitis in Chiang Mai Valley, Thailand. V. Animal infections, *Am. J. Epidemiol.*, 100, 57, 1974.

118. **Shankar, S. K., Vasudev Rao, T., Mruthyunjayanna, B. P., Gourie Devi, M., and Deshpande, M. G.,** Autopsy study of brains during an epidemic of Japanese encephalitis in Karnatuka (South India), *Indian J. Med. Res.*, 78, 431, 1983.

119. **Liu, Y. F., Teng, C. L., and Liu, K.,** Cerebral cysticercosis as a factor aggravating Japanese B encephalitis, *Chin. Med. J.*, 75, 1010, 1957.

120. **Burke, D. S. and Nisalak, A.,** unpublished observations.

121. **Hammon, W. McD.**, Observations on dengue fever, benign protector and killer: a Dr. Jekyll and Mr. Hyde, *Am. J. Trop. Med. Hyg.*, 18, 159, 1969.

122. **Sather, G. and Hammon, W. McD.**, Protection against St. Louis encephalitis and West Nile arboviruses by previous dengue virus (type 1-4) infection, *Proc. Soc. Exp. Biol. Med.*, 135, 573, 1970.

123. **Carey, D. E., Myers, R. M., and Pavri, K. M.**, Japanese encephalitis studies in Vellore, South India. II. Antibody response of patients, *Indian J. Med. Sci.*, 56, 1319, 1968.

124. **Halstead, S. B. and Russ, S. B.**, Subclinical Japanese encephalitis. II. Antibody responses of Americans to a single exposure to JE virus, *Am. J. Trop. Med. Hyg.*, 75, 202, 1962.

125. **Southam, C. M.**, Serologic studies of encephalitis in Japan. I. Hemagglutination-inhibiting, complement-fixing, and neutralizing antibody following overt Japanese B encephalitis, *J. Infect. Dis.*, 99, 155, 1956.

126. **Scherer, W. F., Kitaoka, M., Grossberg, S. E., Okuno, T., Ogata, T., and Chanock, R. M.**, Immunologic studies of Japanese encephalitis virus in Japan. II. Antibody responses following inapparent human infection, *J. Immunol.*, 83, 594, 1959.

127. **Southam, C. M.**, Serologic studies of encephalitis in Japan. II. Inapparent infections of Japanese B encephalitis virus, *J. Infect. Dis.*, 99, 163, 1956.

128. **Lee, H. W. and Scherer, W. F.**, The anamnestic antibody response to Japanese encephalitis virus in monkeys and its implications concerning naturally acquired immunity in man, *J. Immunol.*, 86, 151, 1961.

129. **Igarashi, A., Fukai, K., Ahandrik, S., and Tuchinda, P.**, Antibody against Japanese encephalitis virus in sera of dengue hemorrhagic fever patients in Thailand, *Biken J.*, 11, 41, 1968.

130. **Chanyasanha, C., Bundo, K., and Igarashi, A.**, IgG-ELISA antibody titers against Japanese encephalitis (JE) and dengue virus type 1 among healthy people in JE-endemic areas in Japan and Thailand, *Trop. Med.*, 26, 1, 1984.

131. **Fukunaga, T., Igarashi, A., Okuno, Y., Ishimine, T., Tadano, M., Okamoto, Y., and Fukai, K.**, A seroepidemiological study of Japanese encephalitis and dengue virus infections in the Chiang Mai area of Thailand, *Biken J.*, 27, 9, 1984.

132. **Carey, D. E. and Myers, R. M.**, Japanese encephalitis studies in Vellore, South India. III. Neutralizing activity of human survey sera, *Indian J. Med. Res.*, 56, 1330, 1968.

133. **Hale, J. H. and Lee, J. H.**, Serological evidence of the incidence of Japanese B encephalitis virus infection in Malaysia, *Ann. Trop. Med. Parasitol.*, 49, 293, 1955.

134. **Larson, E. W., Dominik, J. W., and Slone, T. W.**, Aerosol stability and respiratory infectivity of Japanese B encephalitis virus, *Infect. Immun.*, 30, 397, 1980.

135. **Leake, C. J., Pudney, M., and Varma, M. G. R.**, Studies on arboviruses in established tick cell lines, in *Invertebrate Systems In Vitro*, Kurstak, E., Maramorosch, K., and Dubendorfer, A., Eds., Elsevier/North Holland, Amsterdam, 1980, 327.

136. **Newson, H. D., Blakeslee, T. E., Toshioka, S., Sakai, M., Wheeler, C. M., Shimada, T., and Akiyama, J.**, A preliminary report on the laboratory colonization of the mosquito *Culex tritaeniorhynchus* Giles, *Mosq. News*, 16, 282, 1956.

137. **Gresser, J. L., Hardy, J. L., Hu, S. M. K., and Scherer, W. F.**, Factors influencing transmission of Japanese B encephalitis virus by a colonized strain of *Culex tritaeniorhynchus* Giles, from infected pigs and chicks to susceptible pigs and birds, *Am. J. Trop. Med. Hyg.*, 7, 365, 1958.

138. **Takahashi, M.**, Variation in susceptibility among colony strains of *Culex tritaeniorhynchus* to Japanese encephalitis virus infection, *Jpn. J. Med. Sci. Biol.*, 33, 321, 1980.

139. **Takahashi, M.**, Differential transmission efficiency for Japanese encephalitis virus among colonised strains of *Culex tritaeniorhynchus*, *Jpn. J. Sanit. Zool.*, 33, 325, 1982.

140. **Akhter, R., Hayes, C. G., Bagar, S., and Reisen, W. K.**, West Nile virus in Pakistan. III. Comparative vector capability of *Culex tritaeniorhynchus* and eight other species of mosquitoes, *Trans. R. Soc. Trop. Med. Hyg.*, 76, 449, 1982.

141. **Takahashi, M., Yabe, S., and Okada, T.**, Effects of various passages on some properties of an attenuated strain of Japanese encephalitis with special regard to mosquito infectivity, *Jpn. J. Sci. Biol.*, 22, 163, 1969.

142. **Chen, B. Q. and Beaty, B. J.**, Japanese encephalitis vaccine (2-8 strain) and parent (SA 14 strain) viruses in *Culex tritaeniorhynchus* mosquitoes, *Am. J. Trop. Med. Hyg.*, 31, 403, 1982.

143. **Scherer, W. F., Moyer, J. T., Izumi, T., Gresser, I., and McCown, J.**, Ecological studies of Japanese encephalitis in Japan. VI. Swine infection, *Am. J. Trop. Med. Hyg.*, 8, 665, 1959.

144. **Komada, K., Sasaki, N., and Inoue, Y. K.**, Studies of live attenuated Japanese encephalitis vaccine in swine, *J. Immunol.*, 100, 194, 1968.

145. **Maeda, O., Takenokuma, K., Karoji, Y., Kuroda, A., Sasaki, O., Karaki, T., and Ishii, T.**, Epidemiological studies on Japanese encephalitis in Kyoto city area, Japan. IV. Natural infection in sentinel pigs, *Jpn. J. Med. Sci. Biol.*, 31, 317, 1978.

146. **Scherer, W. F., Buescher, E. L., Southam, C. M., Flemings, M. B., and Noguchi, A.**, Ecologic studies of Japanese encephalitis virus in Japan. VIII. Survey for infection of wild rodents, *Am. J. Trop. Med. Hyg.*, 8, 716, 1959.

147. **Mathur, A., Arora, K. L., and Chaturvedi, U. C.,** Transplacental Japanese encephalitis virus (JEV) infection in mice during consecutive pregnancies, *J. Gen. Virol.,* 59, 213, 1982.

148. **Mathur, A., Arora, K. L., Rawat, S., and Chaturvedi, V. C.,** Japanese encephalitis virus latency following congenital infection in mice, *J. Gen. Virol.,* 67, 945, 1986.

149. **Miura, T., Sugamata, M., Ogata, T., and Matsuda, R.,** Japanese encephalitis virus infection in fetal mice at different stages of pregnancy. II. Resistance to Japanese encephalitis virus infection, *Acta Virol. (Engl. Ed.),* 26, 283, 1982.

150. **Gould, D. J., Byrne, R. J., and Hayes, D. E.,** Experimental infection of horses with Japanese encephalitis virus by mosquito bite, *Am. J. Trop. Med. Hyg.,* 1984.

151. **Sulkin, S. E., Allen, R., and Sims, R.,** Studies of arthropod-borne virus infections in chiroptera. I. Susceptibility of insectivorous species to experimental infection with Japanese B and St. Louis encephalitis viruses, *Am. J. Trop. Med. Hyg.,* 12, 800, 1963.

152. **LaMotte, L. C., Jr.,** Japanese B encephalitis in bats during simulated hibernation, *Am. J. Hyg.,* 67, 101, 1958.

153. **Sulkin, S. E., Sims, R., and Allen, R.,** Studies of arthropod-borne virus infections in chiroptera. II. Experiments with Japanese B and St. Louis encephalitis viruses in the gravid bat. Evidence of transplacental transmission, *Am. J. Trop. Med. Hyg.,* 13, 475, 1964.

154. **Mifune, K., Shichijo, A., Ueda, Y., Suenaga, O., and Miyagi, I.,** Low susceptibility of common snakes in Japan of Japanese encephalitis virus, *Trop. Med.,* 11, 27, 1969.

155. **Lee, H. W., Min, B. W., and Lim, Y. W.,** Isolation and serologic studies of Japanese encephalitis virus from snakes in Korea, *J. Korean Med. Assoc.,* 15, 69, 1972.

156. **Oh, S. E., Lee, S. H., and Lee, H. W.,** Overwintering of Japanese encephalitis virus in artificially infected hibernating animals, *Centr. Med. (Korea),* 26, 707, 1974.

157. **Buescher, E. L., Scherer, W. F., McClure, H. E., Moyer, J. T., Rosenberg, M. Z., Yoshii, M., and Okada, Y.,** Ecologic studies of Japanese encephalitis in Japan. IV. Avian infection, *Am. J. Trop. Med. Hyg.,* 8, 678, 1959.

158. **Scherer, W. F., Buescher, E. L., and McClure, H. E.,** Ecologic studies of Japanese encephalitis virus in Japan. V. Avian factors, *Am. J. Trop. Med. Hyg.,* 8, 689, 1959.

159. **Soman, R. S., Rodrigues, F. M., Guttikar, S. N., and Guru, P. Y.,** Experimental viraemia and transmission of Japanese encephalitis virus by mosquitoes in ardeid birds, *Indian J. Med. Res.,* 66, 709, 1977.

160. **Dhanda, V., Banerjee, K., Deshmukh, P. K., and Ilkal, M. A.,** Experimental viraemia and transmission of Japanese encephalitis virus by mosquitoes in domestic ducks, *Indian J. Med. Res.,* 6, 881, 1977.

161. **Chunikhin, S. and Takahashi, M.,** An attempt to establish the chronic infection of pigeons with Japanese encephalitis virus, *Jpn. J. Sanit. Zool.,* 22, 155, 1971.

162. **Reeves, W. C.,** Overwintering of arboviruses, *Prog. Med. Virol.,* 17, 193, 1974.

163. **Oda, T., Wada, Y., and Hayashi, K.,** Considerations on the possibility of overwintering of Japanese encephalitis virus in *Culex tritaeniorhynchus* females, *Trop. Med.,* 20, 153, 1978.

164. **Hayashi, K., Mifune, K., Shichjo, A., Suzuki, H., Matsuo, S., Makino, Y., Akashi, M., Wade, Y., Oda, T., Mogi, M., and Mori, A.,** Ecology of Japanese encephalitis virus in Japan. III. The results of investigations in Amami Island, southern part of Japan from 1973—1975, *Trop. Med.,* 17, 129, 1975.

165. **Ura, M.,** Ecology of Japanese encephalitis virus in Okinawa, Japan. I. The investigation on pig and mosquito infection of the virus in Okinawa island from 1966 to 1976, *Trop. Med.,* 18, 151, 1976.

166. **Lee, H. W.,** Study on overwintering mechanisms of Japanese encephalitis virus in Korea, *J. Korean Med. Assoc.,* 14, 65, 1971.

167. **Mifune, K.,** Transmission of Japanese encephalitis virus by mosquito of *Culex tritaeniorhynchus* after experimental hibernation, *Endemic Dis. Bull. (Nagasaki),* 7, 1978, 1965.

168. **Rosen, L., Tesh, R. B., Lien, J. C., and Cross, J. H.,** Transovarial transmission of Japanese encephalitis virus by mosquitoes, *Science,* 199, 909, 1978.

169. **Mitamura, T., Kitaoka, M., Watanabe, S., Hosoi, T., Tenjin, S., Seki, O., Tagahata, K., Jo, K., and Shimizu, M.,** Weitere untersuchungen Uberdie uber tragung der Japanishen epidemischen Enzephalitis durch mucken, *Trans. Soc. Pathol. Jpn.,* 29, 92, 1939.

170. **Leake, C. J.,** Transovarial transmission of arboviruses by mosquitoes, in *Vectors in Virus Biology,* Mayo, M. A. and Harrap, K. A., Eds., Academic Press, New York, 1984, 63.

171. **Rosen, L., Shroyer, D. A., and Lien, J. C.,** Transovarial transmission of Japanese encephalitis virus by *Culex tritaeniorhynchus* mosquitoes, *Am. J. Trop. Med. Hyg.,* 29, 711, 1980.

172. **Tesh, R., Rosen, L., Beaty, B. J., and Aitken, T. H. G.,** Studies of transovarial transmission of yellow fever and Japanese encephalitis viruses in *Aedes* mosquitoes and their implications for the epidemiology of dengue, *Dengue in the Caribbean,* PAHO Sci. Publ. No. 375, Pan American Health Organization, Washington, D.C., 1979, 179.

173. **Soman, R. S. and Mourya, D. T.,** Transovarial transmission of Japanese encephalitis virus in *Culex bitaeniorhynchus* mosquitoes, *Indian J. Med. Res.,* 81, 257, 1985.

174. **Mogi, M.,** Relationship between number of human Japanese encephalitis cases and summer meterological conditions in Nagasaki, Japan, *Am. J. Trop. Med. Hyg.,* 32, 170, 1983.

175. **Cates, M. D. and Huang, W. C.,** The effect of photoperiod on transmission efficiency of Japanese encephalitis virus by *Culex tritaeniorhynchus summorosus* Dyar, *Mosq. News,* 29, 620, 1969.

176. **Eldridge, B. F.,** The influence of daily photoperiod on blood feeding activity of *Culex tritaeniorhynchus* Giles, *Am. J. Hyg.,* 77, 49, 1963.

177. **Shichijo, A., Mifune, K., and Hayashi, K.,** Experimental infection of *Culex tritaeniorhynchus summorosus* mosquitoes reared in biotron with Japanese encephalitis virus, *Trop. Med.,* 14, 218, 1972.

178. **Reisen, W. K., Aslamkhan, M., and Basio, R. G.,** The effects of climatic patterns and agricultural practices on the population dynamics of *Culex tritaeniorhynchus* in Asia, *Southeast Asian J. Trop. Med. Public Health,* 7, 61, 1976.

179. **Heathcote, O. H. U.,** Japanese encephalitis in Sarawak: studies on juvenile mosquito populations, *Trans. R. Soc. Trop. Med. Hyg.,* 64, 483, 1970.

180. **Buei, K. and Ito, S.,** The age-composition of field populations and the survival rates in *Culex tritaeniorhynchus* Giles, *Jpn. J. Sanit. Zool.,* 33, 21, 1982.

181. **Hill, M. N.,** Japanese encephalitis in Sarawak: studies on adult mosquito populations, *Trans. R. Soc. Trop. Med. Hyg.,* 64, 489, 1970.

182. **Wada, Y., Kawai, S., Ito, S., Oda, T., Nishigaki, J., Suengaga, O., and Omori, N.,** Ecology of vector mosquitoes of Japanese encephalitis, especially of *Culex tritaeniorhynchus.* II. Nocturnal activity and host preference based on all-night catches by different methods in 1965 and 1966 near Nagasaki city, *Trop. Med.,* 12, 79, 1970.

183. **Scherer, W. F., Buescher, E. L., Flemings, M. B., Noguchi, A., and Scanlon, J.,** Ecologic studies of Japanese encephalitis in Japan. III. Mosquito factors. Zootropism and vertical flight of *Culex tritaeniorhynchus* with observations on variations in collections from animal-baited traps in different habitats, *Am. J. Trop. Med. Hyg.,* 8, 665, 1959.

184. **Burke, D. S., Ussery, M. A., Elwell, M. R., Nisalak, A., Leake, C. J., and Laorakpongse, T.,** Isolation of Japanese encephalitis virus strains from sentinel pigs in North Thailand, 1982, *Trans. R. Soc. Trop. Med. Hyg.,* 79, 420, 1985.

185. **Takahashi, M.,** The effects of environmental and physiological conditions of *Culex tritaeniorhynchus* on the pattern of transmission of Japanese encephalitis virus, *J. Med. Entomol.,* 13, 275, 1976.

186. **Takahashi, M. and Suzuki, K.,** Japanese encephalitis virus in mosquito salivary glands, *Am. J. Trop. Med. Hyg.,* 28, 122, 1979.

187. **Huang, C. H., Liang, H. C., and Jia, F. L.,** Beneficial role of a nonpathogenic orbi-like virus. Studies on the interesting effect of M14 virus in mice and mosquitoes infected with Japanese encephalitis virus, *Intervirology,* 24, 147, 1985.

188. **Wada, Y., Kawai, S., Oda, T., Miyagi, I., Suenaga, O., Nishigaki, J., and Omori, N.,** Dispersal experiment of *Culex tritaeniorhynchus* in Nagasaki area, *Trop. Med.,* 11, 37, 1969.

189. **Bailey, C. L. and Gould, D. L.,** Flight and dispersal of Japanese encephalitis vectors in northern Thailand, *Mosq. News,* 35, 172, 1975.

190. **Sellers, R. F.,** Weather, host and vector — their interplay in the spread of insect-borne animal virus diseases, *J. Hyg. (Cambridge),* 85, 65, 1980.

191. **Top, F. H., Gunakasen, P., Chantrasri, C., Supavadee, J.,** Serologic diagnosis of dengue hemorrhagic fever using filter paper discs and one dengue antigen, *Southeast Asian J. Trop. Med. Public Health,* 6, 18, 1975.

192. **Burke, D. S., Chatiyanonda, K., Anandrik, S., Nakornsri, S., Nisalak, A., and Hoke, C. H.,** Improved surveillance of Japanese encephalitis by detection of virus-specific IgM in desiccated blood specimens, *Bull. WHO,* 63, 1037, 1985.

193. **Burke, D. S., Nisalak, A., and Hoke, C. H.,** Field trial of a Japanese encephalitis diagnostic kit, *J. Med. Virol.,* 18, 41, 1986.

194. **Leake, C. J., Hoke, C. H., Andre, R. G., Fleischer, P., and Gingrich, J.,** unpublished data, 1984.

195. **Self, L. S., Ree, H. I., Lofgren, C. S., Shim, J. C., Chow, C. Y., Shin, H. K., and Kim, K. H.,** Aerial applications of ultra-low-volume insecticides to control the vector of Japanese encephalitis in Korea, *Bull. WHO,* 49, 353, 1973.

196. **Hemingway, J.,** personal communication, 1985.

197. **Bang, F. B.,** Genetics of resistance of animals to viruses. I. Introduction and studies in mice, *Adv. Virus Res.,* 23, 270, 1978.

198. **Center for Disease Control (U.S.A.),** Japanese encephalitis: report of a World Health Organization Working Group, *Morb. Mortal. Wkly. Rep.,* 33, 119, 1984.

199. **Burke, D. S.,** Prospects for immunizing against Japanese encephalitis, Appendix D.6 in *Issues and Priorities for New Vaccine Development,* National Academy of Sciences, Washington, D.C., in press.

200. **Takaku, K., Yamashita, T., Osanai, T., Yoshida, I., Kato, M., Goda, H., Takagi, M., Amano, T., Fukai, K., Kunita, N., Inoue, K., Shoji, K., Igarashi, A., and Ito, T.,** Inactivated vaccine from Japanese encephalitis infected mouse brains (ultracentrifuge purified vaccine) in *Immunization for Japanese Encephalitis,* Hammon, W. McD, Kitaoka, M., and Downs, W. G., Eds., Igaku Shoin Ltd., Tokyo, 1971, 59.

201. **Bhamrapravati, N.,** personal communication, 1986.

202. **Pavri, K.,** personal communication, 1985.

203. **Goto, H.,** Efficacy of Japanese encephalitis vaccine in horses, *Equine Vet. J.,* 8, 126, 1976.

204. **Han, G. S., Chen, B. Q., and Huang, C. H.,** Studies on attenuated Japanese B encephalitis virus vaccine. II. Safety, epidemiological and serological evaluation of attenuated 2-8 strain vaccine after immunization of horses, *Acta Microbiol. Sin.,* 14, 185, 1974.

Chapter 29

KYASANUR FOREST DISEASE*

Kalyan Banerjee

TABLE OF CONTENTS

* This paper is dedicated to the memory of Dr. C. R. Anderson, Dr. T. R. Rao, and Dr. J. Boshell.

I. HISTORICAL BACKGROUND

A. Discovery of Agent and Vectors

During January 1957, a number of human cases of fever were reported at the Primary Health Center at Ulvi in the Shimoga district of Karnataka (then Mysore) State. The signs and symptoms of the disease appeared to be somewhat similar to that of typhoid fever. In the preceding year, similar cases occurred in the same area and at the same time of the year. Serological tests like Widal and Weil-Felix were completely negative in these cases.[1]

Preceding the human epidemic, a large number of sick and dead monkeys were noticed in the nearby forest area. The first reaction of the medical scientists when informed about the epidemic associated with monkey mortality was to think of sylvan yellow fever, a disease unknown in the Indian subcontinent. The unlettered villagers, however, shrewdly observed that the disease had occurred among those who had smelled or seen a dead monkey. The Virus Research Centre (VRC), now the National Institute of Virology (NIV), dispatched a team of scientists in March 1957, and by April 1957, the virus was isolated from the post-mortem specimens of blood and visceral organs collected from a langur (*Presbytis entellus*) and two bonnet (*Macaca radiata*) monkeys. The animal found dead near the Kyasanur State Forest yielded the virus and hence the name.[2] Following the isolation of a filterable virus from the monkeys, the virus was also isolated from patients of the nearby villages. Arthropods including culicine mosquitoes, *Haemaphysalis* ticks, and trombiculid mites were collected

Table 1
GENERAL GEOGRAPHIC DISTRIBUTION OF THE SIX
ANTIGENICALLY DISTINCT TYPES OF TICK-BORNE
VIRUSES

Virus	Region
Louping-ill	U.K.
Central European tick-borne	Central Europe and western U.S.S.R.
Omsk hemorrhagic fever	Central U.S.S.R
Kyasanur Forest disease	India
Russian spring-summer encephalitis	Eastern U.S.S.R.
Langat (TP 21)	Malaya

and attempts were made to isolate the virus from them. The virus was isolated only from ticks.

B. History of Epidemics

Since the first record of the disease in 1957, epidemics of Kyasanur Forest Disease (KFD) have occurred every year. The number of cases average about 400 to 500/year, with the low incidence in 1961 (less than 40 suspected cases) and highest incidence (more than 1000 cases) in 1976, 1977, and 1983. No deaths were recorded in 1961, 1964, and 1965; peak mortality occurred in 1983 and 1984, with 150 and 160 deaths, respectively.

C. Social and Economic Impact

The cases occur among the poor villagers who are obliged to work in the forest area collecting firewood and other forest products. The prolonged illness or death of a person, who is often a breadwinner, causes untold hardships to the family, often leading to the brink of starvation. However, the villagers bear this with a considerable degree of fortitude. In 1966, the bamboo forests bloomed in the endemic area of KFD. To collect the bamboo seeds, a large number of persons moved into the forests, and there was a spurt in the number of cases. In 1983, a patch of virgin forest was being cleared in the South Kanara district. A large number of lumbermen contracted the disease, a number of villagers also fell ill, and ultimately the clearing of the forest had to be suspended by the authorities.

II. THE VIRUS

The KFD virus is a typical flavivirus measuring about 40 nm in diameter.[3]

A. Antigenic Relationships

The virus strains isolated from monkeys, humans, and ticks were found identical.[4]

Several tick-borne viruses belonging to group B arboviruses have been isolated from different parts of the world. Clarke[5] reported the antigenic analysis of the different strains of the tick-borne arboviruses. The studies were done with hemagglutination inhibition (HI) and agar gel precipitation techniques using unabsorbed and virus-absorbed immune sera against different viruses. Six antigenically and geographically distinct viruses of the tick-borne virus complex were demonstrated, KFD virus being one of them (Table 1).

Danes[6] carried out cross-neutralization tests employing immune sera produced against the B3 strain of TBE (tick-borne encephalitis), KFD, and RSSE (Russian spring-summer encephalitis). Though immune sera against all these viruses neutralized heterologous viruses to a certain extent, there were significant differences in the neutralization indexes. Formalinized TBE virus vaccine did not render mice resistant to challenge with KFD virus and vice versa.[7] However, Shah et al.[8] reported that mice immunized with formalinized KFD

Table 2
**RESULTS OF N TESTS WITH
SERA OF GUINEA PIGS AND
RABBITS IMMUNIZED WITH 3
DOSES OF LIVE VIRUS
INTRAMUSCULARLY**

| | | Virus | |
Serum	B_3	RSSE	KFD
B_3 rabbit	6.8	4.8	3.7
KFD rabbit I	4.0	5.5	6.2
KFD rabbit II	3.0	4.5	6.3
KFD guinea pig	2.5	2.8	6.5
B_3^a mouse	2.0	2.0	0.8

Note: Results given in logs of neutralization indexes.

a   Mice immunized with 3 doses per 0.5 mℓ of formalinized vaccine intraperitoneally in 3- and 7-day intervals.

vaccine showed no resistance to challenge with RSSE; on the other hand, mice immunized with RSSE showed a considerable degree of protection when challenged with KFD virus (Table 2).

B. Host Range

KFD virus has been isolated from humans, the black-faced langur monkey (*Presbytis entellus*), and the South Indian macaque (*Macaca radiata*).[9] It has also been isolated from the common house shrew (*Suncus murinus*), Blanford's rat (*Rattus blanfordi*), white-bellied rat (*Rattus rattus wroughtoni*),[10-13] and an insectivorous bat (*Rhinolophus rouxi*).[14] Infant mice inoculated by any of the routes succumb to illness within 3 to 4 days of infection. Weanling mice are also susceptible by intracerebral (i.c.) as well as peripheral routes. However, the incubation period is about 5 to 6 days by the peripheral route. Squirrels (*Funambulus tristriatus tristriatus*) when infected by the subcutaneous (s.c.) route developed high-titered viremia and succumbed to infection.[10] Baby hamsters, when inoculated by the i.c. route, succumbed to the infection, whereas adult rabbits and guinea pigs inoculated by the s.c. or the intraperitoneal (i.p.) route developed antibodies but did not show any susceptibility. Day-old chickens infected by the s.c. or intramuscular (i.m.) route developed high-titered viremia with about 60 to 70% mortality.

Virus inoculated into the chorioallantoic membrane of fertile hen's eggs produced hemorrhage around the blood vessels and death of the embryos within 3 to 4 days.

Man circulates the virus for at least 5 to 6 days in blood; consequently, virus isolation is easily accomplished. The virus has been isolated from many patients.

On experimental infection, the monkeys circulate virus in very high titers for 6 to 7 days. The langur monkey usually succumbs to the infection; among the bonnet monkeys there are often survivors.

The virus also multiplies in primary chick embryo cell cultures, primary monkey kidney cell culture, VERO and BSC-1 cell cultures. The virus multiplies in a cell line obtained from *H. spinigera* ticks, but does not produce any significant cytopathic effect.[15] The virus growth can be assayed by complement-fixation (CF) tests on tissue-culture fluid.

C. Strain Variation

Few studies have been undertaken on the question of strain variation of KFD virus. Six strains of KFD obtained from different hosts, different periods of time, and different areas of Shimoga district were tested in HI, kinetic HI, precipitation in gel, and neutralization (N) and cross protection in mice. No significant differences were noticed among these strains.[4] The P9605 strain of KFD virus was passaged serially in primary monkey kidney cell cultures (*M. radiata*). The virus in the 169th passage showed reduced virulence in mice, but the same strain, when further passaged in primary chick kidney cultures after 23 passages, showed increased virulence in mice.[16]

The plaques produced by the tissue-culture-adopted variant in primary monkey kidney epithelial cells were considerably smaller than those produced by the low mouse passage virus. The two variants also showed a difference in their reproducing capacities at 37 and 40°C. The low mouse brain passage virus grew equally well at both temperatures, while the multiplication of the tissue-culture-adapted strain was significantly low at 40°C as compared with its multiplication at 37°C.[17]

D. Methods of Assay

The KFD virus is easily assayed in infant as well as weanling Swiss albino mice. The virus may be inoculated by the i.c. as well as the i.p. routes. However, in the N tests, highest N indexes were obtained when mice of 21 to 25 days of age were inoculated i.p.[18] The virus can be assayed in tissue culture systems like primary monkey kidney cell culture, VERO, or BSC-1 cell cultures either by cytopathic effect or plaque formation or by immunofluorescence.

III. DISEASE ASSOCIATIONS

A. Humans

1. Clinical Features

The incubation period of KFD in man is estimated to be about 2 to 7 days. The onset is sudden, with chills followed by severe frontal headache. The onset is so sudden that most patients can accurately state the time when they fell ill. Fever soon follows headache and rapidly rises to 104°F. The temperature is continuous and lasts for 5 to 12 days, or even longer. Myalgia is of a severe degree and is reminiscent of dengue.[19] Body pains are cramp-like in nature and are of high intensity at the nape of the neck, lumbar region, and calf muscles. Cough and pain in the abdomen are not uncommon symptoms. Diarrhea and vomiting occur by the 3rd or 4th day of illness. Bleeding from the nose, gums, and intestines begins as early as the 3rd day, but the majority of cases run a full course without any hemorrhagic symptoms. Gastrointestinal bleeding is evidenced by hematemesis or fresh blood in the stools. Such bleeding continues in some cases even after the fever has subsided. Some patients have persistent cough with blood-tinged sputum and occasionally substantial hemoptysis.

Physical examinations during the first few days of illness reveal an acutely ill, febrile patient with a severe degree of prostration. There is usually suffusion of the conjunctiva and photophobia. The cervical lymph nodes are usually palpable, as are the axillary epitrochlear lymph nodes in some cases. Neck rigidity is common during the early phase of disease, probably due to guarding of the painful neck and back muscles. A very constant feature is the appearance of papulovesicular lesions on the soft palate, but no skin eruption has been noted.

The pulse is often slow as compared to the rise in temperature. Bradycardia of 48 to 66 beats per minute occurs by the 9th day when the fever subsides, and lasts until about the 19th day. In an occasional case the pulse rate is less than 48/min even after exercise, which is suggestive of heart block. Blood pressure tends to fall with the fall in temperature by about the 5th to 12th day. In the series of cases reported by Webb and Lakshmana Rao,[19] the average lowest BP was 85/54 mmHg. According to the description of Work,[9] hepatomegaly was not common, although the spleen became palpable, while Webb and Lakshmana Rao[19] noted hepatosplenomagaly in 2 of 28 cases and splenomegaly in 2 others. In the acute phase of the disease, marked dehydration is present. Usually there are no localizing neurological abnormalities. However, mental confusion, drowsiness, and occasionally transient disorientation are found. In two cases among NIV personnel, perversion or loss of taste sensation were noted. On admission to a hospital 3 days after onset of the disease, an electroencephalogram of one case showed diffuse slow activity in all leads, usually in the theta range, which was considered abnormal; the EEG repeated 5 days later revealed similar widespread changes. EEG recordings in another case showed only a mild excess of theta activity.[20] Behavioral changes after KFD infection were observed in another case. Most patients recover their mental capacities completely when the fever subsides unless hemorrhagic complications ensue. Persistent high fever, slurred speech, listlessness, tremors, and dyspnea are ominous signs which are followed by coma, 2 to 12 hr before death.[21] The terminal event is usually due to pulmonary edema. The case fatality rate is 8 to 10%.

The convalescent phase of the disease is prolonged. Often, the disease runs a biphasic course; the second phase occurs after an afebrile period of 1 to 2 weeks, i.e., between the 3rd and 4th weeks after the onset of illness. The fever lasts from 2 to 12 days. It is initiated by headache, and by this time central nervous system abnormalities are generally present. Neck stiffness, mental disturbance, coarse tremors, giddiness, and abnormality of reflexes are noted. Whatever the course, recovery is slow and patients need 1 month or more of convalescence. There is extreme weakness of the muscles, and any physical effort is difficult and results in tremors.

Clinically, KFD resembles Omsk hemorrhagic fever (OHF) which occurs in the Omsk Oblast in Siberia.[19] The viruses are also closely related antigenically. The other tick-borne virus disease antigenically related to KFD and OHF is tick-borne encephalitis. In Karnataka, the differential diagnosis of KFD should include consideration of influenza in very mild cases, and the typhoid and rickettsial group of fevers, e.g., Q fever and mite-borne typhus, in moderate to severe cases. Malaria and leptospirosis should also be differentiated.

2. Laboratory Findings

Work et al.[22] reported persistent leukopenia in their series of cases, leukocyte counts returning to normal with improvement in the clinical condition. Lowest levels of the WBC were recorded in patients between the 3rd and 14th days of fever.[23] In most cases, the neutrophil count was below 2000 mm^3 (range 320 to 1872, mean 1276). There was also a significant degree of lymphopenia during the acute phase, whereas lymphocytosis was seen between the 3rd and 5th weeks. Significant eosinopenia was also present during the 1st week or early in the 2nd week. Turk cells were present in the peripheral blood as early as the 2nd day of illness and persisting throughout the course of the disease. The erythrocyte sedimentation rate ranged from 23 to 54 mm (Wintrobe) per hour. Leukoagglutinins and platelet agglutinins were found between the 3rd and 39th days of illness. No cold agglutinins against red cells were observed, and heterophil titers against sheep cells were low in all cases. Significant reductions in platelet count were observed in most of the cases, particularly in the 1st and 2nd weeks of illness, after which the count returned to normal.

The cellularity of bone marrow was not decreased, and in about half the cases it was hypercellular. Active erythropoiesis and granulopoiesis were seen. In a few cases, metamyelocytes were enlarged, resembling giant metamyelocytes. Some increase was observed in plasma cells and reticulum cells. The megakaryocytes were adequate in number, but showed reduced platelet formation in thrombocytopenic cases.

Bleeding time and mean coagulation time were prolonged in some cases; however, prothrombin time and fibrinogen content were normal in all those studied. Alterations in liver function tests are present in a majority of the cases. The common abnormal pattern was low serum albumin, slightly raised gamma globulin, raised serum alkaline phosphatase and transaminases, slightly increased bilirubin, and elevated zinc sulfate turbidity. Blood urea nitrogen, nonprotein nitrogen, and serum chloride values were usually normal.

The cerebral spinal fluid (CSF) was usually clear and under slight tension; cells were 3 to 9 mm^3; proteins were about 10 to 20 mg%; sugar 60 to 77 mg%; and chlorides were within normal range.

3. Pathology

There are few reports of pathological examinations of fatal human cases. Iyer et al.[24] reported the autopsy findings of three human cases. Gross examination revealed hemorrhage and consolidation of the lungs and massive hemorrhage in the gastrointestinal tract. The brains and meninges were hyperemic, but no gross abnormalities were present.

Microscopically, the liver showed general preservation of the architecture. Liver cell cytoplasm often contained brownish-yellow pigment. There was evidence of focal hepatic necrosis, and there was moderate to marked prominence of Kupffer cells, which were sometimes multinucleated and contained evidence of erythrophagocytosis. There was moderate prominence of the Glisson's capsule with a variable number of mononuclear cells in the portal radicals.

In the kidneys, some of the glomeruli were congested, filling the Bowman's capsule. Varying degrees of degenerative changes were seen in the cortical convoluted tubules, including loss of cytoplasmic outlines, fragmentation of the tubular cytoplasm, and sloughing of cytoplasmic debris in the lumens. The basement membrane was usually undisturbed. Changes in the medulla were less prominent, consisting of focal eosinophilic staining and nuclear pyknosis of the cells of the collecting tubules.

The lungs showed patchy consolidation of the parenchyma with dilation of the capillaries. There was marked exudation of plasma, erythrocytes, and leukocytes in the alveoli. Desquamation of the lining of some bronchioles was also present.

The brain showed general preservation of architecture and cellular integrity. There was no specific evidence of encephalitis in any of the cases. The virus was isolated from blood and different organs of the autopsied cases. It was present in the liver, spleen, kidney, lungs, heart, and skeletal muscles. The virus was not isolated from the brains of these three cases.

In patients suffering from KFD, the virus circulated in the blood from the 2nd day onward, sometimes up to the 13th day of illness. Mean viremia levels were highest between the 3rd and 6th days, ranging from 2.6 to 3.1 dex. The highest recorded viremia level was 5.5 dex.[25,26] Though there is a high level of viremia in man, the virus is neither excreted through the milk nor does it infect suckling babies.[27] The virus is normally not excreted in urine.[28]

B. Domestic Animals

Among the domestic animals, cattle do not develop any significant levels of viremia, nor do they show any signs of disease. No disease manifestations have been recorded in other domestic animals, e.g., sheep, goat, dog, or cat.[96]

C. Wild Animals

1. Monkeys

a. Symptomatology

Langur (*Presbytis entellus*) and bonnet (*Macaca radiata*) monkeys show obvious signs of illness, become listless, and stop feeding. Mortality is high among the langurs.[29] Langur and bonnet monkeys constitute about 85 and 15% of the monkeys found dead in the forest.

However, no comment can be made about the natural population of these monkeys. Antibodies to KFD virus have been demonstrated in wild-caught monkeys. Bonnet monkeys also fall ill, and some of them survive even under laboratory conditions.

b. Pathology

Pathological studies were carried out by Iyer et al.[30] in monkeys dying of KFD in nature. The main pathologic alterations were consistent with the picture seen in humans. Liver showed marked to moderate proliferation of Kupffer cells. Round acidophilic inclusions in the liver cell cytoplasm or lying free in the parenchyma were present in some specimens. Renal tubular damage was frequent, and the cortex was relatively more severely affected than the medulla. Small or moderate necrotic foci were found in the ventricular myocardium. The spleen showed erythrophagocytosis and pronouncement of littoral cells. The central nervous system was infrequently involved. Two of 18 monkeys showed distinctive histologic changes consistent with viral encephalitis.

Experimentally infected monkeys developed high titered viremias ranging from 7.5 to 9.0 dex/mℓ between the 5th and 7th postinoculation days. Post-mortem findings showed fatty changes with occasional areas of focal necrosis and slight inflammation of portal radicles. Many reticuloendothelial cells showed red-cell fragments and remnants of leukocytes. This phenomenon was seen in liver, spleen, and lymph nodes. Focal collections of inflammatory cells and histiocytes were seen in the myocardium. In some cases, the brain showed chromatolysis of large neurons, prominent glial cells, and circumscribed areas of demyelination. Occasional perivascular inflammatory cells were observed. The brain changes were present in animals which died in the 3rd week of the disease. Hematologic investigations revealed findings similar to those in humans. There was marked panleukopenia which was maximal by the 7th day (5th day of fever, the day following the peak of viremia) and thrombocytopenia, platelet counts being half of preinoculation values. The monkeys also developed transient agglutinins against red cells, leukocytes, and platelets. The experimentally infected monkeys (*Macaca radiata*) also suffered from bradycardia and high fever ranging from 102 to 105°F.[31] A large amount of nuclear material was seen engulfed by the RE system; erythrophagocytosis was also marked. Neurological damage was found in monkeys dying between the 19th and 27th days of infection. Monkeys dying before the 12th day of inoculation of virus did not develop encephalitis. In monkeys which died later, lesions in the spinal cord and brain were somewhat similar to those in RSSE virus infection in man.[31,32] Encephalitis was present in 3 of 31 monkeys examined, including those infected both in nature and in the laboratory. The histological picture consisted of chromatolytic changes in the pyramidal cells of the cortex, inflammatory changes in pons, midbrain and medulla, neuronophagia, karyolysis, and degeneration of neurons. Perivascular cuffing was present around the small vessels. In the experimentally infected monkeys, the anterior horn cells of the cervical spinal cord were affected, especially the anteromedial group of cells. In a few experimentally infected monkeys examined, SGOT and SGPT levels were moderately elevated 2 to 3 days before the monkeys became ill.[33]

Apart from the monkeys which died after 12 days of infection, the general pathological picture in man and monkeys seems to be nonspecific. The virus was present in high titers in different organs in the monkeys experimentally infected. Similar pathological findings are made in persons with Omsk hemorrhagic fever (OHF). According to Russian workers, the pathology of OHF is due to lesions in the capillaries; this manifests itself by dilation and increased permeability. The absence of overt destructive or inflammatory lesions in the vessels points to a neurogenic capillaropathy probably due to the effect of the virus on the autonomic nervous system.[24] Whether a similar phenomenon can explain the pathology of KFD remains to be seen.

2. Birds

The role of birds has been extensively studied. Chickens 1 or 2 days old circulate the virus in blood in sufficiently high titer to infect ticks.[34] Rodrigues[35] studied viremia in 29 species of birds. He observed effective viremia in two bird species, i.e., red spur fowl and jungle fowl. Ghosh et al.[36] observed only a few sera positive for KFD among hundreds of bird sera collected from the KFD endemic area.

3. Mice

Pathological changes in mice experimentally infected by KFD virus have been described.[37] In adult mice, the essential lesion was vascular necrosis and infiltration of the vessel wall and perivascular region by mononuclear cells. This was followed by degeneration of neurons and glial proliferation leading to a vacuolation and spongiform appearance. In infant mice, inflammatory lesions with perivascular cuffing were not observed. The brain showed areas with varying degrees of degeneration. Interstitial pneumonitis and hemorrhagic foci and degenerative changes in the liver were noticed.

Studies on the pathogenesis of KFD have been fragmentary. In mice, viremia appears within 24 hr; high virus titers are found in the liver, brain, and lungs, whereas the spleen and other viscera have relatively low virus titers.

Weanling mice can be infected intranasally and even by the oral route. In experimentally infected hamster brains, the virus has been visualized by the electron microscope in the cytoplasm of the neurons and capillary pericytes.[3] The incidence of KFD in some laboratory workers who had no history of wound or tick bite strongly suggests that the virus can infect man through aerosols. TBE virus, a close-antigenic-relative virus of KFD, can also infect through the oral route; human infections occur by the consumption of goat's milk, the virus being excreted in the milk. There is no evidence of KFD being spread in this manner.

D. Applicable Diagnostic Procedures

The virus is readily isolated from the blood of patients or from the blood or viscera of monkeys by inoculation into mice. During the convalescent phase, patients develop high-titer antibodies which persist for long periods.[38] The antibodies are easily detected in HI, CF, NT, or gel precipitation tests. KFD antibodies have also been detected by the ELISA test.[39] Raised SGOT and SGPT levels also can help in the differential diagnosis of clinically suspected cases.

IV. EPIDEMIOLOGY

A. Geographic Distribution and Ecology

Clinically and virologically proved cases of KFD have so far been limited to the Karnataka State. It was first recognized in the village around Kyasanur State Forest (Shigga village in the Ulvi Primary Health Centre) situated in the Sorab taluk (a subdivision of a district which is an administrative unit) of the Shimoga District (Figure 1). The first proved cases occurred in 1957, although there were cases of similar description in 1956. No earlier history of similar cases could be traced.

In the initial stages of the discovery, the known activity of the virus was limited within 100 km² of Sagar and Sorab taluks of the Shimoga district. During the following years, the area spread like a smoldering forest fire to Sagar, Sorab, and Shikaripur taluks. For a long time the activity of the virus remained within this area. However, in 1972 a new focus of activity occurred around Gadgeri village in the Sirsi taluk of North Kanara District, northwest of Shimoga. This locality was 8 km from the nearest spot of KFD in the earlier years, and is separated from the Shimoga district by the wide Varda River Valley. A third focus appeared in 1972 around Aramanekoppa village in Hosanagar taluk, Shimoga district, 22 km from the old focus in Sagar. This focus has been expanding centripetally.

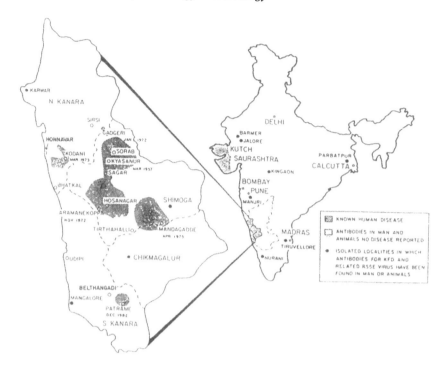

FIGURE 1.   Kyasanur Forest disease distribution in India.

A fourth focus appeared at Kodani in the Honnavar taluk of North Kanara district in March 1973, approximately 50 km from the old theater.

A fifth focus appeared in April 1975 around Mandagadde village in Tirthahalli taluk, Shimoga district, approximately 80 km east of the Aramanekoppa focus. The sixth focus appeared during December 1982, about 80 km south of Mandagadde focus, near Patrame village, Beltangadi taluk of the South Kanara district. The infection appeared to have flared up around the village bordering the Nidle State Forest. This forest clads a spur of the Western Ghats Mountains. It appears that an area of forest of 400 ha was being cleared for a cashew nut plantation (Figure 1). The altered ecosystem seems to have become conducive for the flaring up of KFD.

The foci of KFD in the Shimoga district and of the Sirsi taluk of North Kanara district are situated on the eastern slope or the spurs of the Western Ghats Mountains. They are interspersed with river valleys. The focus on the Honnavar taluk is on the western slope of the Western Ghats, almost reaching the coastal plains. The focus at Beltangadi taluk is also on the western slope, but is not near the sea. The average annual rainfall in the Shimoga district is about 80 in. and the annual range of mean temperature varies from 70°F in January to 80 to 83°F in May. Most of the rain falls between June and September. The rainfall in the South Kanara district is higher (100 to 120 in.). The area is covered with forest comprised of a mosaic of evergreen and deciduous vegetation and dense undergrowth including thickets of lantana, a plant not of indigenous origin, which shelters small mammals and is now the last refuge of jungle fowl. The forest, which was rich in big game up to the early 1950s, has practically none today. Wild pigs and porcupines are present and the monkey population is high. The avifauna is still very rich.

There has been relentless deforestation for agricultural purposes and planting of teak, eucalyptus, and cashew trees. The majority of the people of the area are engaged in agriculture from May to December, growing rice, sugarcane, areca nuts, and banana. During the dry season they are engaged in felling trees, collecting firewood and other forest products, and taking cattle to the forest for grazing.

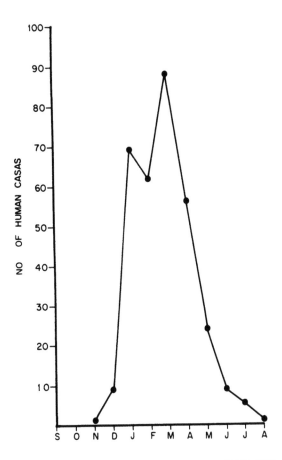

FIGURE 2. Mensal distribution of human cases of KFD, 1959 to 1966 (September to August). All cases were proved either by virus isolation or by serology.

B. Incidence

An average of about 400 to 500 cases are reported annually. During the earlier years when the geographic area affected was small, the number of cases was relatively low. With the increase in number of focuses the incidence has increased. The highest recorded incidence was in 1983, with 1555 cases and 150 deaths.

C. Seasonal Distribution

During the period 1959 to 1966, 322 cases proved as KFD either by virus isolation or through serology were investigated. The monthly distribution of these cases is shown in Figure 2. The epidemic period begins in November or December, peaks from January to April, and declines by May and June.[40] The monsoon or rainy season is from June to September, though a few showers may occur in the months of May and October. The dry period is from November to May.

The largest number of cases occurs in the drier part of the year, i.e., December to June. Monkey epizootics also show a similar pattern. This period corresponds to a high preponderance of nymphal stages of the ticks in the forest. This also corresponds to the human activity in the forest for gathering firewood and other products.

D. Risk Factors

The largest number of cases occurs between 10 and 40 years of age. The male to female ratio is 2.6 to 1, and the sex distribution of deaths roughly parallels that of the cases. The age and sex distribution reflects the occupational exposure in forested areas. The case fatality rate in 1957 to 1958 was about 8 to 10%. Thereafter it declined to 4%. However, in 1983, when the disease flared up in a new area, the case fatality rate again rose to 10%. During the study period of 1957 to 1966, it appeared that the mortality rate was lower in persons below 35 years of age as compared with older persons.

Almost all cases have a history of forest exposure. A few of them did not give a history of actually going to the forest. Nearly all affected villages are surrounded by forest, and infected ticks occur at the forest edge. It is also possible that cattle bring infected ticks from the forest into the villages. Small children who normally do not go out of the immediate vicinity of their households have not often been found infected.

The close association of seeing a dead monkey and human infection which was observed by the villagers is probably due to contracting the infection through infected ticks shed by an infected or dead monkey. The ticks are unable to move themselves more than a few inches. They remain in the vicinity of a dead monkey.

Many laboratory workers have been infected with KFD virus. Cases have been reported from New York, Washington, D.C., Poona, and Shimoga.[41-43] In Shimoga, a number of workers contracted the infection while working in the field, some while conducting autopsies on infected monkeys, others probably by infected tick bites. At Poona, a few workers contracted the infection while handling infected ticks. A number of persons were infected while handling the virus in the laboratory, but others did not have any history of handling the virus personally. These persons seem to have contracted the infection through the inhalation of aerosol.[43]

E. Serological Epidemiology

In the KFD endemic area, blood samples from 1200 persons were collected in 1957—1958, and tested for HI and CF antibodies against KFD and a number of other arboviruses. Of the 123 sera positive in HI against any one or more of the arbovirus antigens, 51 contained CF antibodies only to KFD virus, and 14 had antibodies to KFD and one or more other antigens. The results could be explained by the cumulative effect of the infection in the preceding years.

Sera from squirrels (*Funambulus tristriatus tristriatus*), shrews (*Suncus murinus*), and forest rats (*Rattus rattus wroughtoni*) caught in the KFD area showed neutralizing antibodies. Sera from jungle fowl (*Gallus sonneratii*) and golden woodpecker (*Brachypternus benghalensis*) from the KFD area also showed neutralizing antibodies.[9] On one occasion, serum from *Mus booduga* (a forest mouse) collected from Arehalli (100 km south of Sagar) was found to contain neutralizing antibodies. Antibodies to KFD have also been found in bats in a cave near Poona.[44] However, it appeared that some of the bat sera might have contained nonspecific inhibitors.[45] An extensive serological survey was carried out in bats collected from the KFD area.[46] Sera numbering 727, collected from 17 species of bats, were tested for neutralizing antibodies against KFD virus. Neutralizing antibodies were contained in 138 sera collected from seven species, namely, *Rousettus leschenaulti*, *Eonycteris spelaea*, *Cynopterus sphinx*, *Rhinolophus rouxi*, *Hipposideros speoris*, *Hipposideros lankadiva*, and *Miniopterus schreibersi*. The high incidence of antibodies in certain species suggested activity of the virus in bats. In addition, KFD virus isolations have been made from bats (*R. rouxi*) and the soft ticks (*Ornithodoros chiropterphila*) infesting them.[14] However, the ixodid ticks (vectors of KFD) are rarely found on bats. This poses an unsolved question regarding the role of bats in the epidemiology of KFD.

Cattle sera collected from the KFD area and vicinity had neutralizing antibodies to KFD.[9] However, in the absence of viremia, cattle do not seem to play any significant role as amplifiers of the virus.

Sera from monkeys, from langurs as well as from bonnet monkeys, have shown antibodies

Table 3
NEUTRALIZING ANTIBODIES TO KFD VIRUS
IN ANIMAL SERA FROM KFD ENDEMIC
AREA, KARNATAKA

Hosts	No. positive / tested	% Positive
Rattus rattus wroughtoni	13/236	5.5
Rattus rattus rufescens	0/26	0
Rattus blanfordi	2/76	2.6
Mus booduga	0/13	0
Funambulus t. tristriatus	11/31	35
Tatera indica hardwickei	1/5	20
Suncus murinus	9/59	15
Presbytis entellus	5/25	20
Macaca radiata	1/12	8
Cattle	13/43	30

Table 4
NEUTRALIZING ANTIBODIES IN
HUMAN / ANIMAL SERA IN
KUTCH AREA

Hosts	No. positive / tested	% Positive
Human	28/280	10.0
Camel	7/50	14.0
Cattle	1/12	8.3
Buffalo	0/2	0
Goat	1/23	4.3
Sheep	0/26	0
Donkey	7/35	20.0
Horse	11/12	91.7

to KFD,[96] signifying that not all monkeys in nature succumb to the infection. Table 3 summarizes the serological data from the KFD endemic area in Karnataka.

A few sera collected in 1952 from Kutiyana in Sourashtra (Figure 1), prior to the discovery of KFD, showed the presence of neutralizing antibodies to RSSE virus. A survey carried out in the Bhuj and Banni areas of Sourashtra in 1960 showed the presence of neutralizing antibodies against KFD virus in sera from humans, equines, and camels, and also from cattle and goats.[96] Table 4 shows the results of a serological survey in the Kutch area. It is therefore intriguing that in Sourashtra and Kutch, where the climate varies from arid to semiarid and the areas are geographically removed from Karnataka, should have KFD virus or at least a virus closely related to KFD.

V. TRANSMISSION CYCLES AND ECOLOGICAL DYNAMICS

A. Vectors
1. Virus Isolation from Ticks

In the KFD area 36 species of ticks have so far been recorded. There are 15 species of *Haemaphysalis*, 2 species of *Ixodes*, 3 species of *Amblyomma*, 3 species of *Rhipicephalus*, 2 species of *Boophilus*, 4 species of *Hyalomma*, 2 species of *Ornithodorus*, and 1 species each of *Argas*, *Aponomma*, *Nosomma*, and *Dermacentor*. Of the 15 species of *Haemaphysalis* present in the area, the virus has been isolated from 10. According to the frequency of isolation, they are *H. spinigera*, *H. turturis*, *H. kinneari*, *H. kyasanuresis*, *H. wellingtoni*,

H. bispinosa, H. minuta, H. cuspidata, H. intermedia, and *H. aculeata.* The virus has also been isolated from *Ixodes ceylonensis* and *Ix. petauristae,* and one species each of *Dermacentor, Amblyomma, Rhipicephalus,* and *Ornithodoros.*[14,47,48] A few thousand virus isolations have been made from ticks; *H. spinigera* contributed about 95% of the isolations. It is also the predominant tick species on the forest floor. The infection rate in the *H. spinigera,* caught from the forest floor, varies in different years and localities. The range is from 1 in 10 to 1 in 900 in the areas showing evidence of human cases or monkey deaths.

2. Hosts for Ticks

The life cycle of a tick is briefly as follows. The adults of *H. spinigera,* both male and female, feed upon large animals like cattle or a few wild animals like spotted deer, sambar deer, and Indian bison. Fed females drop on the forest floor and lay their eggs. Under the favorable conditions of temperature and humidity, larvae hatch out. The unfed larvae climb up leaves of small shrubs and plants to lie in wait for passing small mammals or birds. The larvae have their blood meal, usually from ground birds or small mammals like rats, squirrels, or porcupines, and molt to nymphal stage. It is the nymph which seems to be important in passing the virus to man. After the blood meal, the nymphs molt into adults. Thus, *H. spinigera* is a three-host tick and each stage, i.e., larva, nymph, and adult, feeds on different animals.[49-51]

a. Humans

During a study on the infestation of ticks in a sample of 4668 persons, 10% of the persons examined were found infested, with an average of 2.5 ticks per infested person. From this study it was concluded that *H. spinigera* nymphs are the most important virus vector to humans.[52]

b. Monkeys

The monkeys in the KFD area are not entirely arboreal; they spend enough time on the ground to become infested with ticks. In a study by Trapido et al.[53] on dead or shot monkeys, 78% and 20% of the ticks collected were *Haemaphysalis* spp. and *Dermacentor* spp., respectively. Of all *Haemaphysalis,* 90% were *H. spinigera*; the remaining *Haemaphysalis* species were *H. turturis, H. kinneari, H. wellingtoni, H. minuta, H. bispinosa, H. aculeata,* and *H. cuspidata.* The other genera of ticks were represented by *Amblyomma, Ixodes,* and *Rhipicephalus.*

Larval ticks were found on monkeys in all months except April and the monsoon months June to September, with a distinct peak during October and November. Nymphs are found on monkeys principally from November to March.

The tick infestation rate was found to be too low to explain the high mortality of monkeys in the forest.[53] A detailed study of the tick infestation rate of monkeys was carried out later.[54] It was found that the monkeys had a much larger number of larvae and nymphs of *H. spinigera* and *H. turturis* on them. There was a marked preference of *H. turturis* adults for *Presbytis entellus* as compared with *Macaca radiata.*

c. Domestic Animals

Cattle are the most numerous domestic animals having forest exposure. Adult *H. spinigera* is one of the most common bovine ectoparasites, and cattle thus play a principal role in tick reproduction and population density. Adult ticks are usually not found on small mammals or birds. Therefore, cattle rearing can be considered as the most important man-made factor that favors high vector density at the very places frequented by man. Cattle also carry all stages of other *Haemaphysalis* species.

Before 1970, goats and sheep were not considered important in the endemic area. Since

1970, however, goats have been introduced in the KFD area in substantial numbers. All stages of *H. intermedia* parasitize goats and the larvae and nymphs also parasitize small mammals. KFD virus has been isolated from *H. intermedia* on a few occasions. Recent data indicate the presence of *H. intermedia* in substantial numbers in forested habitat.[55]

d. Small Mammals

Tick infestation studies of small mammals showed that the larvae and nymphs parasitize these animals, including *Rattus rattus wroughtoni*, *R. blanfordi*, *R. r. rufescens*, *Golunda ellioti*, *Funambulus tristriatus tristriatus*, *Mus booduga*, *Vandeleuria oleracea*, and *Suncus murinus*.[56] *R. r. wroughtoni* is the most abundant small mammal followed by *R. blanfordi* and *Suncus murinus*, *R. r. rufescens* and *G. ellioti* are comparatively rare in the forest; the former is predominantly peridomestic. *Haemaphysalis* and *Ixodes* constitute the principal tick genera found on captured small mammals. The host predilection studies have shown that *Ix. ceylonensis* is associated with the shrew, *H. kinneari* with *R. r. wroughtoni*, and *H. spinigera* with the squirrel. However, *R. r. wroughtoni*, *R. blanfordi*, and *F. t. tristriatus* also harbor larvae and nymphs of *H. spinigera* and *Ixodes* spp.

KFD virus has been isolated from the larvae and nymphs collected from trapped small mammals. These included *Ixodes* spp. and *H. kinneari* on *R. blanfordi*; *Ixodes* spp. on *Suncus murinus*; *H. kinneari* on *Rattus r. wroughtoni*; and *H. spinigera* and *H. kinneari* on *Funambulus t. tristriatus*.[48]

Special mention must be made of the porcupine (*Hystrix indica*). It is infested with immature stages of *H. spinigera*, all the stages of *H. turturis* and *H. kyasanurensis*, and immature stages of *Dermacentor auratus*, *Amblyomma integrum*, and *A. javanense*. Porcupines are known to be present in the KFD area in significant numbers and are infested with different stages of ticks in large numbers. Porcupines circulate KFD virus in blood at high titers and for prolonged periods; it appears to be an ideal host for amplifying infection and disseminating virus-infected ticks on the forest floor.[57]

Ticks feeding on bats (*Ornithodoros* ticks) are highly species specific, and it seems that they do not feed on other mammals. Therefore, though KFD virus can remain in bat colonies through the bite of their tick parasites, it is unlikely that the bats can act as a source of infection to man.

e. Large Mammals in Forest (Other than Monkeys)

As has been mentioned earlier, larger mammals in the forest have become rare. Wild pigs are still encountered, and *H. kinneari* adults are specifically associated with them.

f. Birds

The association of birds with ticks varies according to the habits of each avian species. Ground-dwelling birds, e.g., *Phasianidae*, have high infestation rates. They usually harbor the larval and nymphal stages, and adult forms are rarely found. Certain other species, e.g., crow pheasant (*Centropus sinensis*), Magpie robin (*Copsychus saularis*), and some babblers also have heavy tick loads. The peafowl (*Pavo cristatus*) and the jungle fowl (*Gallus sonneratii*) are heavily infested with ticks, predominantly *H. wellingtoni* and *H. minuta*. These ticks are also found on small mammals and monkeys, though in smaller numbers. It appears that ticks from birds are rarely infected with KFD virus. Only on one occasion was KFD virus isolated from a pool of nymphs of *H. spinigera* collected from spur fowls (*Galloperdix spadicea*).

B. Laboratory Experiments with Vectors

Transmission of the virus from one vertebrate host to another has been demonstrated with *H. spinigera*,[58] *H. turturis*, *H. kinneari*, *H. minuta*,[59] *H. kyasanurensis*,[60] *H. wellingtoni*,[61]

Ix. petauristae,[62,63] *Ix. ceylonensis,*[63] *Rhipicephalus haemaphysaloides,*[64] *Dermacentor auratus,*[65] and *H. cuspidata.*[96] All these ticks have been found in the KFD area. In addition, virus transmission has been demonstrated by some species of ticks not found in the endemic area, namely, *H. obesa,*[66] *Ornithodoros crossi,*[67] *Argas persicus,*[68] *O. savignyi,* and *Argas arboreus.*[96]

Transstadial transmission of the virus was easily demonstrated in all the experimental studies.[58-62] Transovarial transmission of the virus was demonstrated in the larval progeny of 12 out of 16 female *Ix. petauristae* infected through feeding on viremic squirrels.[69] In the case of *Argas persicus,* virus was transmitted to 8 out of 58 nymphal progeny.[68] However, *H. spinigera* does not transmit the virus transovarially.[70]

C. Vector Dynamics

The monsoon (rainy) season in the KFD area is from June to September. The larval population of *Haemaphysalis* starts building up in September, remains high through October and November, and starts diminishing in December; there is a small second peak in March. The nymphs appear on the forest floor (as evidenced by flag dragging) as well as on captured animals and birds in December and January, and reach their peak population in February and March. The nymphal population starts to dwindle by April and May and tapers off during the rains by the middle of June. Adults increase in numbers as the nymphs decrease and reach a peak during the rainy months, but last throughout the fall and up to the dry season in January to February of the following year. During the period of February to May, the adults are found in flag drags as well as on animals such as monkeys.

Ix. petauristae has a phase fluctuation roughly alternating with *Haemaphysalis.* The larvae appear by mid-June in flag drags as well as on trapped animals, whereas the nymphs have a peak in August and September.

Ixodes ceylonensis has a phase fluctuation similar to *Haemaphysalis.* The larvae have their upsurge in September and last until the end of the year, the nymphs being prevalent during January and February. The life span of these species is generally restricted to 1 year. However, in certain species, namely, *H. kinneari, Dermacentor auratus, Amblyomma* spp., and *R. haemaphysaloides,* overlapping of two adult populations has been observed.

D. Overmonsooning of KFD Virus

Maintenance of KFD virus through the monsoon poses a problem similar to that for overwintering of other arboviruses in the temperate zones. The *Haemaphysalis* larvae cannot be collected in large numbers during the heavy monsoon, which is a season of adult abundance.[71] The virus has not been isolated from unfed larvae in nature. Transovarial transmission of KFD virus in *Haemaphysalis* does not occur, or if at all, at an insignificant rate.

During the monsoon an occasional *Haemaphysalis* nymph has yielded virus,[72] but a more important feature is that larvae of *Ixodes* species are highly prevalent during the monsoon. *Ixodes* species have yielded virus and have also been shown to transmit the virus in the laboratory. It is therefore surmised that *Ixodes* species and small mammals maintain the virus in nature during the monsoon months. It can also be added that *H. kinneari* adults are long-lived, and some of the adults of the previous generation persist in the period of the next generation. An infected larva or nymph of *H. kinneari* can become an infected adult and remain active in the next season.

It is apparent that the epizootic in monkeys and the epidemic in man correspond with the nymphal season of *Haemaphysalis* ticks.

The distribution of infected larvae and nymphs of *H. spinigera* in the KFD area is scattered and focal-geographic concordance has been observed in such foci between the presence of infected nymphs of *H. spinigera* and *H. turturis* and the death of a monkey.[73] It appears that the nymphs came off a sick or dead monkey, hence the creation of foci of infection.

E. Experimental Infections in Animals

Several species of small mammals circulate the virus in sufficiently high titers to infect ticks, especially Blanford's rat (*Rattus blanfordi*), the jungle striped squirrel (*Funambulus tristriatus tristriatus*), and the common house shrew (*Suncus murinus*). Viremia levels in the shrews were the highest (about $10^{7.5}$ LD$_{50}$) and lasted for 11 days.[74] The giant flying squirrel (*Petaurista petaurista philippensis*) is a common nocturnal animal in the forests of the Shimoga district. When experimentally infected, viremia was observed for up to 6 to 7 days (in one animal for 10 days). The peak viremia was on the 3rd to 4th day, ranging up to $10^{7.5}$ LD$_{50}$.[75]

Mus platythrix (a forest mouse) also circulated virus, but at low titer. Some of the animals had the virus persisting in the brain for very long periods (936 days in one animal) in spite of having high antibody titer in the blood.[76] *Mus booduga*[74] and *Vandeleuria oleracea*[77] (forest mice) also develop viremia on experimental infection. However, the levels of viremia in these animals did not exceed $10^{3.5}$ LD$_{50}$. The black-naped hare (*Lepus nigricollis*) is a common lagomorph in the KFD area. When infected experimentally through the bite of infected ticks, low-grade viremia occurred.[78]

Felix chaus (jungle cat) did not circulate the virus when infected experimentally.[96] Frugivorous bats, namely, *Rousettus leschenaulti* and *Cynopterus sphinx*, circulate the virus when experimentally infected. *R. leschenaulti* did not demonstrate any signs of illness and there was no mortality among them, while *C. sphinx* developed illness and some of the bats died.[44,79] Insectivorous bats (*Rhinolophus rouxi*) were also found to develop viremia after experimental infection. These bats could not be kept more than 4 days in the laboratory. However, viremia was seen from the first postinfection day ($10^{2.5}$ to $10^{4.5}$ LD$_{50}$) to the 4th day postinfection (up to $10^{6.2}$ LD$_{50}$).[80] The Indian crested porcupine (*Hystrix indica*) developed viremia which lasted up to the 9th day postinfection; peak viremia was on the 5th to 6th day and ranged up to $10^{7.5}$ LD$_{50}$.[57] All small mammals surviving in captivity after experimental infection developed antibodies to KFD virus.

Cattle constitute the largest number of domestic mammals in the KFD area and the principal host for *Haemaphysalis spinigera*. Fourteen calves without neutralizing antibodies to KFD virus were infected by tick bite. Only one calf had barely detectable virus in the blood on the 1st day of exposure. Eight of the 14 calves developed neutralizing antibodies. Thus, it appears that cattle do not act as amplifiers of the virus.[81]

The role of birds in the natural cycle of KFD has also been studied.[35,82] Rodrigues[35] studied viremia in 29 species of birds. Viremia of some epidemiological significance was found in red spur fowl (*Galloperdix spadicea*) and jungle fowl (*Gallus sonneratii*). Though viremia was detected in a number of other species, the level in them was either too low or the birds did not carry ticks on them which may act as transmitters. Figure 3 shows the natural cycle of KFD virus, as surmised from the studies carried out in the field and the laboratory.

VI. SURVEILLANCE

A. Clinical Hosts

Soon after the discovery of Kyasanur Forest disease, the Public Health Department of the Government of Karnataka established a Virus Diagnostic Laboratory (VDL) at Shimoga. Cases are reported by the primary health centers and are also discovered during tours by the staff of the VDL. Blood samples are collected and virus isolation attempts carried out by inoculating patients' serum into mice. The infected mouse brain is employed as an antigen in the CF test.[83] This surveillance system gives a fairly accurate idea of disease incidence in the various districts.

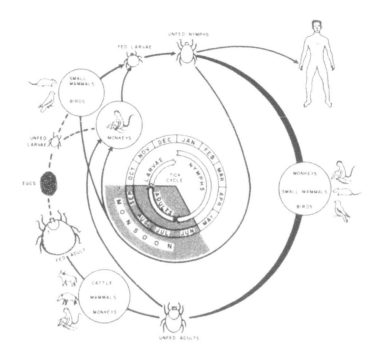

FIGURE 3.   Kyasanur Forest disease, natural cycle.

B. Wild Vertebrates

Up to 1975, the National Institute of Virology had a system of giving a monetary incentive to any villager who reported a sick or dead monkey. A team from the NIV autopsied these animals on the spot, collected ticks from the body of the monkey, and cremated the remains after autopsy. Virus isolation attempts were made on both the monkey organs and the ticks. This method gave very accurate information about the infected foci in the forest.

C. Vectors

Up to 1975, ticks were collected from the forest floor by dragging in the forests where a monkey death had been reported, and virus isolation attempts were made.

VII. INVESTIGATION OF EPIDEMICS AND UNSOLVED QUESTIONS

In the earlier years, epidemics were regularly investigated. Although many aspects of KFD epidemiology and clinical presentation were answered by these studies, other questions remain. Involvement of the heart, especially regarding development and evolution of heart block, neurological signs, derangement of the blood coagulation mechanism, and changes in the electrolyte balance need to be studied in detail. The primary health centers are, however, not equipped to carry out such investigations.

A. Origin of KFD

No history of the existence of KFD could be traced before 1956. The question whether the virus was imported from some other place has not been satisfactorily answered. In Asia, viruses resembling KFD include Omsk hemorrhagic fever virus in Siberia, a human pathogen, and langat virus in Malaysia, which does not produce serious disease in humans.

If we presume that the KFD is a variant of OHF virus introduced from Siberia to Karnataka, the most likely carriers would have been ticks on migratory birds. Of thousands of migratory birds examined for ticks, only one bird had a tick of genus *Hyalomma*, which is found both within and outside India. Therefore, the transfer of infected ticks by birds is not likely. Also, the question of the degree of similarity between KFD, OHF, and Langat viruses has not been settled through modern molecular techniques. The other alternative is that the KFD virus has long been in Karnataka, circulating in a silent forest cycle involving ticks and small mammals; due either to a mutation rendering the virus virulent or to ecological changes rendering humans and monkeys more susceptible to exposure, the disease has become evident in the last 4 to 5 decades.

Banerjee and Bhat[84] defined a concept of the intensity of infection in the ticks, and showed that a direct relationship existed between the intensity of infection of *H. spinigera* and the number of human cases in a particular year. The relationship could not be established with *H. turturis*. This is in accordance with the observed fact that humans are infected only by *H. spinigera*. It was shown that below a critical level of the intensity of infection in *H. spinigera*, no humans cases would occur. This was probably the case before 1956.[84]

H. spinigera predominates in the deciduous forests, while in the evergreen forests *H. spinigera* population is sparse.[55] Therefore, it appears that destruction of the evergreen forests and their replacement with deciduous plantations, creation of farmlands, and increase in the cattle population created a situation conducive to the upsurge of the population of *H. spinigera*, the principal vector of KFD.

B. Spread of KFD Infection in Karnataka

In the Sorab-Sagar track of KFD, the infection spread centripetally, though a few cases did occur in the old area for a few years. Groups of monkeys have limited range of movement, though there is some degree of overlap between the territory of one group with another. The mobility of the ticks corresponds essentially to the mobility of their hosts. Small mammals, except for the porcupine, have very limited ranges of movement. The death of a large number of monkeys in a group would reduce the number of infected ticks in the old area and the infection may pass on to a fresh band of monkeys. This can explain the movement of infection by contiguity, but it cannot explain leapfrogging of the infection by several kilometers, and particularly the recent flare-up at Beltangadi. It is possible that cattle may carry infected ticks to a new area, but in practice this does not seem to be common. The mobility of cattle in the KFD area is not great, particularly since the development of hydroelectric projects in Karnataka beginning in 1962; rivers which the cattle could ford earlier cannot be crossed now. The probability of KFD virus crossing river valleys through ticks on cattle, monkeys, or other animals seems low.

C. The Presence of Antibodies in Kutch

Except for the presence of antibodies to KFD in animals (Section IV.E) not much is known about the virus in this semiarid region. Whether the antibodies found in animals are truly antibodies to KFD or are due to infection with another virus resembling KFD needs to be examined.

VIII. PREVENTION AND CONTROL

A. Vector Control

Infected nymphs and larvae are shed in the forest mainly by the monkeys, rats, shrews, porcupines, squirrels, and probably a few birds, forming foci of infection. Destruction of infected ticks would necessitate control of ticks throughout the entire forested area, and is not technically or economically feasible. As there has been a constant association of humans becoming infected in the vicinity of dead monkeys, use of spray insecticides has been recommended in a 50-m radius around a dead monkey.

Experiments have been carried out to determine the susceptibility of ticks to different insecticides. The relative effectiveness of the insecticides tested was an isomer of benzene hexachloride (lindane) followed in order by carbaryl and malathion. In experimental plots, different concentrations of lindane were sprayed and the relative abundance of ticks on different days was compared with the controls. Use of lindane of 1.12 kg/ha resulted in a significant reduction of the tick population (*H. spinigera* and *H. kinneari*) for at least 12 weeks.[85]

As cattle constitute the most abundant animal to harbor the tick vectors, the effect of insecticides on cattle was also studied. Cattle were sprayed or dusted with the most effective concentrations of insecticide, i.e., 0.5% malathion, 0.25 and 0.5% carbaryl, and 0.01 to 0.03% lindane, resulting in significant control of male *H. bispinosa* for 3 to 7 days and of females for 2 to 3 days.[85] With most treatments, ticks started reappearing on cattle after 1 to 2 days.

Though recommendations have been made for the spraying of insecticides around the place of monkey death, technically it is difficult in certain inaccessible areas to transport large volumes of water needed for the spray. The economics and logistics make the implementation of a regular insecticide spray program difficult. Under these circumstances, the prevention of tick bites by the use of repellents should be considered. The effectiveness of different repellents, e.g., dimethylphthalate (DMP), dibutylphthalate (DBP), *N,N*-diethyl-*m*-toluamide (DEET), a formulation of DEET and DMP (from Boots Co. Ltd., Nottingham, England), "Mylol" (a formulation of DBP and DMP, from Boots Co. India Ltd.,), and pyrethrum (Pyrect extract 2%) was studied on eight species of ticks. These included *H. spinigera, H. turturis, H. kinneari, H. intermedia, H. obesa,* and *H. wellingtoni,* and two species of argasids (*A. persicus* and *A. arboreus*). The *Haemaphysalis* species were much more sensitive to the repellents than *Argas* species. The relative efficacy of the repellents was in the following order: DEET > pyrethrum > DEET + DMP > Mylol > DMP > DBP. The effectiveness of DBP and DMP were much less than DEET or pyrethrum.[86] Mylol was found to be an effective repellent on langur monkeys against ticks when they were "walked" in the forest.[87]

B. Vaccines

A vaccination campaign was organized in 1958—1959 with the formolized RSSE vaccine. Notwithstanding the difficulties in organization and evaluation of an ideal vaccine trial,[88,89] serological studies with the sera from vaccinated persons failed to show development of significant levels of antibodies against KFD virus in most of the persons.[90]

A formolized tissue culture (FTC) vaccine of KFD virus was developed at the Haffkine Institute, Bombay.[91] After proper safety trials, the vaccine was tested in monkeys and in a small group of human volunteers. Six langur monkeys were inoculated s.c. with two doses of 1 mℓ of the FTC vaccine at an interval of 1 month. Serological studies showed development of neutralizing antibodies, although HI and CF antibody responses were poor. The monkeys successfully withstood challenges with live KFD virus 1 to 15 months after vaccination.[92]

Two doses of the FTC vaccine were administered to 87 human volunteers who had no history of KFD. Twelve persons had antibodies to KFD virus in their prevaccination sera, indicating the KFD virus could cause subclinical infection in man. The vaccine evoked the production of neutralizing antibodies in about 72% of the persons; HI and CF antibody development was poor.[93] In 1970, a larger quantity of FTC vaccine was produced at the NIV, Poona. Based upon the previous experience of the prevalence of the disease, villagers from Sagar and Sorab taluks of the Shimoga district were vaccinated. After randomization, 1405 persons were administered two doses of the vaccine, while 584 persons received a single dose. There was no untoward reaction in any of the vaccinees. The vaccinees were followed for 4 years. There was only one case of KFD among the vaccinees and 32 cases among the unvaccinated persons. The difference was statistically significant.[94] Paired serum samples from 214 vaccinated persons and 204 matched controls were tested for the development of HI, CF, and neutralizing antibodies. Seroconversion was demonstrable in 59% of the vaccinees who had received two doses of the vaccine. The presence of antibodies to the other flaviviruses, especially of West Nile virus, seemed to interfere with the efficacy of the formolized KFD vaccine.[95] This was also seen in the study among the laboratory volunteers.[93] However, work on better vaccines is required.

ACKNOWLEDGMENTS

My sincerest thanks are due to Dr. Mrs. Shobhana Kelkar and Dr. G. Geevarghese for their invaluable help in the preparation of the manuscript. I am also indebted to Dr. H. R. Bhat for discussions and providing certain unpublished data.

REFERENCES

1. **Seshagiri Rau, S.,** Preliminary report on epidemic of continuous fever in human beings in some villages of Sorab taluk, Shimoga district (Malnad area) in Mysore, *Indian J. Public Health,* 1, 195, 1957.
2. **Work, T. H. and Trapido, H.,** Kyasanur Forest disease, a new virus disease in India, *Indian J. Med. Sci.,* 11, 341, 1957.
3. **Jelinkova, A. A., Danes, L., and Novak, M.,** Electron microscopic detection of Kyasanur Forest Disease virus in olfactory bulb and tract of intranasally inoculated hamsters, *Acta Virol. (Engl. Ed.),* 18, 254, 1974.
4. **D'Lima, L. V. and Pavri, K. M.,** Studies on antigenicity of six Kyasanur Forest disease virus strains isolated from various sources, *Indian J. Med. Res.,* 57, 1832, 1969.
5. **Clarke, D. H.,** Antigenic relationships among viruses of the tick-borne encephalitis complex as studied by antibody absorption and agar gel precipitin techniques, in *Symposia CSAV Biology of Viruses of The Tick-Borne Encephalitis Complex,* Libikova, H., Ed., Czechoslovak Academy of Sciences, Praha, 1962, 67.
6. **Danes, L.,** Contribution to the study of antigenic relationship between tick-borne encephalitis and Kyasanur Forest disease viruses, in *Symposia CSAV Biology of Viruses of the Tick-Borne Encephalitis Complex,* Libikova, H., Ed., Czechoslovak Academy of Sciences, Praha, 1962, 81.
7. **Benda, R. and Danes, L.,** Evaluation of the immunogenic efficiency of a tick-borne encephalitis virus vaccine, in *Symposia CSAV Biology of Viruses of the Tick-Brone Encephalitis Complex,* Libikova, H., Ed., Czechoslovak Academy of Sciences, Praha, 1962, 354.
8. **Shah, K. V., Buescher, A., et al.,** Discussion in, *Symposia CSAV Biology of Viruses of the Tick-Borne Encephalitis Complex,* Libikova, H., ED., Czechoslovak Academy of Sciences, Praha, 1962, 85.
9. **Work, T. H.,** Russian spring-summer virus in India. Kyasanur Forest disease, *Prog. Med. Virol.,* 1, 248, 1958.
10. **Webb, H. E.,** Kyasanur Forest disease virus in three species of rodents, *Trans. R. Soc. Trop. Med. Hyg.,* 59, 205, 1965.
11. **Boshell, M. J., Rajagopalan, P. K., Goverdhan, M. K., and Pavri, K. M.,** The isolation of Kyasanur Forest disease virus from small mammals of Sagar-Sorab forests, Mysore State, India, 1961—1964, *Indian J. Med. Res.,* 56, (Suppl.), 569, 1968.
12. **Boshell, M. J.,** Kyasanur Forest disease: ecological considerations, *Am. J. Trop. Med. Hyg.,* 18, 67, 1969.
13. **Rajagopalan, P. K., Paul, S. D., and Sreenivasan, M. A.,** Involvement of *Rattus blanfordi* (Rodentia:Muridae) in the natural cycle of Kyasanur Forest disease virus, *Indian J. Med. Res.,* 57, 999, 1969.
14. **Rajagopalan, P. K., Paul, S. D., and Sreenivasan, M. A.,** Isolation of Kyasanur Forest disease virus from the insectivorous bat, *Rhinolophus rouxi* and from *Ornithodorus* ticks, *Indian J. Med. Res.,* 57, 805, 1969.
15. **Banerjee, K., Guru, P. Y., Dhanda, V.,** Growth of arboviruses in cell cultures derived from the tick *Haemaphysalis spinigera, Indian J. Med. Res.,* 66, 530, 1977.
16. **Bhatt, P. N. and Anderson, C. R.,** Attenuation of a strain of Kyasanur Forest disease virus for mice, *Indian J. Med. Res.,* 59, 199, 1971.
17. **Paul, Sharada Devi,** Some biological properties of two variants of Kyasanur Forest disease virus, *Indian J. Med. Res.,* 54, 419, 1966.
18. **Anderson, C. R.,** Serum neutralization of Kyasanur Forest disease virus in mice, *Indian J. Med. Res.,* 58, 1584, 1970.

19. **Webb, H. E. and Lakshmana Rao, R.,** Kyasanur Forest disease: a general clinical study in which some cases with neurological complications were observed, *Trans. R. Soc. Trop. Med. Hyg.,* 55, 284, 1961.

20. **Wadia, R. S.,** Neurological involvement in Kyasanur Forest disease, *Neurol. India,* 23, 115, 1975.

21. **Lakshmana Rao, R.,** Clinical observations on Kyasanur Forest disease cases, *J. Indian Med. Assoc.,* 31, 113, 1958.

22. **Work, T. H., Trapido, H., Narasimha Murthy, D. P., Lakshmana Rao, R., Bhatt, P. N., and Kulkarni, K. G.,** Kyasanur Forest disease. III. A preliminary report on the nature of the infection and clinical manifestations in humans beings, *Indian J. Med. Sci.,* 11, 619, 1957.

23. **Chatterjea, J. B., Swarup, S., Pain, S. K., and Lakshmana Rao, R.,** Haematological and biochemical studies in Kyasanur Forest disease, *Indian J. Med. Res.,* 51, 419, 1963.

24. **Iyer, C. G. S., Lakshmana Rao, R., Work, T. H., and Narasimha Murthy, D. P.,** Kyasanur Forest disease. VI. Pathological findings in three fatal human cases of Kyasanur Forest disease, *Indian J. Med. Sci.,* 13, 1011, 1959.

25. **Upadhyaya, S., Narasimha Murthy, D. P., and Yashodhara Murthy, B. K.,** Viraemia studies on the Kyasanur Forest disease human cases of 1966, *Indian J. Med. Res.,* 63, 950, 1975.

26. **Haldane, J. B. S.,** "Dex" or "Order of Magnitude", *Nature (London),* 187, 879, 1960.

27. **Shah, K. V. and Narasimha Murthy, D. P.,** Investigation of the possibility of transmission of Kyasanur Forest disease virus from mother to child by milk or across the placenta, *Acta Virol. (Engl. Ed.),* 4, 329, 1960.

28. **Banerjee, K. and Bhat, H. R.,** Absence of viruria in Kyasanur Forest disease patients, *Acta Virol. (Engl. Ed.),* 21, 174, 1977.

29. **Goverdhan, M. K., Rajagopalan, P. K., Narasimha Murthy, D. P., Upadhyaya, S., Boshell, M. J., Trapido, H., and Ramachandra Rao, T.,** Epizootiology of Kyasanur Forest disease in wild monkeys of Shimoga district, Mysore State (1957—1964), *Indian J. Med. Res.,* 62, 497, 1974.

30. **Iyer, C. G. S., Work, T. H., Narasimha Murthy, D. P., Trapido, H., and Rajagopalan, P. K.,** Kyasanur Forest disease. VII. Pathological findings in monkeys, *Presbytis entellus* and *Macaca radiata,* found dead in the forest, *Indian J. Med. Res.,* 48, 276, 1960.

31. **Webb, H. E. and Burston, J.,** Clinical and pathological observations with special reference to the nervous system in *Macaca radiata* infected with Kyasanur Forest disease virus, *Trans. R. Soc. Trop. Med. Hyg.,* 60, 325, 1966.

32. **Webb, H. E. and Chatterjea, J. B.,** Clinical-pathological observations on monkeys infected with Kyasanur Forest disease virus, with special reference to the haemopoietic system, *Br. J. Haematol.,* 8, 401, 1962.

33. **Banerjee, K.,** Serum transaminases in Kyasanur Forest disease: a preliminary report, *Indian J. Med. Res.,* 67, 1, 1978.

34. **Singh, K. R. P. and Anderson, C. R.,** Relation of *Haemaphysalis spinigera* larval infection rates and host viremia levels of Kyasanur Forest disease virus, *Indian J. Med. Res.,* 56, 137, 1968.

35. **Rodrigues, F. M.,** A study on the susceptibility of wild birds of Shimoga district, Mysore State, to infection with Kyasanur Forest disease virus, with a view to determining their role in the epidemiology of the disease, Ph.D. thesis, University of Poona, 1968.

36. **Ghosh, S. N., Rajagopalan, P. K., Singh, G. K., and Bhat, H. R.,** Serological evidence of arbovirus activity in birds of KFD epizootic-epidemic area, Shimoga district, Karnataka, India, *Indian J. Med. Res.,* 63, 1327, 1975.

37. **Nayar, M.,** Histological changes in mice infected with Kyasanur Forest disease virus, *Indian J. Med. Res.,* 60, 1421, 1972.

38. **Achar, T. R., Patil, A. P., and Jayadevaiah, M. S.,** Persistence of humoral immunity in Kyasanur Forest disease, *Indian J. Med. Res.,* 73, 1, 1981.

39. **Geetha, P. B., Ghosh, S. N., Gupta, N. P., Shaikh, B. H., and Dandawate, C. N.,** Enzyme linked immunosorbent assay (Elisa) using β-lactamase for the detection of antibodies to KFD virus, *Indian J. Med. Res.,* 71, 329, 1980.

40. **Upadhyay, S., Narasimha Murthy, D. P., and Anderson, C. R.,** Kyasanur Forest diseae in the human population of Shimoga district, Mysore State, 1959—1966, *Indian J. Med. Res.,* 63, 1556, 1975.

41. **Morse, L. J., Russ, S. B., Needy, C. F., and Buescher, E. L.,** Studies of viruses of the tick-borne complex. II. Disease and immune response in man following accidental infection with Kyasanur Forest disease virus, *J. Immunol.,* 88, 240, 1962.

42. **Hanson, R. P., Sulkin, S. E., Buescher, E. L., Hammon, W. McD., McKinney, R. W., and Work, T. H.,** Arbovirus infections of laboratory workers: extent of problems emphasizes the need for more effective measures to reduce hazards, *Science,* 158, 1283, 1967.

43. **Banerjee, K., Gupta, N. P., and Goverdhan, M. K.,** Viral infections in laboratory personnel, *Indian J. Med. Res.,* 69, 363, 1979.

44. **Pavri, K. M. and Singh, K. R. P.,** Demonstration of antibodies against the virus of Kyasanur Forest disease (KFD) in the frugivorous bat *Rousettus leschenaulti* near Poona, Indian, *Indian J. Med. Res.,* 53, 956, 1965.

45. **Gadkari, D. A., Banerjee, K., and Bhat, H. R.,** Critical evaluation of Kyasanur Forest disease virus neutralizing antibodies found in bats (a preliminary report), *Indian J. Med. Res.,* 64, 64, 1976.

46. **Bhat, H. R., Sreenivasan, M. A., Goverdhan, M. K., Naik, S. V., and Banerjee, K.,** Antibodies to Kyasanur Forest disease virus in bats in the epizootic-epidemic area and neighbourhood, *Indian J. Med. Res.,* 68, 387, 1978.

47. **Trapido, H., Rajagopalan, P. K., Work, T. H., and Varma, M. G. R.,** Kyasanur Forest disease. VIII. Isolation of Kyasanur Forest disease virus from naturally infected ticks of the genus *Haemaphysalis, Indian J. Med. Res.,* 47, 133, 1959.

48. **Boshell, M. J., Rajagopalan, P. K., Patil, A. P., and Pavri, K. M.,** Isolation of Kyasanur Forest disease virus from Ixodid ticks: 1961—1964, *Indian J. Med. Res.,* 56(Suppl.) 541, 1968.

49. **Bhat, H. R.,** Infestation of cattle with *Haemaphysalis spinigera* Neumann, 1897 (Acarina:Ixodidae) in Kyasanur Forest disease area Shimoga district, Karnataka, *Indian J. Anim. Sci.,* 44, 750, 1974.

50. **Bhat, H. R.,** Life history of *Haemaphysalis spinigera* Neumann, 1897 (Acarina:Ixodidae), *Indian J. Anim. Sci.,* 49, 517, 1979.

51. **Bhat, H. R.,** Observations on the biology of *Haemaphysalis spinigera* Neumann, 1897 (Acarina:Ixodidae) under natural conditions of KFD area, *J. Bombay Nat. Hist. Soc.,* 82, 548, 1985.

52. Indian Council of Medical Research, Kyasanur Forest Disease 1957—1964, Virus Research Centre, Poona, 1964.

53. **Trapido, H., Goverdhan, M. K., Rajagopalan, P. K., and Rebello, M. J.,** Ticks ectoparasitic on monkeys in the Kyasanur Forest disease area of Shimoga district, Mysore State, India, *Am. J. Trop. Med. Hyg.* 13, 763, 1964.

54. **Rajagopalan, P. K. and Anderson, C. R.,** Further studies on ticks of wild monkeys of Kyasanur Forest disease area Shimoga district, *Indian J. Med. Res.,* 59, 847, 1971.

55. **Bhat, H. R.,** personal communication, 1986.

56. **Rajagopalan, P. K., Patil, A. P., and Boshell, M. J.,** Ixodid ticks on their mammalian hosts in the Kyasanur Forest disease area of Mysore State, India, 1961—64, *Indian J. Med. Res.,* 56, 510, 1968.

57. **Bhat, H. R., Sreenivasan, M. A., Goverdhan, M. K., Naik, S. V., and Banerjee, K.,** Susceptibility of *Hystrix indica* Kerr, 1792, Indian crested porcupine (Rodentia, Hystricidae) to KFD virus, *Indian J. Med. Res.,* 64, 1566, 1976.

58. **Varma, M. G. R., Webb, H. E., and Pavri, K. M.,** Studies on the transmission of Kyasanur Forest disease virus by *Haemaphysalis spinigera* Neumann, *Trans. R. Soc. Trop. Med. Hyg.,* 54, 509, 1960.

59. **Singh, K. R. P., Pavri, K. M., and Anderson, C. R.,** Transmission of Kyasanur Forest disease virus by *Haemaphysalis turturis, Haemaphysalis papuana kinneari* and *Haemaphysalis minuta, Indian J. Med. Res.,* 52, 566, 1964.

60. **Bhat, H. R., Sreenivasan, M. A., Goverdhan, M. K., and Naik, S. V.,** Transmission of Kyasanur Forest disease virus by *Haemaphysalis kyasanurensis* Trapido, Hoogstraal and Rajagopalan, 1964, (Acrina:Ixodidae), *Indian J. Med. Res.,* 63, 879, 1975.

61. **Bhat, H. R. and Naik, S. V.,** Transmission of Kyasanur Forest disease virus by *Haemaphysalis wellingtoni* Nuttall and Warburton, 1907 (Acarina:Ixodidae), *Indian J. Med. Res.,* 67, 697, 1978.

62. **Boshell, M. J. and Rajagopalan, P. K.,** Preliminary studies on experimental transmission of Kyasanur Forest disease virus by nymphs of *Ixodes petauristae* Warburton, 1933, infected as larvae on *Suncus murinus* and *Rattus blanfordi, Indian J. Med. Res.,* 56(Suppl.), 589, 1968.

63. **Singh, K. R. P., Goverdhan, M. K., and Ramachandra Rao, T.,** Experimental transmission of Kyasanur Forest disease virus to small mammals by *Ixodes petauristae, I. ceylonensis* and *Haemaphysalis spinigera, Indian J. Med. Res.,* 56(Suppl.), 594, 1968.

64. **Bhat, H. R., Naik, S. V., Ilkal, M. A., and Banerjee, K.,** Transmission of Kyasanur Forest disease virus by *Rhipicephalus haemaphysaloides* ticks, *Acta Virol. (Engl. Ed.),* 22, 241, 1978.

65. **Sreenivasan, M. A., Bhat, H. R., and Naik, S. V.,** Experimental transmission of Kyasanur Forest disease virus by *Dermacentor auratus* Supino, *Indian J. Med. Res.,* 69, 701, 1979.

66. **Ilkal, M. A., Dhanda, V., Goverdhan, M. K., and Deshmukh, P. K.,** Life history of *Haemaphysalis obesa* Larrousse (Acarina:Ixodidae) and its susceptibility to Kyasanur Forest disease virus, *Indian J. Parasitol.,* 5, 27, 1981.

67. **Bhat, U. K. M. and Goverdhan, M. K.,** Transmission of Kyasanur Forest disease virus by the soft tick *Ornithodoros crossi, Acta Virol. (Engl. Ed.),* 17, 337, 1973.

68. **Singh, K. R. P., Goverdhan, M. K., and Bhat, U. K. M.,** Transmission of Kyasanur Forest disease virus by soft tick *Argas persicus* (Ixodoidae:Argasidae) *Indian J. Med. Res.,* 59, 213, 1971.

69. **Singh, K. R. P., Goverdhan, M. K., and Bhat, H. R.,** Transovarial transmission of Kyasanur Forest disease virus by *Ixodes petauristae, Indian J. Med. Res.,* 56(Suppl.), 628, 1968.

70. **Singh, K. R. P., Pavri, K. M., and Anderson, C. R.,** Experimental transovarial transmission of Kyasanur Forest disease virus in *Haemaphysalis spinigera, Nature (London),* 199, 513, 1963.

71. **Rajagopalan, P. K., Patil, A. P., and Boshell, M. J.,** Studies on Ixodid tick populations on the forest floor in the Kyasanur Forest disease area (1961—64), *Indian J. Med. Res.,* 56, 497, 1968.

72. **Rajagoplan, P. K. and Anderson, C. R.,** Trans-monsoonal persistence of Kyasanur Forest disease virus in *Haemaphysalis* nymphs infected in nature, *Indian J. Med. Res.,* 58, 1184, 1970.

73. **Sreenivasan, M. A., Rajagopalan, P. K., and Bhat, H. R.,** Spatial distribution of infected *Haemaphysalis* nymphs in the epizootic localities of Kyasanur Forest disease, *Indian J. Med. Res.,* 78, 531, 1983.

74. **Boshell, M. J., Goverdhan, M. K., and Rajagopalan, P. K.,** Preliminary studies on the susceptibility of wild rodents and shrews to KFD virus, *Indian J. Med. Res.,* 56(Suppl.), 614, 1968.

75. **Bhat, H. R., Sreenivasan, M. A., and Naik, S. V.,** Susceptibility of common giant flying squirrel to experimental infection with KFD virus, *Indian J. Med. Res.,* 69, 697, 1979.

76. **Goverdhan, M. K. and Anderson, C. R.,** The reaction of *Mus platythrix* to Kyasanur Forest disease virus, *Indian J. Med. Res.,* 60, 1002, 1972.

77. **Sreenivasan, M. A. and Bhat, H. R.,** Susceptibility of *Vandeleuria oleracea* Bennet, 1832, (Rodentia:Muridae) to experimental infection with Kyasanur Forest disease virus, *Indian J. Med. Res.,* 64, 568, 1976.

78. **Sreenivasan, M. A. and Bhat, H. R.,** Susceptibilty of *Lepus nigricollis* Cuvier to experimental infection with Kyasanur Forest disease virus, *Indian J. Med. Res.,* 65, 17, 1977.

79. **Pavri, K. M. and Singh, K. R. P.,** Kyasanur Forest disease virus infection in the frugivorous bat, *Cynopterus sphinx, Indian J. Med. Res.,* 56, 1202, 1968.

80. **Sreenivasan, M. A. and Bhat, H. R.,** Susceptibility of an insectivorous bat *Rhinolophus rouxi* to experimental infection with Kyasanur Forest disease virus, *Curr. Sci.,* 46, 268, 1977.

81. **Anderson, C. R. and Singh, K. R. P.,** The reaction of cattle to Kyasanur Forest disease virus, *Indian J. Med. Res.,* 59, 195, 1971.

82. **Pavri, K. M., Rodrigues, F. M., and Rajagopalan, P. K.,** Role of birds in the ecology of arboviruses in India, in *Transcontinental Connections of Migratory Birds and Their Role in the Distribution of Arboviruses,* Proc. 5th Symp. Study of Role of Migratory Birds in Distribution of Arboviruses 1969, Nauka, Moscow, 1972, 193.

83. **Pavri, K. M. and Shaikh, B. H.,** A rapid method of specific identification of Japanese encephalitis-West Nile subgroup of arboviruses, *Curr. Sci.,* 35, 455, 1966.

84. **Banerjee, K. and Bhat, H. R.,** Correlation between the number of persons suffering from Kyasanur Forest disease and the intensity of infection in the tick population, *Indian J. Med. Res.,* 66, 175, 1977.

85. **Drummond, R. O., Rajagopalan, P. K., Sreenivasan, M. A., and Menon, P. K. B.,** Tests with Ixodicides for the control of the tick vectors of Kyasanur Forest disease, *J. Med. Entomol.,* 6, 245, 1969.

86. **Kulkarni, S. M. and Naik, V. M.,** Laboratory evaluation of six repellents against some Indian ticks, *Indian J. Med. Res.,* 82, 14, 1985.

87. **Dhanda, V. and Sreenivasan, M. A.,** Field studies on repellents against ticks in a forested area in Shimoga district, Karnataka, in *Proc. All India Symp. Acarology,* Vol. 1, 1979, 104.

88. **Shah, K. V., Anikar, S. P., Narasimha Murthy, D. P., Rodrigues, F. M., Jayadevaiah, M. S., and Prasanna, H. A.,** Evaluation of the field experience with formalin inactivated mouse brain vaccine of Russian spring summer encephalitis virus against Kyasanur Forest disease, *Indian J. Med. Res.,* 50, 162, 1962.

89. **Aniker, S. P., Work, T. H., Chandrasekharaiya, T., Narasimha Murthy, D. P., Rodrigues, F. M., Ahmed, R., Kulkarni, K. G., Rahman, S. H., Mansharamani, H., and Prasanna, H. A.,** The administration of formalin-inactivated RSSE virus vaccine in the Kyasanur Forest disease area of Shimoga district, Mysore State, *Indian J. Med. Res.,* 50, 147, 1962.

90. **Pavri, K. M., Gokhale, T., and Shah, K. V.,** Serological response to Russian spring-summer encephalitis virus vaccine as measured with Kyasanur Forest disease virus, *Indian J. Med. Res.,* 50, 153, 1962.

91. **Mansharamani, H. J., Dandawate, C. N., and Krishnamurthy, B. G.,** Experimental vaccine against Kyasanur Forest diesase (KFD) virus from tissue culture source. I. Some data on the preparation and antigenicity tests of vaccines, *Indian J. Pathol. Bacteriol.,* 10, 9, 1967.

92. **Bhatt, P. N. and Dandawate, C. N.,** Studies on antibody response of a formalin inactivated Kyasanur Forest disease virus vaccine in langurs, *Presbytis entellus, Indian J. Med. Res.,* 62, 820, 1974.

93. **Banerjee, K., Dandawate, C. N., Bhatt, P. N., and Ramachandra Rao, T.,** Serological response in humans to a formolized Kyasanur Forest disease vaccine, *Indian J. Med. Res.,* 57, 969, 1969.

94. **Upadhyaya, S., Dandawate, C. N., and Banerjee, K.,** Surveillance of formalized KFD virus vaccine administration in Sagar-Sorab taluks of Shimoga district, *Indian J. Med. Res.,* 69, 714, 1979.

95. **Dandawate, C. N., Upadhyaya, S., and Banerjee, K.,** Serological response to formalized Kyasanur Forest disease virus vaccine in humans at Sagar and Sorab taluks of Shimoga district, *J. Biol. Stand.,* 8, 1, 1980.

96. National Institute of Virology, unpublished data.

Chapter 30

LOUPING-ILL

Hugh W. Reid

TABLE OF CONTENTS

I. HISTORY

A. Discovery of Agent

A clinical disease of hill sheep known as louping-ill has been present in southern Scotland for at least two centuries.[1,2] It was, however, an ill-defined condition, and not until specific histopathological lesions in the central nervous system (CNS) were described by MacGowan and Rettie[4] in 1913 was a distinct entity defined. However, the prevalence of another infection, the then-unrecognized, tick-transmitted infection, tick-borne fever (*Cytocaetes phagocytophila*), which frequently parasitizes neutrophils of sheep affected with louping-ill, confounded subsequent attempts to determine the cause of the condition.[5,6] Thus, it was not until 1931 that the etiological agent was established as a filterable virus transmitted by the sheep tick *Ixodes ricinus*.[7,8] Louping-ill virus is now recognized to be a member of a closely related complex of tick-transmitted flaviviruses known as the tick-borne encephalitides (TBE).[9]

B. History of Epidemics

The earliest description of louping-ill refers to a disease prevalent in the Scottish border counties but unknown in the northern uplands.[1,10] However, by the time the disease was investigated in the 1930s, it was recognized as a problem throughout much of the hill sheep farming areas of Scotland and northern England.[7] Subsequently, the disease was recognized in Ireland, Wales, and in 1978 was also identified as a disease problem in southwestern England.[11,12,13] Encephalitis in sheep caused by viruses considered to be of the louping-ill group has also been described in Bulgaria and Turkey.[14,15] In addition, encephalomyelitis affecting sheep in Norway and Spain has recently been recognized and found to be associated with a high incidence of antibody to louping-ill virus in affected flocks.[13] Furthermore, a virus isolated from affected sheep in Norway could not be differentiated from louping-ill virus by plaque morphology or by standard hemagglutination and neutralization tests. The conventional view that louping-ill is restricted to the British Isles should therefore be accepted with caution until critical comparison of viruses associated with encephalomyelitis of sheep in different geographical areas has been made. The following consideration is, however, restricted to analysis of louping-ill in the British Isles.

II. THE VIRUS

A. Antigenic Relationships

Louping-ill, the only flavivirus that has been identified as existing naturally in the British Isles, is a member of a group of viruses distributed primarily throughout the northern temperate latitudes and known as the TBE complex. This group includes Central European encephalitis (CEE) virus, Russian spring-summer encephalitis virus, Kyasanur Forest disease virus, Omsk hemorrhagic fever virus, Langat virus, Negishi virus, and Powassan virus.[9] Differentiation by conventional serological techniques is not readily achieved, but analysis using a panel of 16 monoclonal antibodies suggests that louping-ill virus belongs to an antigenically distinct western subtype.[16]

B. Host Range

1. Arthropods

The epizootiology of louping-ill implicates *Ix. ricinus* as the exclusive virus vector. However, experimentally both *Rhipicephalus appendiculatus* and *Hyalomma anatolicum* were also found to transmit virus transstadially, and it is probable that most species of *Ixodoidea* can be vectors.[17,18]

2. Vertebrates

The catholic host range of *Ix. ricinus* ensures that in enzootic regions most terrestrial vertebrates encounter infection. Thus, serum antibody has been detected and/or virus isolated from a number of native wild species, including red grouse (*Lagopus lagopus scoticus*), common shrew (*Sorex araneus*), wood mouse (*Apodemus sylvaticus*), blue hare (*Lepus timidus*), badger (*Meles meles*), roe deer (*Capreolus capreolus*), red deer (*Cervus elaphus*), and feral goat.[13,19-24] There is similar evidence that in enzootic areas man and most domestic species, including pig, sheep, cattle, farmed red deer, horses, and dogs, become infected.[25-33] The susceptibility of laboratory animals to experimental infection with louping-ill virus is summarized in Appendix 1, while information or replication in cell culture systems is given in Appendix 2.

C. Strain Variation

Hemagglutinin (HA) could be demonstrated in only one of two isolates of virus from red grouse, suggesting that strain variations might exist.[20] Likewise, Timoney[34] detected HA in only 28% of 55 isolates employing "crude" infected mouse brains as antigens. However, all of 21 isolates from both field-caught pools of *Ix. ricinus* and a variety of vertebrates from different geographical localities appeared to be identical to the reference strain Li31 M5 when compared by plaque morphology, neutralization, and hemagglutination test.[13,35] In this latter study, the source of HA was acetone extracts of infected BHK-21 cell supernatant fluids, and it is now thought that failure to demonstrate HA from some isolates in the earlier studies may have been due to the techniques employed, rather than intrinsic characteristics of the virus isolates.

In addition, the pathogenicity of 5 of the 21 isolates examined in the above study was determined for 3-week-old mice and 1-day-old domestic chickens. For each isolate a group of five mice and five chickens was inoculated subcutaneously with 10^4 plaque forming units (PFU), and the magnitude and duration of the ensuing viremias measured. No significant differences between the isolates was detected.

It is therefore concluded from the available evidence that isolates of louping-ill made in the British Isles are essentially identical, and that no strain diversity has been established.

D. Methods of Assay

The transmission of louping-ill virus to laboratory mice by intracerebral (i.c.) inoculation provided the first practical method of assay for both virus and antibody to the virus.[36] Thus, many of the early investigations of louping-ill relied on this method of quantification, but have largely been superseded in contemporary studies by in vitro techniques. Madrid and Porterfield[37] described a method for plaquing louping-ill virus under an overlay of carboxymethyl cellulose using the pig kidney cell line PS, while Reid and Doherty[38] reported the use of the pig-kidney cell line IB/RS2 clone 60. This latter method was demonstrated to be at least as sensitive as i.c. inoculation of mice when assaying virus from tissues collected under experimental conditions, though experience has shown that for isolation of virus from field material the i.c. inoculation of mice is more sensitive.[21,39]

III. DISEASE ASSOCIATION

A. Humans

The description by Rivers and Schwentker[40] of severe encephalitis in three laboratory workers who had been working with louping-ill virus provided the first evidence of its pathogenicity for man. Retrospective serology indicated that all three had developed neutralizing antibody as had two other men who had been working with the virus. Numerous other laboratory-acquired infections have been reported, and in some cases virus has been recovered during the initial systemic phase.[41-43]

Apart from laboratory workers, a veterinary surgeon, two farmers, and an abattoir worker were shown by retrospective serology in 1948 and 1949 to have been infected.[25,44,45] Since then, cases have been identified only in surveys of sera collected from patients with aseptic meningitis or encephalitis of unknown etiology. These included 5/35 patients in Ireland who developed antibody to Russian spring-summer encephalitis virus, while examination of 775 sera from Scottish patients identified one case of louping-ill virus encephalitis in a farmer and one fatal case in a slaughterman.[26,46,47]

However, serological surveys of abattoir workers indicated that 8% had experienced infection, which is surprising in view of the fact that there have been only two cases of clinical disease reported in this occupational category.[45,48]

B. Domestic Animals

Most categories of domestic livestock become infected in endemic areas. Apart from sheep, which is the main species at risk, fatal infections have been described in cattle, goats, horses, pigs, farmed deer, and dogs.[27-33]

1. Sheep

Traditionally, the disease in sheep is associated with lambs during the 1st year and female animals that are retained as replacement breeding stock.[7,49] However, field observations suggest that the occurrence of clinical disease is governed by a complex interaction of numerous factors.

The incidence of louping-ill infection in a sheep flock may either be relatively stable or fluctuate with time. When the challenge is constant over a number of years, losses in sheep are frequently unrecognized. This can be explained on the basis of the variability of the clinical signs, low case mortality, the extensive nature of the grazings of much of the enzootic areas, and an accepted death rate of 10% or greater due to other causes. In addition, immune ewes transfer antibody to their lambs very efficiently, and this ensures the complete protection of lambs during their first exposure to tick activity.[50] Furthermore, lambs that are susceptible because of inadequate colostrum intake are prone to succumb to other neonatal diseases, and they are therefore not likely to die from louping-ill.[51,52]

As colostral immunity completely prevents infection during their first season of exposure to tick activity, lambs tend to be fully susceptible to infection after their first season of exposure. The majority of lambs are, however, removed for slaughter, and only the ewe lambs retained for breeding are liable to experience disease. It is in this category of animal that louping-ill is most frequently recognized, although in some endemic areas where antibody may be detected in all mature animals, the incidence of disease is sporadic and insufficient to attract the attention of stock owners.

Losses may be high when susceptible sheep are introduced into enzootic areas, and the case mortality can exceed 60%.[53,54] In addition, severe outbreaks of disease may occur following the introduction of virus to sheep grazings where ticks are numerous but from which louping-ill has been absent. When this occurs, few cases are recognized during the first 1 or 2 years, after which there is rapid increase in the incidence of disease affecting all age categories before infection becomes enzootic.

2. Other Species

Cattle would appear to be less susceptible than sheep to clinical disease and are more likely to recover following the development of clinical signs.[28,55] Likewise, there are few reports of clinical disease in horses, although fatal encephalomyelitis may develop following infection.[31] The only reported incident in pigs involved 16 piglets, 10 of which developed severe disease, having probably acquired infection through ingesting lambs that had died from louping-ill.[27] Natural disease in goats has not been reported, and experimental infection resulted in only mild transient disease in 1/6 subcutaneously (s.c.) inoculated animals.[56] However, five kids that acquired infection through ingesting infected milk all developed severe disease, indicating the potential susceptibility of this species. In dogs, clinical disease has been recognized principally in working sheep dogs, although the occasional gun dog has also succumbed to infection.[32]

The single report of clinical louping-ill in a farmed red deer indicates that this species can develop clinical disease, although observations of infection in free-living populations and experimentally inoculated animals suggests that infection is normally subclinical.[24,29,57]

C. Wildlife

In endemic areas, infection is prevalent in wildlife, but with one exception there is no evidence that it results in clinical disease. The exception is the red grouse. Louping-ill virus was isolated from 23/31 young grouse found dead or dying in Speyside, Scotland.[13,19-21] These observations were augmented by a field study of free-living populations of red grouse in which louping-ill virus infection was identified as responsible for 95% of the mortality in enzootic foci.[58]

D. Diagnostic Procedures

The clinical manifestations of louping-ill are varied and therefore provide an unreliable basis for diagnosis. Laboratory confirmation may be sought by serology, virus isolation, and histopathological examination.

1. Serology

Antibody to louping-ill virus has been detected by gel diffusion, complement fixation, hemagglutination inhibition (HI), neutralization, and fluorescent antibody tests.[38,59-62] However, the HI test has proved of greatest utility and is regularly employed in serodiagnosis in all species. Suitable virus antigen will agglutinate trypsinized human O, rooster, and pigeon red blood cells, but for large-scale testing, gander cells have proved to be most practical.[13,59] Antigen may be prepared from either infected suckling mouse brains or from BHK-21 culture supernatants, which for best results should be extracted with cold acetone.[63]

Both acetone and kaolin have been employed to remove nonspecific inhibitors from sera, although the latter method has proved the best, being both convenient and economical.[39]

From approximately 6 days after infection, antibody titers rise rapidly, and depending on the species involved may reach maximum HI titers of between 1/640 and 1/40,960.[38,64] In general, high titers are indicative of recent exposure, although residual titers of 1/640 have been recorded in sheep 3 years after a single infection.[39]

2. Virus Isolation

Only in man has the isolation of virus during the viremic phase proved to be of diagnostic value.[41,42] Thus, in laboratory personnel following suspected accidental infection, virus isolation has been successful during the initial illness when the presenting signs are systemic and nonspecific. However, in other species confirmation is sought only after death or the development of neurological disease, in which case virus can be isolated only from the central nervous tissues. Homogenates of cerebellum and brain stem are inoculated onto cell cultures or i.c. into mice.[60] Isolates may be identified by preparing HA, by neutralization in a plaque assay, or by immunofluorescence.

3. Histopathology

Identification of typical neurohistopathological lesions may be useful in reaching a presumptive diagnoses. A series of paraffin-embedded blocks are prepared from at least five levels of brain which has been fixed in 10% formal saline and sections cut and stained with hematoxylin and eosin. The presence of a characteristic nonsuppurative meningoencephalomyelitis can assist in reaching a diagnosis, particularly when the animal is from an area where other encephalitogenic agents are unlikely to be encountered.[65-68] Viral antigen may be detected by direct immunofluorescence with specific antisera, but this has not been adopted for routine application.[60]

IV. EPIDEMIOLOGY

A. Geographical Distribution

Within the U.K., louping-ill is restricted to the rough, upland grazings and unimproved pastures of the western seaboard that are largely devoted to sheep rearing. Louping-ill may also be encountered over much of rural Ireland where the rainfall is relatively evenly distributed throughout the year and tick infestations extend to improved pastures. However, there are quite extensive areas where ticks are present but from which louping-ill virus is absent.

B. Prevalence

In enzootic areas the prevalence of infection in sheep flocks as assessed by the detection of HI antibody can vary widely, from all animals over 1 year of age having antibody to only a few positive animals in the older age categories.[13] Similarly, in sera from cattle, deer, and grouse the prevalence of antibody may vary widely between different locations.[13,24,58] The factors determining the prevalence of infection within enzootic areas are not fully understood, but clearly, vector density, access of animals to pasture during periods of tick activity (Section VI.A), and in particular the intervention of vaccination regimes will all play a part.

C. Seasonal Distribution

The activity of the vector tick follows a distinct seasonal pattern. *Ix. ricinus* becomes active in the late spring and early summer, after which there is a period of inactivity through the summer months. This is followed in the west of the country by a recrudescence of tick activity from late August to October.

MacLeod[69,70] determined that the commencement of tick activity in the spring correlated with the average day temperature exceeding 7°C, a temperature which agrees with laboratory studies for the lower temperature limit for tick activity. The point at which this threshold temperature is reached varies from early April in the southwest to mid-June in the northeast highlands above 300 m. However, in all areas the period of peak infestation is followed by a rapid decline to low levels. This decline in activity may be attributed to depletion due to ticks successfully attaching to a host, and the death of ticks due to desiccation and exhaustion of energy stores after failing to attach to a host.[71,72] Through the summer months, engorged larvae and nymphs metamorphose to the next stage while females deposit eggs. Emerging ticks, however, are reluctant to feed before December and thus do not actively quest until the following spring.[71]

Campbell[71] established that the autumn-feeding ticks represented a separate population which overwintered in the engorged state, with egg laying and metamorphosis occurring during the following summer, but which did not actively quest until late August.

The seasonal incidence of clinical louping-ill closely parallels this seasonal pattern of tick activity.

D. Risk Factors

The variability of clinical disease following peripheral challenge with louping-ill virus has confounded many studies of the disease in sheep.[8,49,73,74] It was long considered that factors which facilitated virus invasion of the CNS were critical to the outcome, and Smith et al.[49] suggested that important factors were age, nutritional status, concurrent disease, and cold. Subsequent studies have indicated that viral invasion of the CNS following exposure to virus invariably occurred, irrespective of the clinical course of infection.[60,66,75,76] There was, however, a correlation between the intensity of the viremia, later appearance of antibody, and the development of fatal encephalitis.[38] A similar correlation was found in studies of the experimental disease in cattle and red grouse.[55,64]

Concurrent infection with *Cytocaetes phagocytophila*, the cause of tick-borne fever (TBF), has been implicated frequently as a contributory factor to precipitation of clinical louping-ill, but only recently has this possibility been examined critically.[49,73,77-79] Compared to sheep experimentally infected with louping-ill virus alone, those infected 5 days previously with *C. phagocytophila* developed greater viremias, delayed immune responses, and consequently experienced a dramatically increased mortality.[77] The significance of this interaction under field conditions is not clear. The incidence of *C. phagocytophila* in ticks is high, and maternal antibody does not provide protection from TBF.[79] Thus, during their first year, lambs will become infected with *C. phagocytophila* while protected from louping-ill by colostral antibody. Concomitant infection will thus be unusual in flocks resident in enzootic areas. The most severe losses from louping-ill have always been considered to occur in unacclimatized sheep when introduced to infected pasture, when it is possible that animals could be exposed to both infections concurrently. This has been confirmed by a field trial reported by O'Reilly et al.[80] in which all eight control animals that became infected with louping-ill during the first 2 months died, whereas none of seven animals that were infected during the subsequent 3 months developed clinical disease. No explanation for this is offered, but it is possible that concurrent infection with TBF and louping-ill virus during the first 2 months could have contributed to the high mortality rate, whereas those animals that were exposed to louping-ill virus subsequently were immune to TBF, which they had encountered during the initial part of the trial.

Likewise, mice chronically infected with *Trypanosoma brucei* and mice and lambs infected with *Toxoplasma gondii* were much more susceptible to louping-ill virus challenge than were animals infected with virus alone.[81-85] Furthermore, although the proportion dying was greater in the dually infected animals, deaths were also delayed, and it was concluded that the increased mortality was largely due to the immunosuppressive effect of the protozoan infections.

V. TRANSMISSION CYCLES

A. Field Studies

Since the original observation that louping-ill virus was transmitted by the sheep tick, field investigations have been perfunctory.

1. Vector

The two reported attempts to establish incidence indicate that louping-ill virus is present only in a small proportion of ticks. Varma and Smith[86] made two isolates of virus from a collection of 2000 questing ticks from Ayrshire, while Swanepoel[18] made a single virus isolate from 1284 nymphs collected in Argyll.

2. Vertebrates

It has been assumed that louping-ill virus is maintained in a cycle between wild vertebrates and the vector tick *Ix. ricinus*.[20,22] However, investigations to establish the role of the native fauna have been unrewarding. Virus has been recovered from 2 of 26 wood mice and 1 of 68 common shrews examined by Smith et al.[22] although the level of infestation by *Ix. ricinus* on those rodents was very low.[86] Milne[87] reported that a proportion of engorged ticks was destroyed, and provided evidence to suggest that rodents, in particular the common shrew, were the responsible predators. In view of this, it is tempting to conclude that as louping-ill virus can readily infect through the oral route, infection in rodent species could arise through ingestion of engorged ticks rather than through parasitism by the tick.

In some areas, red grouse frequently become infected (Section III.C). However, studies have indicated that the proportion of grouse that survive in endemic areas is inadequate to provide sufficient numbers of replacement birds to sustain the population.[58] Hence, grouse disappear and the species should not be regarded as an essential maintenance host.

In an extensive investigation of red deer sera for antibody to both louping-ill virus and the protozoan parasite *Babesia*, it became apparent that all deer populations examined were exposed to *Babesia*, which indicated that *Ix. ricinus* was present in all the areas from which the sera were collected, but antibody to louping-ill virus was absent from a proportion of these sites, one of which is of particular interest.[24] Louping-ill was known to have been present on the island of Rhum prior to the removal of all the sheep stock approximately 15 years before the survey was carried out.[88] The absence of antibody to louping-ill virus in the deer sera suggested that while deer and the native fauna were able to support a viable tick population as evidenced by the high incidence of antibody to *Babesia*, louping-ill virus had not persisted.

B. Experimental Studies

1. Vector

Compared to infection of insect vectors, infection of ticks is complex. Blood is ingested in two separate phases for several days, during which time the titer of viremia in the vertebrate alters. Furthermore, the life cycle of ticks involves three distinct instars, each of which requires only one blood meal. Thus, to ensure transmission, virus must translocate from the ingested blood to the salivary glands of the succeeding instar. These variables have not been exhaustively investigated, but the work reported by Beasley et al.[89] provides a basis of understanding. When larvae were fed on viremic, 1-day-old chicks, virus could be recovered from 100% of freshly engorged larvae, but from only 10% of the resultant nymphs. In subsequent experiments, it was shown that the concentration of virus in the blood during

Table 1
VIREMIA IN EXPERIMENTAL LOUPING-ILL VIRUS INFECTION OF SOME VERTEBRATES

| Species | No. examined | Mean no. of days[a] when infective for | | Max titer,[b] mean (range) | Ref. |
		Larvae	Nymphs		
Red fox *(Vulpes vulpes)*	8	0	0	1.6 (1.4—2.4)	35
Wood mouse *(Apodemus sylvaticus)*	28	0	0	2.2 (1.6—2.9)	35
Bank vole *(Clethrionomys glareolus)*	25	0	0	1.6 (0.0—2.8)	35
Field vole *(Microtus agrestis)*	59	0.07	0.46	2.2 (0.0—4.4)	35
Brown rat *(Rattus norvegicus)*	3	0	0	0.43 (0.0—1.3)	35
Blue hare *(Lepus timidus)*	3	0	0	0.43 (0.0—1.3)	35
Cattle *(Bos taurus)*	6	0	0.17	1.3 (0.1—3.0)	55
Goat *(Capra aegagrus)*	7	0.14	0.42	2.6 (1.6—4.0)	56
Sheep *(Ovis aries)*	33	2.25	3.09	5.6 (3.2—7.1)	38
Horse *(Equus caballus)*	7	0	1.29	3.4[c] (2.9—3.9)	93
Red deer *(Cervus elaphus)*	4	0	0	1.2 (0.8—1.8)	57
Roe deer *(Capreolus capreolus)*	3	0	0.33	2.6 (2.3—3.2)	57
Red grouse *(Lagopus lagopus scoticus)*	27	3.48	5.04	5.3 (3.7—6.9)	64
Ptarmigan *(L. mutus)*	7	5.71	7.43	6.2 (5.7—7.4)	91
Willow grouse *(L. lagopus)*	5	4.20	5.60	4.8 (4.3—5.7)	91
Black grouse *(Tetrao tetrix)*	4	0	0	2.5 (2.4—2.6)	92
Capercaillie *(T. urogallus)*	7	0	0.29	2.2 (0.7—3.5)	91
Pheasant *(Phasianus colchicus)*	5	0	0	1.5 (0.0—2.7)	35

[a] Total number of days when viremia exceeded threshold titer for larvae and nymph, respectively.
[b] PFU/0.2 mℓ of whole blood or plasma.
[c] Calculated on the basis of one 50% mouse infectious dose = 1 PFU.

the initial feeding phase was critical, with virus establishing in approximately 40% of ticks when the ingested blood contained 10^6 PFU/0.2 mℓ. It was concluded that the threshold titer of viremia for larvae was approximately 10^4 PFU/0.2 mℓ while the concentration of virus in blood required to infect nymphs was approximately tenfold less. As louping-ill virus does not appear to be transmitted transovarially, infection acquired by adults is probably of limited epidemiological significance.

Thus, these experimental data indicate that infection of the vector is a relatively inefficient process, an observation which is in agreement with the small proportion of field-caught ticks that are infected.[18,86]

Though there are 15 other species of *Ixodes* as well as *Dermacentor reticulatus* and *Haemaphysalis punctata* in the U.K., there is no evidence to suggest that any of these ticks is involved in the epidemiology of louping-ill.[90]

2. Vertebrates

The potential of 18 vertebrates species to transmit louping-ill virus has been assessed by examining the intensity and duration of viremia following experimental subcutaneous inoculation.[35,38,55-58,64,91-93] Viremia was detected in all species, but titers were generally low and only in a few species approached threshold titers of infection for nymphs or larvae (Table 1). Of eight native species of mammals examined, viremias that exceeded threshold titers were detected in only two. The viremia detected in most of the 59 field voles was transient and of low intensity, but in a few animals it was more prolonged and in five reached titers in excess of 10^4 PFU/0.2 mℓ. The other species to achieve a threshold titer was roe deer, in which one of three animals had a titer of $10^{3.2}$ PFU/0.2 mℓ on 1 day only.

Table 2
CUMULATIVE MORTALITY IN SIX SPECIES OF BIRDS FOLLOWING
SUBCUTANEOUS INOCULATION WITH LOUPING-ILL VIRUS

Species	Total infected	Time after inoculation (days)						
		2	4	6	8	10	12	14(%)
Lagopus lagopus scoticus[64]	37	0	4	14	22	27	28	29(78)
L. lagopus[35]	5	0	0	0	3	3	4	5(100)
L. mutus[35]	7	0	0	0	2	5	6	7(100)
Phasianus colchicus[91]	5	0	0	0	0	0	0	0(0)
Tetrao urogalus[35]	7	0	0	0	0	0	0	0(0)
T. tetrix[92]	4	0	0	0	0	0	0	0(0)

In contrast, all four domestic species examined developed viremias that exceeded threshold titers, but this was consistently achieved only by sheep, all of which maintained viremias that would have infected ticks for 2 to 3 days.

The six avian species examined fell into two distinct categories; red grouse, ptarmigan, and willow grouse developed high and sustained viremias that exceeded threshold titers for up to 8 days, while black grouse, capercaillie, and pheasants developed viremias of low intensity. No evidence of clinical disease was detected in these latter three species, but between 78 and 100% of the three species of *Lagopus* died (Table 2).

C. Conclusion

Of the mammalian species experimentally infected, only sheep consistently developed viremias of a sufficient intensity to infect the vector and are thus the only species that is likely to have a significant role in the maintenance of louping-ill virus.

However, three of the avian species examined developed substantial viremias following experimental infection. Ptarmigan inhabit the mountain tundra of the northern U.K. above the altitude of the habitat of the vector *Ix. ricinus*, whereas willow grouse, a subspecies found in the tundra of northern Europe, is not present in the U.K. Neither is therefore likely to encounter infection and can be excluded as potential amplifier hosts. In contrast, the habitat of the red grouse is suitable for the vector of louping-ill, but the role of the red grouse in the epidemiology of louping-ill can be only transitory. Both field and experimental studies indicate that the mortality that occurs following infection is such that residual numbers of surviving birds are too small to maintain a stable population.[58,64] Thus, in louping-ill endemic areas the grouse population will decline and disappear and can therefore have only a temporary amplifying effect in the epidemiology of louping-ill virus.

VI. ECOLOGICAL DYNAMICS

A. Vector

During the 3-year life cycle of *Ix. ricinus* only approximately 3 weeks are spent parasitizing vertebrate hosts.[72] The remainder of the life cycle is spent in the mat of decaying vegetation close to the soil, where survival is dependent on the maintenance of a relative humidity close to saturation throughout the year.[69,94] Thus, the availability of suitable microclimatic conditions is considered to determine the distribution of *Ix. ricinus*, but not critically to dictate the intensity of infestation which is governed primarily by the availability of vertebrate hosts, in particular the large mammalian species to which the adult ticks are restricted.[95,96]

At least one stage of *Ix. ricinus* has been recovered from 29 mammal and 39 avian species, and Arthur[90] states that any vertebrate above Amphibia which either feeds or nests on the ground may be parasitized. Tick infestations found on wild vertebrates in the north of England

and in southwestern Scotland have been shown to be insignificant when compared to those on sheep, and it is concluded that between 94 and 99% of the tick population is supported on sheep.[95,96] This is in accord with the observation of MacLeod[97] that tick infestation declined dramatically in an area from which sheep had been excluded.

In limited areas, red deer may be important maintenance hosts of *Ix. ricinus*, while in others blue hares can be heavily parasitized and may represent the principal host of the adult tick.[98,99] However, throughout the louping-ill enzootic region in Scotland there can be little doubt that the 300,000 domestic cattle and 2.5 million sheep are the principal hosts of the adult stage, and this is probably true also in other regions of the U.K.[100,101] However, as cattle tend to be restricted to the more fertile areas where the microhabitat is less likely to be suitable for survival of the nonparasitic stages of the tick, their role is less significant. It is thus apparent that provided the nature of the vegetation and soil maintain a suitable microclimate for survival of the tick, the principal factor governing vector density is the sheep stocking rate.

B. Vertebrates

Within enzootic areas all species of vertebrates may become infected with louping-ill virus, but it would appear that generally the intensity of the viremia is inadequate to transmit infection to the vector, and such infections are of no ecological relevance. Only red grouse and sheep have so far been identified as likely to have a role (Section V.B.2).

Both red grouse and louping-ill occur naturally and exclusively in the U.K.[102] Thus, the susceptibility of grouse to infection cannot be explained in terms of the recent introduction of either to the U.K. However, it is possible that the distribution of either the grouse or the virus has altered so that contact has occurred only recently. Red grouse are a species highly specialized to utilize the heather (*Caluna vulgaris*) dominated moorlands, and it is probable that they have been restricted to this habitat for many centuries. The density of grouse populations may have increased due to moorland management policies initiated in the 19th century and designed to increase the numbers of birds available for shooting.[103] It is unlikely, however, that increased numbers alone could be responsible for new foci of virus.

Land utilization in louping-ill enzootic areas, particularly in Scotland, has changed markedly over the last two centuries.[104] Sheep farming in the uplands of southern Scotland has been practiced for hundreds of years, but north of the Forth-Clyde valley it has been introduced in comparatively recent times. Formerly, intensive subsistence farming was practiced in the low-lying areas with the hill pasture being used only for a few months during the summer by equal numbers of sheep and cattle. However, through the 19th century this type of farming was progressively replaced by large-scale sheep ranching.

Such changes in agricultural practices would have greatly increased the host potential of the area for *Ix. ricinus*, the principal factor governing the abundance of ticks within suitable habitats.[94,95] Thus, the change to extensive sheep farming would not only have increased the abundance of ticks, but would have provided for the first time in this habitat a substantial number of vertebrates able to transmit louping-ill virus. This conclusion is supported by Stevenson,[10] who described louping-ill in 1807 as a disease not infrequently observed in the south of Scotland but scarcely known in the Highlands. It is therefore probable that louping-ill originally was absent from the habitat of the red grouse, and that this host-parasite relationship has evolved only recently.

It has further been suggested that the relative resistance to infection with louping-ill virus of the three forest species of birds examined, compared to the moorland and tundra species (Section V.B.2), reflects the extent of exposure to louping-ill or related viruses during their evolution.[91,92] Thus, while the ancestral origins of louping-ill virus may have been the forest habitat, the current ecological climax is dictated by agricultural practices with domestic sheep representing the single essential vertebrate species. Aspects of sheep management thus have a preeminent role in the ecology of louping-ill virus.

VII. SURVEILLANCE

A. Disease Detection

Throughout much of the louping-ill enzootic regions, disease surveillance of domestic animals is inadequate. The nature of the extensive grazings and the traditionally high levels of mortality that are accepted, together with meager veterinary resources, have rendered it impossible to assess the impact of infection in enzootic areas. Furthermore, the relatively low value of affected animals and the cost of veterinary and virological investigations mitigate against effective surveillance.

Generally, disease incidents are investigated only where exceptional losses are encountered and veterinary advice is sought. Ideally, brain from affected animals is examined both for histological evidence of encephalitis and for infective virus, but not infrequently diagnosis relies on the detection of serum HI antibody. Subsequent serological investigations may then be carried out to determine the advisability of adopting a vaccination regime. To establish the immune status of a flock, sera should be collected from approximately 20 mature ewes, 20 animals that have been exposed for 2 seasons, and 20 animals of approximately 5 to 6 months of age. Analysis of the incidence of HI antibody in these sera provide a measure of the annual challenge rate and indicate the appropriate vaccination regimen to be adopted.

Disease in other species is generally investigated only sporadically or in association with research projects.

B. Surveillance by Other Means

Because sheep are the dominant large mammalian species present throughout much of the enzootic region and blood samples can readily be collected from them, there are few instances when they are not the preferred source of material for qualitative or quantitative investigations. Occasionally, concern for red grouse stock may prompt land owners to submit brains of dead young grouse or blood collected during shoots. This may provide useful information for establishing that louping-ill is causing mortality in the grouse, but it is not the preferred method of surveillance.

Examination of sera from hunter-shot red deer has proved to be of value in establishing the absence of louping-ill virus in areas where no sheep have been present.[24] In some preliminary studies, examination of sera from shot blue hares was found of value in detecting the presence or absence of louping-ill on an estate, but only in exceptional circumstances does such an approach have any advantage over the examination of sheep serum.[13]

Likewise, attempts to establish the infection rate in the vector tick have been unrewarding because it is so low, and collecting significant numbers of unengorged ticks is a very slow process.[18,86]

The concept of deploying sentinel animals is generally inappropriate because the catholic host preference of the vector ensures that available resident vertebrates can supply adequate data on the prevalence of louping-ill virus. However, occasionally where the principal concern has been to investigate deaths in red grouse, a flock of sheep has been introduced into an area and monitored for seroconversion. In addition, to establish the challenge rate in vaccine trials, equal numbers of vaccinated and control sheep have been exposed to the same environment.[54,80]

VIII. PREVENTION AND CONTROL

A. Vector Control

The major role that sheep have in the maintenance of *Ix. ricinus* (Section VI.A) suggests that the tick would be vulnerable to control by the systematic application of insecticidal sheep dips. No such study has been made, although it is the opinion of many sheep farmers that ticks have become a much greater problem since sheep dips incorporating dieldrin have been withdrawn. Most sheep dips currently available are effective against ticks for only a limited period.[105] In addition, as the period of maximum tick activity alters according to weather, latitude, and altitude, the most appropriate time to dip for the control of ticks varies from year to year and between regions. Furthermore, problems with sheep management, such as abortion and mismothering of lambs following dipping, can render strategic dipping impractical. Thus, the use of sheep dips has only a limited role in tick control.

The reduction of tick infestation in an area following the removal of sheep has already been discussed (Section VI.A). A system of sheep management designed to provide areas of improved pasture and the use of hill grazing only during midsummer and winter, the periods when there is minimal tick activity, can succeed in dramatically reducing tick infestation.[106] In addition, the application of insecticide (chlorpyrifos) directly to pasture has been reported to reduce tick infestation markedly, but the economic considerations and environmental implications of this procedure make such a strategy both unfeasible and unacceptable.[107]

B. Vaccination

Shortly after the discovery of the viral etiology of louping-ill, a vaccine consisting of homogenized, formalin-treated brains from experimentally infected sheep was developed and widely deployed.[108] It was considered that while this vaccine failed to induce a detectable immune reaction, it sensitized animals to viral antigens so that on exposure to natural challenge the immune response was accelerated and provided protection from clinical encephalitis.[109] It was, however, postulated that the systemic phase of virus infection following challenge after vaccination was essentially unaltered, and thus the capacity of vaccinated animals to transmit virus was unaffected.[49] In this way it was proposed that vaccination had little effect on the epidemiology of the virus.

This crude vaccine was subsequently replaced by one prepared from virus propagated in BHK-21 cells, inactivated with formalin, and incorporated in an oil-based adjuvant.[110,111] This vaccine induced substantial titers of HI and neutralizing antibody and provided complete protection both from disease and from the establishment of infection. The systematic administration of such a vaccine to sheep, the one essential maintenance host, should interrupt the transmission of louping-ill virus, and over a period of 2 to 3 years eliminate it from an area. Militating against such a stragegy is the possible introduction of virus-infected ticks from adjacent properties carried by wild or domestic animals and difficulties in ensuring that all sheep on an extensive grazing are vaccinated. Thus, to overcome these problems, a test site was selected on an island to which the lateral spread of ticks was considered to be insignificant.[13]

Following 3 successive years when all available sheep were vaccinated, the bovine calves employed as sentinels did not show evidence of infection. However, in the following years some cattle did seroconvert, indicating that virus had survived. It was considered that virus persistence on the island was due at least in part to problems with the shelf life of the vaccine which were experienced at the time, and thus all sheep may not have been fully protected.

The feasibility of eradicating louping-ill from an area by systematic vaccination of sheep thus remains unresolved.

IX. FUTURE RESEARCH

The proposed epidemiology of louping-ill virus relies largely on data generated within the laboratory and remains to be examined critically in controlled field trials. The logistics of such investigations and the limited resources available for surveillance and research into animal disease precludes such studies at the present time. Studies on free-living populations of red grouse have been limited to one area of Scotland.[58] Initial observations suggest that louping-ill virus infection may have an effect on grouse populations elsewhere, but its full significance is not known. In the absence of such information, the relevance of developing control strategies specifically to protect commercial interest in grouse remains a matter for conjecture.

Studies of the bionomics of the vector tick have been made primarily in the relatively dry eastern seaboards of the U.K. Critical examination of *Ix. ricinus* populations in other areas, particularly the wetter regions of Scotland, is required. The application of field observation to a predictive model of the life cycle of the tick suggests that even limited but specific observations could be of value in determining the validity of extrapolating the currently held view of the bionomics of *Ix. ricinus* throughout its distribution.[112]

The recent observations in mainland Europe of a disease of sheep similar to louping-ill require further investigation. Monoclonal antibodies generated to CEE virus suggest that the western subtype of virus including louping-ill may be particularly closely related. A panel of monoclonal antibodies to louping-ill virus should therefore be developed to compare isolates of louping-ill virus from the U.K. and isolates associated with meningoencephalitis in sheep in other areas of Europe. Only in this way will the relationship of these viruses be revealed.

APPENDIX 1
SUSCEPTIBILITY OF LABORATORY ANIMALS TO LOUPING-ILL VIRUS INFECTION

Species[a]	Route of inoculation	Clinical disease	Ref.
Monkey *(Macacus rhesus)*	Intracerebral	+	113
	Intramuscular	−	113
Mouse	Intracerebral	+	36,114
	Intraperitoneal and subcutaneous	+	115
Rat	Intracerebral	?	36
	Intranasal	−	116
Rat (suckling)	Intracerebral	+	117
Guinea pig	Intracerebral	−	36,114
Guinea pig (20-day-old)	Intracerebral	±	118
Rabbit	Intracerebral	−	36,114
Hamster	Intracerebral	+	119
Hamster (<10-day-old)	Intraperitoneal	+	119

[a]   Unless otherwise stated animals are assumed to be young adults.

APPENDIX 2
VIRUS REPLICATION IN CULTURED CELLS

Cell type	Cytopathic effect	Ref.
Human cell line		
HeLa	+	120
Detroit-6	+	121
KB	+	121
Hep-2	−	122
Angiosarcoma	+	122
Cynomolgus monkey heart	+	122
Primary pig kidney	+	123
Pig kidney cell line		
PS	+	124
IBRS/2	+	38
PK/15	+	125
Hamster kidney (BHK-21)	+	126
Sheep kidney	−	127
Chick embryo fibroblasts[a]	+	128
Common frog	−	130
Hyalomma dromedarii	−	131
Boophilus microplus	−	131
Rhipicephalus appendiculatus	−	131

[a]   Lesions also produced on the chorioalantoic membrane following inoculation of 6- to 10-day-old chick embryos.[129]

REFERENCES

1. **Duncan, A.,** A treatise on the diseases of sheep: drawn up from original communications presented to the Highland Society of Scotland, *Trans. R. Highl. Agric. Soc. Scotl.,* 3, 339, 1807.
2. **M'Fadzean, J.,** The etiology of louping-ill, *J. Comp. Pathol. Ther.,* 13, 145, 1900.
3. **M'Fadzean, J.,** Louping-ill in sheep, *J. Comp. Pathol. Ther.,* 7, 207, 1894.
4. **M'Gowan, J. P. and Rettie, T.,** Poliomyelitis in sheep suffering from "loupin'-ill", *J. Pathol. Bacteriol.,* 18, 47, 1913.
5. **Stockman, S.,** Louping-ill, *J. Comp. Pathol. Ther.,* 31, 137, 1918.
6. **Stockman, S.,** Louping-ill, *Trans. R. Highl. Agric. Soc. Scotl.,* 36, 1, 1924.
7. **Greig, J. R., Brownlee, A., Wilson, D. R., and Gordon, W. S.,** The nature of louping-ill, *Vet. Rec.,* 11, 325, 1931.
8. **MacLeod, J. and Gordon, W. S.,** Studies in louping-ill (an encephalomyelitis of sheep). II. Transmission by the sheep tick *Ixodes ricinus* L., *J. Comp. Pathol. Ther.,* 45, 240, 1932.
9. **Andrewes, C. and Pereira, H. G.,** Tick-borne flaviviruses, in *Viruses of Vertebrates,* 3rd ed., Bailliere Tindall, London, 1978, 91.
10. **Stevenson, R.,** Cases of sickness, *Trans. R. Highl. Agric. Soc. Scotl.,* 3, 380, 1807.
11. **Walton, G. A. and Kennedy, R. C.,** Tick-borne encephalitis virus in southern Ireland, *Br. Vet. J.,* 122, 427, 1966.
12. **Hughes, L. E. and Kershaw, G. F.,** Louping-ill in mid-Wales, *Vet. Rec.,* 69, 102, 1957.
13. **Reid, H. W.,** unpublished data.
14. **Pavlov, P.,** Studies on tick-borne encephalitis of sheep and their natural foci in Bulgaria, *Zentralbl. Bakteriol. Parasitenkd.,* 206, 360, 1968.
15. **Hartley, W. J., Martin, W. B., Hakioglu, F., and Chifney, S. T. E.,** A viral encephalitis of sheep in Turkey, *Pendik Inst. J.,* 2, 89, 1969.

16. **Stephenson, J. R., Lee, J. M., and Wilton-Smith, D.**, Antigenic variation among members of the tick-borne encephalitis complex, *J. Gen. Virol.*, 65, 81, 1984.

17. **Alexander, R. A. and Neitz, W. O.**, The transmission of louping-ill to sheep by ticks, *Vet. J.*, 89, 320, 1933.

18. **Swanepoel, R.**, Quantitative Studies on the Virus of Louping-Ill in Sheep and Ticks, Ph.D. thesis, University of Edinburgh, 1968.

19. **Watt, J. A., Brotherston, J. G., and Campbell, J.**, Louping-ill in red grouse, *Vet. Rec.*, 75, 1151, 1963.

20. **Williams, H., Thorburn, H., and Ziffo, G. S.**, Isolation of louping-ill from the red grouse, *Nature (London)*, 200, 193, 1963.

21. **Reid, H. W. and Boyce, J. B.**, Louping-ill virus in red grouse in Scotland, *Vet. Rec.*, 95, 150, 1974.

22. **Smith, C. E. G., Varma, M. G. R., and McMahon, D.**, Isolation of louping-ill virus from small mammals in Ayrshire, Scotland, *Nature (London)*, 203, 992, 1964.

23. **Dunn, A. M.**, Louping-ill: the red deer *(Cervus elaphus)* as an alternative host of the virus in Scotland, *Br. Vet. J.*, 116, 284, 1960.

24. **Adam, K. M. G., Beasley, S. J., and Blewett, D. A.**, The occurrence of antibody to babesia and to the virus of louping-ill in deer in Scotland, *Res. Vet. Sci.*, 23, 133, 1977.

25. **Brewis, E. G., Neubauer, C., and Hurst, E. W.**, Another case of louping-ill in man. Isolation of the virus, *Lancet*, 1, 689, 1949.

26. **Williams, H. and Thorburn, H.**, Serum antibodies to louping-ill virus, *Scott. Med. J.*, 7, 353, 1962.

27. **Bannatyne, C. C., Wilson, R. L., Reid, H. W., Buxton, D., and Pow, I.**, Louping-ill virus infection of pigs, *Vet. Rec.*, 106, 13, 1980.

28. **Dunn, A. M.**, Louping-Ill. A Study of the Diseases in Cattle, Ph.D. thesis, University of Edinburgh, 1952.

29. **Reid, H. W., Barlow, R. M., Pow, I., and Maddox, J.**, Isolation of louping-ill virus from red deer *(Cervus elaphus)*, *Vet. Rec.*, 102, 463, 1978.

30. **Fletcher, J. M. and Galloway, I. A.**, Louping-ill in the horse, *Vet. Rec.*, 49, 17, 1937.

31. **Timoney, P. J., Donnelly, W. C., Clements, C., and Fenlon, M.**, Louping-ill infection in the horse, *Vet. Rec.*, 95, 540, 1974.

32. **Mackenzie, C. P.**, Recovery of a dog from louping-ill, *J. Small Anim. Pract.*, 23, 233, 1982.

33. **Mackenzie, C. P., Smith, S. T., and Muir, R. W.**, Louping-ill in a working collie, *Vet. Rec.*, 92, 354, 1973.

34. **Timoney, P. J.**, Variations in the haemagglutination content of crude mouse brain antigens containing louping-ill virus, *Res. Vet. Sci.*, 12, 490, 1971.

35. **Reid, H. W.**, Epidemiology of louping-ill, in *Vectors in Virus Biology*, Mayo, M. A. and Harrap, K. A., Eds., Academic Press, London, 1984, 161.

36. **Alston, J. M. and Gibson, H. J.**, A note on the experimental transmission of "louping-ill" to mice, *Br. J. Exp. Pathol.*, 12, 82, 1931.

37. **deMadrid, A. T. and Porterfield, J. S.**, A simple microculture method for the study of group B arboviruses, *Bull. WHO*, 40, 113, 1969.

38. **Reid, H. W. and Doherty, P. C.**, Louping-ill encephalomyelitis in sheep. I. The relationship of viraemia and the antibody response to susceptibility, *J. Comp. Pathol.*, 81, 521, 1971.

39. **Reid, H. W.**, A Study of the Pathogenesis of Louping-Ill, Ph.D. thesis, University of Edinburgh, 1975.

40. **Rivers, T. M. and Schwentker, F. F.**, Louping-ill in man, *J. Exp. Med.*, 59, 669, 1934.

41. **Edward, D. G.**, Immunisation against louping-ill. Immunisation of man, *Br. J. Exp. Pathol.*, 29, 372, 1948.

42. **Reid, H. W., Gibbs, C. A., Burrells, C., and Doherty, P. C.**, Laboratory infections with louping-ill virus, *Lancet*, 1, 593, 1972.

43. **Webb, H. E., Connolly, J. H., Kane, F. F., O'Reilly, K. J., and Simpson, D. I. H.**, Laboratory infections with louping-ill with associated encephalitis, *Lancet*, 2, 255, 1968.

44. **Davison, G., Neubauer, C. and Hurst, E. W.**, Meningo-encephalitis in man due to louping-ill virus, *Lancet*, 2, 453, 1948.

45. **Lawson, J. H., Manderson, W. G., and Hurst, E. W.**, Louping-ill meningo-encephalitis, a further case and a serological survey, *Lancet*, 2, 696, 1949.

46. **Likar, M. and Dane, D. S.**, An illness resembling acute poliomyelitis caused by a virus of the Russian spring-summer encephalitis/louping-ill group in Northern Ireland, *Lancet*, 1, 456, 1958.

47. **Ross, C. C.**, Louping-ill in the west of Scotland, *Lancet*, 2, 527, 1961.

48. **Schonell, M. E., Brotherston, J. G., Burnett, R. C. S., Campbell, J., Coghlan, J. D., Moffat, M. A. J., Norval, J., and Sutherland, J. A. W.**, Occupational infections in the Edinburgh abattoir, *Br. Med. J.*, 2, 148, 1966.

49. **Smith, C. E. G., McMahon, D. A., O'Reilly, K. J., Wilson, A. L., and Robertson, J. M.,** The epidemiology of louping-ill in Ayrshire the first year of studies in sheep, *J. Hyg. Cambr.,* 62, 53, 1964.
50. **Reid, H. W. and Boyce, J. B.,** The effect of colostrum-derived antibody on louping-ill virus infection in lambs, *J. Hyg. Cambr.,* 77, 349, 1976.
51. **Findlay, C. R.,** Serum immune globulin levels in lambs under a week old, *Vet. Rec.,* 92, 530, 1973.
52. **Harker, D. B.,** Serum immune globulin levels in artificially reared lambs, *Vet. Rec.,* 95, 229, 1974.
53. **Gordon, W. S., Brownlee, A., Wilson, D. R., and Macleod, J.,** The epizootiology of louping-ill and tick-borne fever with observations on the control of these sheep diseases, *Symp. Zool. Soc. London,* 6, 1, 1962.
54. **Brotherston, J. G., Bannatyne, C. C., Mathieson, A. O., and Nicolson, T. B.,** Field trials of an inactivated oil-adjuvant vaccine against louping-ill (Arbovirus group B), *J. Hyg. Cambr.,* 69, 479, 1971.
55. **Reid, H. W., Buxton, D., Pow, I., and Finlayson, J.,** Experimental louping-ill virus infection of cattle, *Vet. Rec.,* 108, 497, 1981.
56. **Reid, H. W., Buxton, D., Pow, I., and Finlayson, J.,** Transmission of louping-ill virus in goat milk, *Vet. Rec.,* 114, 163, 1984.
57. **Reid, H. W., Buxton, D., Pow, I., and Finlayson, J.,** Experimental louping-ill virus infection in two species of British deer, *Vet. Rec.,* 111, 61, 1982.
58. **Reid, H. W., Duncan, J. S., Phillips, J. D. B., Moss, R., and Watson, A.,** Studies on louping-ill virus (Flavivirus group) in wild red grouse *(Lagopus lagopus scoticus), J. Hyg. Cambr.,* 81, 321, 1978.
59. **Clarke, D. H. and Casals, J.,** Techniques for hemagglutination and hemagglutination-inhibition with arthropod-borne viruses, *Am. J. Trop. Med. Hyg.,* 7, 561, 1958.
60. **Doherty, P. C. and Reid, H. W.,** Louping-ill encephalomyelitis in sheep. II. Distribution of virus and lesions in nervous tissue, *J. Comp. Pathol.,* 81, 531, 1971.
61. **Thorburn, H. and Williams, H.,** A serological examination of Scottish strains of louping-ill and their relation to other members of the complex, *Arch. Gesamte Virusforsch.,* 19, 155, 1966.
62. **Williams, H. E.,** Complement-fixation test for louping-ill in sheep, *Am. J. Vet. Res.,* 29, 1619, 1968.
63. **Reid, H. W., Barlow, R. M., Boyce, J. B., and Inglis, D. M.,** Isolation of louping-ill virus from a roe deer *(Capreolus capreolus), Vet. Rec.,* 98, 116, 1976.
64. **Reid, H. W.,** Experimental infection of red grouse with louping-ill virus (flavivirus group). I. The viraemia and antibody response, *J. Comp. Pathol.,* 85, 223, 1975.
65. **Brownlee, A. and Wilson, D. R.,** Studies in the histopathology of louping-ill, *J. Comp. Pathol. Ther.,* 45, 67, 1932.
66. **Doherty, P. C. and Reid, H. W.,** Experimental louping-ill in sheep and lambs. II. Neuropathology, *J. Comp. Pathol.,* 81, 331, 1971.
67. **Doherty, P. C., Smith, W., and Reid, H. W.,** Louping-ill encephalomyelitis in the sheep. V. Histopathogenesis of the fatal disease, *J. Comp. Pathol.,* 83, 481, 1972.
68. **Doherty, P. C., Reid, H. W., and Smith, W.,** Louping-ill encephalomyelitis in the sheep. IV. Nature of the perivascular inflammatory reaction, *J. Comp. Pathol.,* 81, 545, 1971.
69. **MacLeod, J.,** Ixodes ricinus in relation to its physical environment. II. The factors governing survival and activity, *Parasitology,* 27, 123, 1935.
70. **MacLeod, J.,** Ixodes ricinus in relation to its physical environment. IV. An analysis of the ecological complexes controlling distribution and activities, *Parasitology,* 28, 295, 1936.
71. **Campbell, J. A.,** Recent work on the ecology of the pasture-tick *Ixodes ricinus* L. in Britain, *Rep. 14th Int. Vet. Congr., London, (1949),* 2, 113, 1952.
72. **Lees, A. D. and Milne, A.,** The seasonal and diurnal activities of individual sheep ticks *(Ixodes ricinus L.), Parasitology,* 41, 189, 1951.
73. **Gordon, W. C., Brownlee, A., Wilson, D. R., and MacLeod, J.,** Tick-borne fever, *J. Comp. Pathol. Ther.,* 45, 301, 1932.
74. **Pool, W. A., Brownlee, A., and Wilson, D. R.,** The etiology of "louping-ill", *J. Comp. Pathol. Ther.,* 43, 253, 1930.
75. **Zlotnik, I., Keppie, J., and Grant, D. P.,** A method of testing the efficacy of louping-ill vaccines in sheep, *Vet. Rec.,* 86, 659, 1970.
76. **Reid, H. W., Doherty, P. C., and Dawson, A. McL.,** Louping-ill encephalomyelitis in the sheep. III. Immunoglobulins in the cerebrospinal fluid, *J. Comp. Pathol.,* 81, 537, 1971.
77. **Reid, H. W., Buxton, D., Pow, I., Brodie, T. A., Holmes, P. H., and Urquhart, G. M.,** Response of sheep to experimental concurrent infection with tick-borne fever *(Cytoecetes phagocytophila)* and louping-ill virus, *Res. Vet. Sci.,* 41, 56, 1986.
78. **MacLeod, J.,** Ticks and disease in domestic stock in Great Britain, *Symp. Zool. Soc. London,* 6, 29, 1962.
79. **Woldehiwet, Z.,** Tick-borne fever: a review, *Vet. Res. Commun.,* 6, 163, 1983.
80. **O'Reilly, K. J., Smith, C. E. G., McMahon, D. A., Bowen, E. T. W., and White, G.,** A comparison of methods of measuring the persistence of neutralizing and haemagglutination-inhibiting antibodies to louping-ill in experimentally infected sheep, *J. Hyg. Cambr.,* 66, 217, 1968.

81. **Buxton, D., Reid, H. W., Finlayson, J., and Pow, I.,** Immunosuppression in toxoplasmosis: preliminary studies in mice infected with louping-ill virus, *J. Comp. Pathol.,* 90, 331, 1980.

82. **Reid, H. W., Buxton, D., Finlayson, J., and Holmes, P. H.,** Effect of chronic Trypanosoma brucei infection on the course of louping-ill virus infection in mice, *Infect. Immun.,* 23, 192, 1979.

83. **Buxton, D., Reid, H. W., Finlayson, J., Pow, I., and Anderson, I.,** Immunosuppression in toxoplasmosis: studies in sheep with vaccines for chlamydial abortion and louping-ill virus, *Vet. Rec.,* 109, 559, 1981.

84. **Reid, H. W., Buxton, D., Pow, I., and Finlayson, J.,** Immunosuppression in toxoplasmosis further studies on mice infected with louping-ill virus, *J. Med. Microbiol.,* 13, 313, 1980.

85. **Reid, H. W., Buxton, D., Gardiner, A. C., Pow, I., Finlayson, J., and MacLean, M. J.,** Immuno-suppression in toxoplasmosis: studies in lambs and sheep infected with louping-ill virus, *J. Comp. Pathol.,* 92, 181, 1982.

86. **Varma, M. G. R. and Smith, C. E. G.,** The epidemiology of louping-ill in Ayrshire, Scotland. II. Ectoparasites of small mammals (Ixodidae), *Folia Parasitol. (Prague),* 18, 63, 1971.

87. **Milne, A.,** The ecology of the sheep tick, *Ixodes ricinus* L. Spatial distributions, *Parasitology,* 40, 35, 1950.

88. **Dunn, A. M.,** personal communication, 1984.

89. **Beasley, S. J., Campbell, J. A., and Reid, H. W.,** Threshold problems in infection of *Ixodes ricinus* with the virus of louping-ill, in *Tick-Borne Diseases and Their Vectors,* Wilde, J. K. H., Ed., Edinburgh University Press, 1978, 497.

90. **Arthur, D. R.,** *British Ticks,* Butterworths, London, 1963, 1.

91. **Reid, H. W., Moss, R., Pow, I., and Buxton, D.,** The response of three grouse species *(Tetrao urogallus, Lagopus mutus, Lagopus lagopus)* to louping-ill virus, *J. Comp. Pathol.,* 90, 257, 1980.

92. **Reid, H. W., Buxton, D., Pow, I., and Moss, R.,** Experimental louping-ill virus infection of black grouse *(Tetrao tetrix),* *Arch. Virol.,* 78, 299, 1983.

93. **Timoney, P. J.,** Susceptibility of the horse to experimental inoculation with louping-ill virus, *J. Comp. Pathol.,* 90, 73, 1980.

94. **MacLeod, J.,** *Ixodes ricinus* in relation to its physical environment: the influence of climate on development, *Parasitology,* 26, 282, 1934.

95. **Milne, A.,** The ecology of the sheep tick *Ixodes ricinus* L. Host relationships of the tick. II. Observations on hill and moorland grazing in northern England, *Parasitology,* 39, 173, 1949.

96. **Milne, A.,** The ecology of the sheep tick *Ixodes ricinus* L. Host relationships of the tick. I. Review of previous work in Britain, *Parasitology,* 39, 167, 1949.

97. **MacLeod, J.,** The part played by alternative hosts in maintaining the tick population on hill pastures, *J. Anim. Ecol.,* 3, 161, 1934.

98. **Moore, W.,** The sheep or grass tick *Ixodes ricinus* L., *Vet. Rec.,* 50, 1031, 1938.

99. **MacLeod, J.,** The tick problem, *Vet. Rec.,* 50, 1245, 1938.

100. **Cunningham, J. M. M., Smith, A. D. M., and Doney, J. M.,** Trends in livestock population in hill areas of Scotland, in *5th Rep. Hill Farming Research Organisation, 1967— 1970,* Constable, Edinburgh, 1970, 1.

101. **MacLeod, J.,** The bionomics of *Ixodes ricinus* L., the "sheep tick" of Scotland, *Parasitology,* 24, 382, 1932.

102. **Leslie, A. S. and Shipley, A. E.,** *The Group in Health and in Disease,* Smith, Elder, London, 1912, 1.

103. **Douglas-Home, H.,** Great shooting counties twixt Tweed and Lammermuirs. Berwickshire's variety of attractions, *Field,* 1038, 1938.

104. Report of the Committee on Hill Sheep Farming in Scotland, Cmd. 6494, Department of Agriculture for Scotland, 1944, 1.

105. **Platt, N. E.,** An evaluation of amitraz, a new acaricidal sheep dip, against the castor bean tick *Ixodes ricinus* L., in Scotland and Lankashire, in *Tick-Borne Diseases and Their Vectors,* Wilde, J. K. H., Ed., Edinburgh University Press, 1978, 206.

106. **Whitelaw, A.,** personal communication, 1980.

107. **Bevan, W. J. and Sykes, G. B.,** Use of chlorpyrifos for the control of *Ixodes ricinus* on pasture, *Vet. Rec.,* 113, 341, 1983.

108. **Gordon, W. S.,** The control of certain diseases in sheep, *Vet. Rec.,* 14, 1, 1938.

109. **Smith, C. E. G.,** Arbovirus vaccines, *Br. Med. Bull.,* 25, 142, 1969.

110. **Brotherston, J. G. and Boyce, J. B.,** A new vaccine against louping-ill, *Vet. Rec.,* 84, 514, 1969.

111. **Brotherston, J. G. and Boyce, J. B.,** Development of a non-infective protective antigen against louping-ill (Arbovirus group B), *J. Comp. Pathol.,* 80, 377, 1970.

112. **Gardiner, W. P. and Gettinby, G.,** A weather-based prediction model for the life-cycle of the sheep tick, *Ixodes ricinus, Vet. Parasitol.,* 13, 77, 1983.

113. **Hurst, E. W.,** The transmission of louping-ill to the mouse and monkey, *J. Comp. Pathol. Ther.,* 44, 231, 1931.

114. **Greig, J. R., Brownlees, A., Wilson, D. R., and Gordon, W. S.,** The nature of louping-ill, *Vet. Rec.,* 11, 325, 1931.
115. **Doherty, P. C.,** Louping-ill in C57 black mice, *J. Comp. Pathol.,* 79, 571, 1969.
116. **Burnet, F. M.,** Inapparent (subclinical) infection of the rat with louping-ill virus, *J. Pathol. Bacteriol.,* 42, 213, 1936.
117. **Gresikova, M., Albrecht, P., and Ernek, E.,** Comparison of an attenuated and virulent louping-ill strain, *Nature (London),* 190, 508, 1961.
118. **Zlotnik, I., Carter, G. B., and Grant, D. P.,** The persistence of louping-ill virus in immunosuppressed guinea pigs, *Br. J. Exp. Pathol.,* 52, 395, 1971.
119. **Doherty, P. C.,** Effect of age on louping-ill encephalitis in the hamster, *J. Comp. Pathol.,* 79, 431, 1969.
120. **Oker-Blom, N.,** Propagation of louping-ill virus in malignant human epithelial cells, strain HeLa, *Ann. Med. Exp. Biol. Fenn.,* 34, 199, 1956.
121. **von Zeipel, G. and Svedmyr, A.,** Growth of viruses of the Russian spring-summer louping-ill group in tissue culture, *Arch. Gesamte Virusforsch.,* 8, 370, 1958.
122. **Levcovick, E. N. and Karpovich, L. G.,** Study on biological properties of viruses of the tick-borne encephalitis complex in tissue cultures, in *Biology of Viruses of the Tick-Borne Encephalitis Complex,* Czechoslovak Academy of Science, Prague, 1962, 161.
123. **Williams, H. E.,** Growth and titration of louping-ill virus in monolayer tissue culture of pig kidney, *Nature (London),* 181, 497, 1958.
124. **de Madrid, A. T. and Porterfield, J. S.,** A simple microculture method for the study of group B arboviruses, *Bull. WHO,* 40, 113, 1969.
125. **Timoney, P. J., Geraghty, V. P., Harrington, A. M., and Dillon, P. B.,** Microneutralisation test in PK (15) cells for assay of antibodies to louping-ill virus, *J. Clin. Microbiol.,* 20, 128, 1984.
126. **Karabatsos, N. and Buckley, S. M.,** Susceptibility of the baby hamster kidney line (BHK-21) to infection with arboviruses, *Am. J. Trop. Med. Hyg.,* 16, 99, 1967.
127. **Brotherston, J. G. and Boyce, J. B.,** A new vaccine against louping-ill, *Vet. Rec.,* 83, 514, 1969.
128. **Porterfield, J. S.,** Plaque production with yellow fever and related arthropod-borne viruses, *Nature (London),* 183, 1069, 1959.
129. **Burnet, F. M.,** Observations on the effect of louping-ill virus on the developing egg, *Br. J. Exp. Pathol.,* 17, 294, 1936.
130. **Pudney, M. and Varma, M. G. R.,** The growth of some tick-borne arboviruses in cell cultures derived from tadpoles of the common frog *Rana temporaria, J. Gen. Virol.,* 10, 131, 1971.
131. **Pudney, M., Varma, M. G. R., and Lake, C. J.,** The growth of some arboviruses in tick cell lines, in *Tick-Borne Diseases and Their Vectors,* Wilde, J. K. H., Ed., Edinburgh University Press, 1978, 490.

Chapter 31

MAYARO VIRUS DISEASE

Francisco P. Pinheiro and James W. LeDuc

TABLE OF CONTENTS

I. HISTORICAL BACKGROUND

A. Discovery of the Agent

Mayaro virus was first isolated in Trinidad in 1954 from the blood of five sick persons.[1] Four of the cases occurred in adult male forest workers, and the fifth in a young female urban resident. In the subsequent year, the agent was recovered from patients in the Amazon region of Brazil and Bolivia.[2,3] The disease in these episodes consisted of a mild to moderately severe acute febrile illness of short duration with uneventful recovery.

B. History of Epidemics

Three epidemics associated with Mayaro virus have been described in the literature: two in Brazil,[2,4] and one in Bolivia.[3] This small number of epidemics is somewhat intriguing since immunity to Mayaro virus is very widespread among rural populations in northern South America and in Trinidad. The epidemics occurred among persons that had close contact with forests.

The first reported epidemic occurred in 1955 in a community of quarry and forest workers on the Guama River, 120 miles east of Belém in the Amazon region of Brazil. Six strains of Mayaro virus were isolated from the blood of febrile patients. Neutralizing antibodies against Semliki Forest virus (interpreted as induced by Mayaro virus) were detected in 18.9% of the population within a month of the beginning of the outbreak.

The epidemic in Bolivia was also seen in 1955, and affected a community of some 400 Okinawan workers who had recently colonized a forested area east of the Rio Grande, Bolivia.[3] A virus was isolated from the blood of two patients. Initially, the agent was considered as a new group A arbovirus and was named Uruma virus,[5] but subsequently it was found to be antigenically very closely related or identical to Mayaro virus.[6,7] Although almost one half of the Okinawan settlers were affected during the outbreak, serological studies indicated that probably only 10 to 15% of the cases were due to Uruma virus. Fifteen of the cases had a fatal outcome, but none of them could be related to this virus.

The third reported outbreak occurred in the first half of 1978 and affected the village of Belterra, which is situated near the confluence of the Tapajós and Amazon Rivers, Para State, Brazil.[4] A total of 55 cases were confirmed, 43 by virus isolation and serology, and 12 by serology only. The outbreak lasted about 6 months, peaking in April. Serological surveys conducted during the peak and at the end of the outbreak revealed that some 20% of the 4000 inhabitants of Belterra were infected.[8] A very high proportion of persons infected developed overt clinical manifestations. Yellow fever virus was also circulating in Belterra concurrently with the Mayaro virus outbreak and was responsible for five fatalities. The mosquito *Haemagogus janthinomys* was identified as the vector of both viruses.[9]

C. Social and Economic Impact

Although no fatalities due to Mayaro virus infection have been documented, there is evidence suggesting that this disease may cause a significant morbidity among rural populations. The infection is widespread in rural communities throughout northern South America and Trinidad, with up to 60% of the residents immune in some areas[10] and a clinically apparent attack rate of over 80%.[8]

Data obtained during epidemics of Mayaro are also useful to illustrate the impact of this disease. Of the outbreaks investigated, the one in Belterra provided the most complete information to data concerning Mayaro fever in an affected population. Most patients observed during the outbreak in Belterra were too ill to continue their daily routine during the febrile period, and some even became prostrate. Arthralgia was common and was sometimes severe, leading to temporary incapacitation. That outbreak was associated with an epidemic of yellow fever, but clinical, virological, and serological findings enabled differentiation of the two entities. The yellow fever outbreak was quickly controlled through an immunization campaign; however, the Mayaro epidemic continued for 6 months. The outbreaks on the Guama River in Brazil and in Bolivia were also associated with other disease, which may have interfered with an accurate assessment of the impact of Mayaro virus. As 15 to 20% of persons can become infected during outbreaks, the loss of work can be significant. However, a precise estimate of costs due to loss of work, and medical assistance including treatment, remains to be determined.

II. THE VIRUS

A. Antigenic Relationships

Early studies done by hemagglutination inhibition (HI) tests, complement-fixation (CF) tests and neutralization (N) tests showed that Mayaro virus is a group A arbovirus, most closely related to Semliki Forest virus.[11] However, by both HI and CF tests, Mayaro virus was sufficiently distinguishable from Semliki virus to justify its being considered as a new agent. They are closely related by mouse intraperitoneal N test, but different by the plaque-inhibition,[6] microculture,[7] and plaque N tests.[12] The available evidence indicates that the Uruma strain isolated in Bolivia is very closely related or identical to the prototype Mayaro from Trinidad.[6,7] The Trinidadian and Brazilian (Guama River) isolates are identical by HI and CF tests.[11]

Mayaro virus is now recognized as a member of the family Togaviridae, genus *Alphavirus*. Alphaviruses registered in the *International Catalogue of Arboviruses* have been classified in six antigenic complexes.[13] The classification takes into account studies employing HI, CF, and cross-protection tests in mice,[14] plaque-inhibition tests,[6] and N tests in cell cultures.[7,12,15] In this scheme, Mayaro virus is classified in the Semliki Forest complex as a species, which in turn has Mayaro and Una viruses as subtypes.

B. Host Range

Mayaro virus is highly pathogenic for newborn Swiss mice by intracerebral and intraperitoneal routes, but not for weanling or adult mice. The first signs of illness in infant mice consist of irritability and failure to nurse, followed by prostration and death within 1 to 3 days. Some of the animals may develop temporary flaccid paralysis of one or both hind limbs. Newborn, but not adult, hamsters are highly susceptible to the Uruma strain of Mayaro; the same strain causes fever in guinea pigs after intracerebral inoculation and no reaction in rabbits inoculated intraperitoneally.[5] A hemagglutinin active against chicken and goose erythrocytes can be prepared from brains and serum of infected mice; with certain strains, the hemagglutinin can be demonstrated in suckling mouse serum, but not in suckling mouse brains.[11] The hemagglutinin can also be obtained from hamster kidney cell cultures, but only in low titer.[16]

The marmosets *Callithrix argentata* and *C. humeralifer* become viremic after subcutaneous inoculation with Mayaro virus and survive.[9] Two experimentally infected marmosets developed virus titers in blood presumably high enough to infect susceptible vectors. A cebus monkey inoculated subcutaneously with Mayaro virus had leukopenia and viremia, and subsequently developed immunity.[17] Cynomolgus monkeys inoculated intracerebrally developed a febrile reaction and viremia, but when subcutaneously inoculated, no febrile reaction was observed.[18] Experimental studies in rhesus monkeys with chikungunya and Mayaro viruses revealed that when these monkeys were inoculated with either virus, then later challenged with the heterologous virus, none had detectable viremia following challenges, thus indicated cross-protection.[19]

Mayaro virus replicates and induces cytopathic effect and plaques in several cell cultures, including BHK-21, VERO, and HeLa cell lines, and in primary cultures of Peking duck or rhesus monkey kidney cells, and in chicken or mouse embryo fibroblast.[20-24]

C. Strain Variation

Very limited information is available germane to biological variation of Mayaro virus. Two Brazilian isolates of the virus, one from *Haemogogus* mosquitoes and the other from a patient, were both found to produce large and small plaques in overlaid cultures of primary chick embryo cells.[25] Studies with the mosquito isolate revealed that the plaque variants are antigenically very similar by HI and plaque-inhibition tests. Nevertheless, the large plaque variant causes more intense lesions in connective tissue of infant mice, while only the small variant produces a cytopathic effect in chick embryo cultures maintained under fluid medium.

Several interesting biological differences were also observed among two Trinidadian Mayaro isolates, one from a person and one from the mosquito *Mansonia venezuelensis*. Distinct differences were noted in pathogenicity for adult mice, production of hemagglutinating antigen from mouse brain, and growth in chick embryo tissue culture.[26]

Minor antigenic differences were demonstrated by HI and CF tests between a Brazilian Mayaro strain of human origin and the Uruma strain. No differences, however, were seen between prototype Mayaro (Trinidad) and Uruma virus using microneutralization tests in cell cultures.[7]

The fact that arthralgia and exanthema were more conspicuous during the outbreak in Belterra than in cases previously observed led to the suggestion that the strain of Mayaro from Belterra was more virulent, or that the population of the village was more susceptible. A comparison made between a human isolate of Mayaro from Belterra and two strains previously isolated from the Amazon region of Brazil revealed complete identity between them by HI and CF tests, and only minor differences by the plaque-reduction N test.[4]

It is conceivable that sensitive serologic and biological tests, and also molecular analysis, may reveal strain variation of Mayaro virus and their biological significance; however, to date, these analyses have not been attempted.

D. Methods for Assay

Newborn or infant mice and several cell culture systems are useful for assaying Mayaro virus infectivity. Serological tests such as HI, CF, and N tests have been employed for the antigenic characterization of the virus. Other tests such as radioimmunoassay (RIA) and the enzyme-linked immunosorbent assay (ELISA) may also be useful for antigenic analysis, as well as for antigen and antibody detection.

III. DISEASE ASSOCIATIONS

A. Humans

According to early observations, Mayaro virus disease was usually manifested by fever, headache, epigastric pain, backache, sometimes generalized pain, chills, nausea, photophobia, and vertigo.[1-3] One Trinidadian case complained of joint pains and exhibited a swollen

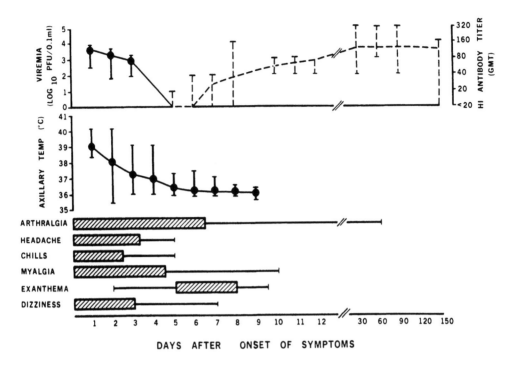

FIGURE 1. Diagrammatic representation of clinical manifestations, viremia curve, and antibody response in Mayaro virus infection as seen in 21 patients with daily clinical examinations during 10 days and periodic follow-up thereafter, and from whom Mayaro virus was isolated; Belterra, Para, Brazil, 1978. (Reproduced with permission of the *Am. J. Trop. Med. Hyg.*)

finger joint on one hand.[1] Slight icterus was observed in three Brazilian patients,[2] and rash was seen in a case of laboratory infection.[4] Polyuria and rash were reported in Bolivian cases.[3] During the outbreak in Belterra, Brazil, fever, arthralgia, and rash were very prominent among 43 laboratory-documented cases of Mayaro.[4] Most of these patients were examined daily. The onset of clinical illness was usually abrupt. More than 80% reported headache, myalgias, and chills. Dizziness, eye pain, nausea, vomiting, photophobia, and diarrhea were referred to less frequently. Axillary temperature was over 39°C in most cases, and reached 40.2°C in one. Arthralgia was present in all cases and was a very prominent part of the clinical picture; many patients referred to it as being quite severe. Wrists, ankles, and toes were predominantly affected, although aches in the elbows and knees were also often reported. Swelling of the affected joints was observed in about 20% of the cases. In some patients the arthralgia was so severe that they became temporarily incapacitated. Inguinal lymphadenopathy, which was not painful at palpation, was observed in half of the patients. Rash was present in about two thirds of patients, and was more commonly observed among children than in older people. It usually appeared on the 5th day of illness and lasted about 3 days. Rash consisted of either small maculopapular or micropapular isolated lesions, which occasionally formed small areas of confluence. Rash was more prominent on the chest, back, arms, and legs, with the face less affected. Occasionally rash was generalized. Clinical manifestations lasted from 2 to 5 days, with the exception of the arthralgia, which persisted in some patients for 2 months, and caused a temporary incapacity for work in some. Leukopenia, with moderate lymphocytosis, was seen in all patients during the 1st week of illness. Platelet counts, and serum bilirubin, glutamic-pyruvic transaminase, and glutamic-oxaloacetic levels were within normal limits or showed minor alterations. Figure 1 presents a generalized schematic summary of the duration of clinical manifestations of infection, as well as the magnitude and duration of viremia and the onset of HI antibody.

Table 1
GEOGRAPHICAL DISTRIBUTION OF MAYARO VIRUS IN THE AMERICAS,
ACCORDING TO VIRUS ISOLATIONS AND IMMUNITY TO THE AGENT

Virus isolation	
Humans	Brazil, Bolivia, Suriname, and Trinidad
Vertebrates	Brazil, U.S.[a]
Arthropods	Brazil, Panama, and Trinidad
Serology	
Humans	Brazil, Colombia, French Guiana, Guyana, Peru, Suriname, Trinidad, Venezuela, Costa Rica, Guatemala, and Panama
Vertebrates	Brazil, Colombia, Guatemala, Honduras, Panama, and Trinidad

Note: No isolations have been obtained outside the Americas.

[a] Brazil: One from a marmoset[9] and two from lizards (considered technically suspect).[29] U.S.: one from a migrating bird.[30]

B. Disease Associated with Domestic Animals
No disease observed was among domestic animals.

C. Disease Associated with Wildlife
No disease was observed.

D. Diagnostic Procedures
The laboratory diagnosis of Mayaro infection can be established by isolation of the agent from blood and/or by demonstration of an antibody rise. Virus can be recovered from blood collected during the first 3 to 4 days after onset by intracerebral inoculation into infant Swiss mice and/or in cell cultures such as VERO, BHK-21, or primary chick embryo fibroblasts. Isolation rates of Mayaro virus from known positive human blood specimens were higher in VERO cell cultures maintained under fluid medium or agar medium than in infant mice.[4] Mayaro virus could not be recovered from throat swabs, feces, or spinal fluid from viremic persons.[4] The HI test has been commonly used to detect antibody rises among paired serum samples collected from patients early during illness and 1 to 3 weeks later.

E. Adverse Effects of Virus on Vector
No effects have been reported.

IV. EPIDEMIOLOGY

A. Geographic Distribution
Mayaro virus activity has been documented only in the Americas. With the exception of a single strain obtained in North America, all other isolations of Mayaro virus came from South America, Trinidad, and Panama. The virus has been recovered from persons in Brazil, Bolivia, Suriname, and Trinidad.[1-4,27-28] Most human strains come from the Amazon region of Brazil, either from sporadic cases or during epidemics. The only isolations of Mayaro virus obtained from wild animals have been from two lizards[29] and a marmoset[9] in Brazil, and from a wild bird captured in the southern U.S.[30] Isolations from arthropods have been made from Brazil, Panama, and Trinidad.[9,26,28,29,31,32] Isolations of Mayaro virus from *Psorophora* mosquitoes obtained in Colombia[28,33] were later reported as being Una virus.[34] The geographical distribution of the agent is summarized in Table 1.

B. Incidence
Attack rates of Mayaro virus infection have been determined during epidemic periods.

The study done in Bolivia by serology suggested that some 10 to 15% of the sick persons were infected by the virus during an outbreak of jungle fever.[3] Serological and clinical surveys made during the epidemic in Belterra, Brazil revealed an attack rate of 20% among villagers over a period of approximately 6 months.[8] A study involving 374 Dutch soldiers stationed in Suriname indicated that 5.3% acquired antibodies to Mayaro virus during their 1-year tour of duty.[35] Immunity to Mayaro developed in 1.6% of a second group of 500 Dutch soliders also after a 1-year stay in Suriname.[36]

C. Seasonal Distribution

Both epidemics of Mayaro reported in Brazil occurred during the rainy season. The epidemic in Belterra began with the onset of the wet season and ended with the onset of the dry season, which corresponded to the rise and fall of abundance of the apparent mosquito vector.[8]

D. Risk Factors

Forest contact seems to be an important risk factor for infection with Mayaro virus. During the Belterra oubreak, the greatest concentration of cases and highest incidence of antibodies were in residents of houses closest to the forest.[8]

Age is also a risk factor related to Mayaro infection. Among persons living in areas where Mayaro virus is endemic, the prevalence of antibodies to the agent among children generally increases with age. In certain population groups, there is a steady progression toward increasing HI antibody prevalence with increasing age, which reaches very high levels in older adults.[37-39] In other communities there seems to be a leveling off of HI antibody prevalence after childhood. This may indicate decreased Mayaro virus activity in the area, or it may be that antibody levels tend to fall below the measurable threshold some years after initial infection.[37]

In general, Mayaro infections are more predominant in males than in females. Antibody rates among males vary from only slightly higher than females, to twofold higher or even greater.[38-40] A survey made during the peak of the Mayaro outbreak in Belterra revealed that seropositive males outnumbered females 2:1; at the end of the outbreak, however, the sex ratio was nearly equal.[8]

No information is available concerning the importance of race and other genetic factors, nutrition, or immunological background as risk factors for Mayaro infection.

E. Serologic Epidemiology

The distribution of immunity in humans to Mayaro virus is very widespread in northern South America, including Brazil, Colombia, French Guiana, Guyana, Peru, Suriname, Trinidad, and Venezuela.[28,29,34-43] The presence of HI antibodies in sera from residents of Costa Rica, Guatemala, and Panama suggests that the infection is probably present throughout Central America.[41,44,45] Human infections seem to be closely associated with forested areas. The presence of neutralizing antibodies to Semliki Forest virus in 9.6% of 551 residents of the Amazon Valley examined in 1953[46] probably represents cross reactivity between this agent and Mayaro virus. Immunity to Mayaro virus was found throughout rural communities of the Amazon basin of Brazil, with HI antibody prevalence rates from 10 to 60%.[29] From 20 to 47% of Brazilian Indians in the Amazon region show immunity to the agent,[37] whereas 20% of Xavante Indians of the Mato Grosso were found immune.[47] In a serosurvey of Brazilian military recruits made in 1964, neutralizing antibodies to Mayaro virus were found among recruits from the Amazon and central regions of the country, but only a single serum (0.5%) from the northeast was found positive, and all sera from the southern regions were negative.[48] HI antibodies were also detected in 16% of inhabitants from rural areas of Goiás state in central Brazil, examined in 1973.[49] Serum surveys conducted among residents of three towns situated along the Trans-Amazon Highway revealed HI antibody prevalence rates from 10 to 26% to alphaviruses, most of which is due to Mayaro virus; in contrast, 2% or less of immigrants from other parts of Brazil, particularly from the northeast, were reactive against alphaviruses.[50] Studies in Trinidad showed high immunity rates concentrated in the forested areas in the southeastern part of the island.[38] In Guyana, high Mayaro immunity rates were found in both the Rupununi Savannah and the Mazaruni River regions, whereas in the in coastal region of the country Mayaro infection was virtually absent.[51] No evidence of human infection was found in Barbados, Grenada, Jamaica, and Tobago.[38] Studies in Suriname showed high Mayaro immunity rates among residents from the interior reaching 81% in adult males and 51% in adult females.[52] Infections with Mayaro were also detected in Dutch military personnel stationed in Suriname.[35,36] Mayaro infections appear to be common in French Guiana.[41] Studies in Venezuela showed that Mayaro infections are common in the southern part of the country near Brazil, and in the Sierra de Perija, west of Lake Maracaibo.[43] High Mayaro immunity rates were found among Indian and mestizo residents of tropical foothill and lowland aras in eastern Peru.[42]

V. TRANSMISSION CYCLES

Mayaro virus is probably maintained in nature by a silent transmission cycle in tropical forests of the Americas. The available evidence suggests that nonhuman primates and *Haemagogus* mosquitoes play an important role in the transmission of Mayaro virus. Other animals, including birds, and certain other vectors have also been implicated in the virus cycle, although the evidence supporting this assumption is less convincing.

A. Evidence from Field Studies

Extensive arbovirus ecological studies carried out in the Amazon region of Brazil over the past 30 years have provided data concerning vectors and vertebrate hosts of Mayaro virus. Similarly, field studies done in other areas of the Americas such as Trinidad, Colombia, and Panama have also been helpful in accumulating knowledge on the virus cycle.

1. Vectors

Mayaro virus has been isolated from at least five genera of mosquitoes, comprising *Haemagogus, Psorophora, Mansonia, Culex, and Sabethes.*[9,26,28,29,31,32] At least 38 isolations originated from *Haemagogus,*[31] whereas only five strains have been recovered from the other four genera. All *Haemagogus* isolates originated from Brazil. The majority of these strains were recovered from *Haemagogus* spp. which were collected during nonepidemic periods. Nine strains of Mayaro were obtained from *Haemagogus janthinomys* captured during the Mayaro epidemic in Belterra, Brazil.[9] The single isolations from *Mansonia venezuelensis, Psorophora ferox,* and *Culex* spp.[26,28,29,32] came from Trinidad, Panama, and Brazil, respectively, and may represent fortuitous findings, as are also the two isolations from mixed Sabethini[28,29] made in Brazil. Four isolations of Mayaro virus from *Psorophora albipes* and *P. ferox* have been reported from Colombia.[28,33] These isolates were identified by neutralization tests, but subsequent studies using the HI technique revealed differences between them and prototype Mayaro virus from Trinidad, and showed that the four stains were closely related to Una virus.[34] The agent was also isolated on one occasion from a pool of *Gigantolaelaps* mites combed from four *Oryzomys* rats captured near Belém, Brazil.[17]

2. Vertebrate Hosts

Evidence from field studies implicate monkeys and marmosets as important vertebrate hosts in the transmission cycle of Mayaro virus. High immunity levels to the virus have been found among nonhuman primates from Brazil, Panama, Trinidad, Colombia, Honduras, and Guatemala.[28,29,41,53] Antibody conversions to Mayaro have been demonstrated in sentinel monkeys in Panama.[54] During the outbreak in Belterra, 27% of *Callithrix* tested were found immune, whereas only 1.3% of wild birds were positive, and all bats, rodents, marsupials, edentates, and carnivores were negative.[9] Low to moderate HI antibody rates were found among wild birds (except in *Columbigallina* spp., which shows a high rate), rodents, marsupials, edentates, and lizards from the Amazon region, but confirmation by neutralization tests was not done.[29] Bats were usually found free of antibodies to Mayaro.[9,53-55]

The only isolations of Mayaro virus from wild vertebrates in the tropics were from two lizards and one marmoset in Brazil.[9,29] Outside the tropics, the virus was recovered from one migrating bird captured in the southern U.S.; however, sera from wild birds captured in the same location and sera from persons living in the vicinity were devoid of neutralizing antibodies to Mayaro virus.[30]

B. Evidence from Experimental Infection

1. Vectors

Mayaro virus has been shown to multiply readily in both *Aedes aegypti* and *Anopheles quadrimaculatus* mosquitoes following experimental inoculation.[1] Following parenteral inoculation, *Ae. scapularis, Ae. serratus, Culex quinquefasciatus, Mansonia arribalzagai, M. venezuelensis, M. wilsoni,* and *Psorophora ferox* were shown to harbor Mayaro virus for at least 12 days; a single transmission by one *Ae. scapularis* was obtained after an incubation period of 7 days, whereas *Ae. serratus, Cx. quinquefasciatus* and *Mansonia wilsoni* failed to transmit the virus.[56] No experimental studies have been reported using *Haemagogus* spp.

2. Vertebrate Hosts

Nonhuman primates from the New and Old World have been shown to be susceptible to Mayaro virus following experimental infection. The marmosets *Callithrix argentata* and *Callithrix humeralifer* developed viremia probably high enough to infect susceptible vectors.[9] A *Cebus* monkey had viremia on post-inoculation days 4 to 6.[17] Rhesus and cynomolgus monkeys also develop viremia when injected with Mayaro virus.[18,19]

C. Maintenance / Overwintering Mechanisms

Mayaro virus appears to be maintained in nature in a cycle involving mammals, and possibly birds, and mosquito vectors. The bulk of the evidence, however, supports the existence of a monkey-*Haemagogus*-monkey transmission cycle, which may serve to maintain the virus. Studies undertaken after the Belterra outbreak failed to demonstrate persistence of the virus in the area, therefore suggesting that transovarial transmission among *Haemagogus janthinomys* may not occur in nature.[9]

VI. ECOLOGICAL DYNAMICS

Based on evidence from immunity surveys among humans and nonhuman primate populations, Mayaro virus occurs in tropical forests from Guatemala in the north to those in central Brazil in the south. Virus circulation occurs in tropical rain forest and possibly in other types of tropical forests. Although there is insufficient knowledge on the maintenance of Mayaro virus, based on the available data it can be postulated that the Mayaro cycle is similar to that of sylvatic yellow fever, a constantly moving wave of virus activity among susceptible vertebrates, primarily monkeys, and transmitted by sylvatic culicine mosquitoes, especially those of the genus *Haemagogus*. Consequently, the microenvironment where transmission occurs is probably similar to yellow fever, where virus transmission between *Haemagogus* and monkeys takes place mainly in the tropical rain forest canopy. In forests where trees are low and spaced apart, the capture rates of *Haemagogus* may be similar for ground and tree canopy; therefore, the transmission cycle may occur at different levels of these forests.[57] In tropical rain forests, the temperature fluctuates within relatively narrow limits and rainfall is fairly abundant and well distributed throughout the year. Because of these climatic conditions, the virus is probably permanently enzootic in these forests, and may periodically become epizootic and cause human epidemics. Factors leading to epizootics are probably dictated by the introduction of the virus into areas with high density of *Haemagogus* and of susceptible monkeys. *Haemagogus* mosquitoes are primarily tree-hole breeders, although they may occasionally be found in bamboo stumps.

Seasonal variations in the density of *Haemagogus* have been demonstrated in an area of small residual forests in Brazil, where high densities were found in the wet season, but very small numbers were captured during the dry season.[58] Similar oscillations have also been observed with *Haemagogus janthinomys* in forests of the Amazon region.[9] The longevity and fecundity of *Haemagogus* in nature is unknown.

Repeated observations have emphasized the strict preference of *Haemagogus* for feeding during the midday period.[59] *Haemagogus* also has a marked preference for the upper levels in tropical forests. These mosquitoes exhibit a marked phototropism, although humidity and temperature may also be critical factors in the microclimate.[57] Nevertheless, *Haemagogus* does not confine its activity to the forest, as they can follow man 200 to 300 yards into open cleared land. They have also been collected biting man both inside and outside houses located close to the forest.[57] The attractiveness of man as a blood source is well known, but little is known about the blood preferences of the genus in nature. Laboratory observations indicate that it is much more difficult to induce feeding on marsupials than on monkeys, but this may be due in part to the relative cleanliness of the laboratory animals.[59] No studies have been undertaken to correlate monkey density and immunity status in relation to Mayaro transmission, although during the Mayaro outbreak in Belterra, nearly 30% of *Calithrix* population became immune, and this, combined with reduced vector abundance due to the onset of the dry season, may have resulted in the cessation of transmission. The vector competence of *Haemagogus* for Mayaro virus has not been studied.

Based on studies on the movement of vectors and monkeys in regard to jungle yellow fever, some speculation can be made regarding Mayaro virus activity. Labeled *Haemagogus* specimens have been recaptured 11.5 km from the point of release, and it is believed that the mosquito can be carried by wind currents for distances of 50 km or more. In contrast, monkeys have a limited range of movement and tend not to migrate from one forest patch to another. Other animals implicated as vertebrate hosts of Mayaro, such as rodents, have even more limited movements. If migrant birds are really involved in the Mayaro cycle, this would allow dissemination of the virus at great distances. Although excursions of Mayaro virus from its enzootic zones to susceptible areas have not been documented, it is conceivable that this phenomenon occurs periodically, as observed with yellow fever.

The influence of humans in Mayaro disease ecology has not been assessed. It can be speculated, however, that in the long run man-made deforestation will have an adverse effect on circulation of the virus in enzootic zones. If *Ae. aegypti* can become infected and transmit the virus, there is a risk of urban epidemics and even urban endemicity as has been observed in Asia with chikungunya virus, another *Alphavirus* antigenically related to Mayaro.

VII. SURVEILLANCE

Humans are the only known hosts that develop significant disease following Mayaro virus infection. Therefore, human surveillance, in conjunction with laboratory diagnosis, appears to be the best approach to detect Mayaro disease. In practice, however, this may be difficult since the disease usually is not severe, and it occurs in forested areas where little medical infrastructure is available for adequate surveillance. The three outbreaks of Mayaro virus described in the literature were all associated with other clinical entities which, due to their severity, were probably responsible for the alarm, rather than Mayaro disease. Several animals have been used extensively as sentinels in areas where Mayaro virus is known to be enzootic, but so far virus infection has been documented in only two sentinel monkeys in Panama.[54]

VIII. INVESTIGATION OF EPIDEMICS

Epidemics of Mayaro will offer the opportunity to investigate clinical, epidemiological, and ecological aspects of the disease. Differences in clinical expression of Mayaro disease have been documented in previous outbreaks; therefore, it will be desirable to conduct detailed studies of clinical manifestations and the course of the disease. The fact that chikungunya infection, which is clinically very similar to Mayaro disease, has been associated with hemorrhagic manifestations and rarely death, prompts a search for rare complications in cases of Mayaro infection. Particular attention should be given to the pathogenesis and treatment of arthralgia, because arthralgia can cause significant physical incapacitation. Studies on the magnitude and persistence of viremia, together with the dynamics of immune response, including IgM antibodies, should also be undertaken. The only data on clinical attack rates comes from the epidemic in Belterra; consequently, more studies on this aspect appear justified. As limited data are available on the impact of Mayaro disease, epidemics will offer the opportunity to attempt to estimate with more precision the impact of the infection, including work loss, school absenteeism, and costs of medical care. Ecological studies should address the role of possible vectors other than *Haemagogus*, and of vertebrate hosts in the Mayaro virus cycle. It will be of particular interest to study the vector competence of vectors, especially of *Ae. aegypti*, which will help to predict the possibility of urbanization of Mayaro virus. Similar studies could be attempted using *Haemagogus* mosquitoes to see if man can be an amplifying host. The insects could be fed on viremic persons, and subsequently the infection and transmission rates could be determined using appropriate methods.

IX. PREVENTION AND CONTROL

The Mayaro virus cycle occurs only in forested areas, where human infections occur inside or near forests. Consequently, the use of vector or vertebrate host control measures is not practical. Similarly, environmental modifications are useless to control or prevent Mayaro infection. An experimental vaccine prepared by formalin inactivation of Mayaro virus grown in human diploid cell cultures has produced a satisfactory antibody response in weanling mice.[60] In the event that a Mayaro vaccine for human use be developed, its use will probably be limited to very selected groups of persons.

X. FUTURE RESEARCH

Areas of future research include detailed studies of the clinical features of Mayaro disease and its pathogenesis, particularly of the arthralgia. Biological and molecular characterization of Mayaro strains isolated from persons in different outbreaks is also needed in order to detect possible strain variations and to evaluate their significance in terms of virulence to man and epidemiological implications. Mechanisms of transmission and maintenance of the virus in known or suspected endemic areas must also be clarified. Antibody studies of vertebrate populations should utilize serological techniques of recognized specificity to minimize cross reactions due to infections caused by related alphaviruses. Additional experimental studies should be done with nonhuman primates in order to define the parameters of infection of ecological importance. Similar studies should be undertaken with other suspected vertebrate hosts, such as wild birds, rodents, and reptiles. It will also be of interest to undertake vector competence studies of known vectors. Most important is the investigation of the vector competence of *Ae. aegypti* and the possible role of man as a source of virus for this mosquito species, in order to assess the potential for urbanization of Mayaro virus. The progressive expansion of urban areas infested with *Ae. aegypti* observed in the Americas in recent years, and the proximity of some of these areas to forests where Mayaro virus is enzootic, highlights the importance of such studies.

REFERENCES

1. **Anderson, C. R., Downs, W. G., Wattley, G. H., Ahin, N. W., and Reese, A. A.,** Mayaro virus: a new human disease agent. II. Isolation from blood of patients in Trinidad, B.W.I., *Am. J. Trop. Med. Hyg.,* 6, 1012, 1957.
2. **Causey, O. R. and Maroja, O. M.,** Mayaro virus: a new human disease agent. III. Investigation of an epidemic of acute febrile illness on the River Guama in Para, Brazil, and isolation of Mayaro virus as causative agent, *Am. J. Trop. Med. Hyg.,* 6, 1017, 1957.
3. **Schaeffer, M., Gajdusek, D. C., Lema, A. B., and Eichenwald, H.,** Epidemic jungle fevers among Okinawan colonists in the Bolivian rain forest. I. Epidemiology, *Am. J. Trop. Med. Hyg.,* 8, 372, 1959.
4. **Pinheiro, F. P., Freitas, R. B., Travassos da Rosa, J. F. S., Gabbay, Y. B., Mello, W. A., and LeDuc, J. W.,** An outbreak of Mayaro virus disease in Belterra, Brazil, I. Clinical and virological findings, *Am. J. Trop. Med. Hyg.,* 30, 674, 1981.
5. **Schmidt, J. R., Gajdusek, D. C., Schaeffer, M., and Gorrie, R. H.,** Epidemic jungle fever among Okinawan colonists in the Bolivian rain forest. II. Isolation and characterization of Uruma virus, a newly recognized human pathogen, *Am. J. Trop. Med. Hyg.,* 8, 479, 1959.
6. **Porterfield, J. S.,** Cross-neutralization studies with group A arthropod-borne viruses, *Bull. WHO,* 24, 735, 1961.
7. **Chanas, A. C., Johnson, B. K., and Simpson, D. I. H.,** Antigenic relationships of alphaviruses by a simple micro-culture cross-neutralization method, *J. Gen. Virol.,* 32, 295, 1976.
8. **LeDuc, J. W., Pinheiro, F. P., and Travassos da Rosa, A. P. A.,** An outbreak of Mayaro virus disease in Belterra, Brazil. II. Epidemiology, *Am. J. Trop. Med. Hyg.,* 30, 682, 1981.
9. **Hoch, A. L., Peterson, N. E., LeDuc, J. W., and Pinheiro, F. P.,** An outbreak of Mayaro virus disease in Belterra, Brazil. III. Entomological and ecological studies, *Am. J. Trop. Med. Hyg.,* 30, 689, 1981.
10. **Pinheiro, F. P.,** Mayaro fever, in *Handbook Series in Zoonoses Section B: Viral Zoonoses,* Vol. 1, Beran, G. W., Ed., CRC Press, Boca Raton, Fla., 1981, 159.
11. **Casals, J. and Whitman, L.,** Mayaro virus: a new human disease agent. I. Relationship to other arboviruses, *Am. J. Trop. Med. Hyg.,* 6, 1004, 1957.
12. **Karabatsos, N.,** Antigenic relationships of group A arboviruses by plaque reduction neutralization testing, *Am. J. Trop. Med. Hyg.,* 24, 527, 1975.
13. **Calisher, C. H., Shope, R. E., Brandt, W., Casals, J., Karabatsos, N., Murphy, F. A., Tesh, R. B., and Wiebe, M. E.,** Proposed antigenic classification of registered arboviruses. I. Togaviridae, *Alphavirus, Intervirology,* 14, 229, 1980.
14. **Casals, J.,** The arthropod-borne group of animal viruses, *Trans. N.Y. Acad. Sci.,* 19(Ser. 2), 219, 1957.
15. **Calisher, C. H., Monath, T. P., Muth, D. J., Lazuick, J. S., Trent, D. W., Francy, D. B., Kemp, G. E., and Chandler, F. W.,** Characterization of Fort Morgan virus, an alphavirus of the western equine encephalitis virus complex in an unusual ecosystem, *Am. J. Trop. Med. Hyg.,* 29, 1428, 1980.
16. **Diercks, F. H., Kundin, W. D., and Porter, T. J.,** Arthropod-borne virus hemagglutinin production by infected hamster kidney-cell cultures, *Am,. J. Hyg.,* 73, 164, 1961.
17. **Causey, O. R.,** cited by Woodall, J. P., Virus research in Amazônia, in *Atas Simpósio Biota Amazônica,* Vol. 6, Lent, H., Ed., Conselho Nacional de Pesquisas, Rio de Janeiro, 1967, 31.
18. **Verlinde, J. D.,** Susceptibility of cynomolgus monkeys to experimental infection with arboviruses of group A (Mayaro and Mucambo), group C (Oriboca and Restan) and an unidentified arbovirus (Kwatta) originating from Surinam, *Trop. Geogr. Med.,* 20, 385, 1968.
19. **Binn, L. N., Harrison, V. R., and Randall, R.,** Patterns of viremia and antibody observed in rhesus monkeys inoculated with Chikungunya and other serologically related group A arboviruses, *Am. J. Trop. Med. Hyg.,* 16, 782, 1967.
20. **Buckley, S. M.,** Applicability of the HeLa (Gey) strain of human malignant epithelial cells to the propagation of arboviruses, *Proc. Soc. Exp. Biol. Med.,* 116, 354, 1964.

21. **Henderson, J. R. and Taylor, R. M.**, Propagation of certain arthropod-borne viruses in avian and primate cell cultures, *J. Immunol.*, 84, 590, 1960.
22. **Bergold, G. H. and Mazzali, R.**, Plaque formation by arboviruses, *J. Gen. Virol.*, 2, 273, 1968.
23. **Sellers, R. F.**, The use of a line of hamster kidney cells (BHK-21) for growth of arthropod-borne viruses, *Trans. R. Soc. Trop. Med. Hyg.*, 57, 433, 1963.
24. **Pinheiro, F. P.**, unpublished data, 1968.
25. **Pinheiro, F. P. and Dias, L. B.**, Virus Mayaro e Una: estudo de variantes produzindo grandes e pequenas placas, in *Atas Simposio Biota Amazonica*, Vol. 6, Lent, H., Ed., Conselho Nacional de Pesquisas, Rio de Janeiro, 1967, 211.
26. **Aitken, T. H. G., Downs, W. G., Anderson, C. R., and Spence, L.**, Mayaro virus isolated from a Trinidadian mosquito, *Mansonia venezuelensis, Science*, 131, 986, 1960.
27. **Metselaar, D.**, Isolation of arboviruses of group A and group C in Surinam, *Trop. Geogr. Med.*, 18, 137, 1966.
28. **Karabatsos, N., Ed.**, *International Catalogue of Arboviruses, Including Certain Other Viruses of Vertebrates*, 3rd ed., American Society for Tropical Medicine and Hygiene, San Antonio, Tex., 1985, 673.
29. **Woodall, J. P.**, Virus research in Amazonia, in *Atas Simposio Biota Amazonica*, Vol. 6, Lent, H., Ed., Conselho Nacional de Pesquisas, Rio de Janeiro, 1967, 31.
30. **Calisher, C. H., Gutiérrez, E., Maness, K. S. C., and Lord, R. D.**, Isolation of Mayaro virus from a migrating bird captured in Louisiana in 1967, *Bull. PAHO*, 8, 243, 1974.
31. **Pinheiro, F. P.**, Situação das arboviroses na Região Amazônica. *Int. Symp. Tropical Arboviruses and Haemorrhagic Fevers*, Pinheiro, F. P., Ed., Academic Brasileira de Ciencias, Rio de Janeiro, 27, 1982.
32. **Galindo, P., Srihongse, S., De Rodaniche, E., and Grayson, M. A.**, An ecological survey for arboviruses in Almirante, Panamá, 1959—1962, *Am. J. Trop. Med. Hyg.*, 15, 385, 1966.
33. **Groot, H., Morales, A., and Vidales, H.**, Virus isolations from forest mosquitoes in San Vicente de Chucurí, Colombia, *Am. J. Trop. Med. Hyg.*, 10, 397, 1961.
34. **Groot, H.**, Estudios sobre virus transmitidos por artropodos en Colombia, *Rev. Acad. Colomb. Cienc. Exactas Fis. Nat.*, 12, 197, 1964.
35. **Karbaat, J., Jonkers, A. H., and Spence, L.**, Arbovirus infections in Dutch military personnel stationed in Surinam. A preliminary study, *Trop. Geogr. Med.*, 4, 370, 1964.
36. **Jonkers, A. H., Spence, L., and Karbaat, J.**, Arbovirus infections in Dutch military personnel stationed in Surinam. Further studies, *Trop. Geogr. Med.*, 20, 251, 1968.
37. **Black, F. L., Hierholzer, W. J., Pinheiro, F. P., Evans, A. S., Woodall, J. P., Opton, E. M., Emmons, J. E., West, B. S., Edsall, G., Downs, W. G., and Wallace, G. D.**, Evidence for persistence of infectious agents in isolated human populations, *Am. J. Epidemiol.*, 100, 230, 1974.
38. **Downs, W. G. and Anderson, C. R.**, Distribution of immunity to Mayaro virus infection in the West Indies, *West Indies Med. J.*, 7, 190, 1958.
39. **Pinheiro, F. P. and Travassos da Rosa, A. P. A.**, unpublished data, 1980.
40. **Dixon, K. E., Llewellyn, C. H., Travassos da Rosa, A. P. A., and Travassos da Rosa, J. F.**, Programa multidisciplinario de vigilancia de las enfermedades infecciosas en zonas colindantes con la carretera transamazonica en Brasil. II. Epidemiología de las infecciones por arbovirus, *Bol. Of. Sanit. Panam.*, 91, 200, 1981.
41. **Theiler, M. and Downs, W. G.**, *The Arthropod-Borne Viruses of Vertebrates*, 1st ed., Yale University Press, New Haven, 1973, 131.
42. **Madalengoitia, J., Flores, W., and Casals, J.**, Arbovirus antibody survey of sera from residents of eastern Peru, *Bull. PAHO*, 7, 25, 1973.
43. **Downs, W. G., Spence, L., and Nuñez-Montiel, O.**, Un estudio serológico sobre la frecuencia de virus de encephalitis transmitido por artropodos (arbovirus) en Venezuela: estudio preliminar, *Rev. Venez. Sanid. Asist. Soc.*, 26, 145, 1961.
44. **Fuentes, L. G. and Mora, J. A.**, Encuesta serológica sobre arbovirus en Costa Rica, *Rev. Latinoam. Microbiol.*, 13, 25, 1971.
45. **Srihongse, S., Stacy, H. G., and Gauld, J. R.**, A survey to assess potential human disease hazards along proposed sea level canal routes in Panama and Colombia. IV. Arbovirus surveillance in man, *Mil. Med.*, 138, 422, 1973.
46. **Causey, O. R. and Theiler, M.**, Virus antibody survey on sera of residents of the Amazon Valley in Brazil, *Am. J. Trop. Med. Hyg.*, 7, 36, 1958.
47. **Neel, J. V., Andrade, A. H. P., Brown, G. E., Eveland, W. E., Goobar, J., Sodeman, W. A., Jr., Stollerman, G. H., Weinstein, E. D., and Wheeler, A. H.**, Further studies of the Xavante Indians. IX. Immunologic status with respect to various diseases and organisms, *Am. J. Trop. Med. Hyg.*, 17, 486, 1968.
48. **Niederman, J. C., Henderson, J. R., Opton, E. M., Black, F. L., and Skvrnova, K.**, A nationwide serum survey of Brazilian military recruits, 1964. II. Antibody patterns with arboviruses, polioviruses, measles and mumps, *Am. J. Epidemiol.*, 86, 319, 1967.

49. **Pinheiro, F. P., Bensabath, G., Andrade, A. H. P., and Moraes, M. A. P.,** Febre amarela no Estado de Goiás, 1973, *Bol. Epidemiol. (Minist. Saude),* 6, 1, 1974.

50. **Pinheiro, F. P., Bensabath, G., Andrade, A. H. P., Lins, A. C., Fraiha, H., Tang, A. T., Lainson, R., Shaw, J. J., and Azevedo, M. C.,** Infectious diseases along Brazil's Trans-Amazon Highway: surveillance and research, *Bull. PAHO,* 8, 111, 1974.

51. **Spence, L. and Downs, W. G.,** Virological investigations in Guyana, 1956—1966, *West Indies Med. J.,* 17, 83, 1968.

52. **Jonkers, A. H., Downs, W. G., Aitken, T. H. G., and Spence, L.,** Arthropod-borne encephalitis viruses in northeastern South America. I. A serological survey of northeastern Surinam, *Am. J. Trop. Med. Hyg.,* 14, 304, 1965.

53. **Seymour, C., Peralta, P. H., and Montgomery, G. G.,** Serologic evidence of natural togavirus infections in Panamanian sloths and other vertebrates, *Am. J. Trop. Med. Hyg.,* 32, 854, 1983.

54. **Srihongse, S., Galindo, P., Eldridge, B. F., Young, D. G., and Gerhardt, R. R.,** A survey to assess potential human disease hazards along proposed sea level canal routes in Panama and Colombia. V. Arbovirus infection in non-human vertebrates, *Mil. Med.,* 139, 449, 1974.

55. **Price, J. L.,** Serological evidence of infection of Tacaribe virus and arboviruses in Trinidadian bats, *Am. J. Trop. Med. Hyg.,* 27, 162, 1978.

56. **Aitken, T. H. G. and Anderson, C. R.,** Virus transmission studies with Trinidadian mosquitoes. II. Further observations, *Am. J. Trop. Med. Hyg.,* 8, 41, 1959.

57. **Pinheiro, F. P., Travassos da Rosa, A. P. A., and Moraes, M. A. P.,** An epidemic of yellow fever in central Brazil, 1972—1973. II. Ecological studies, *Am. J. Trop. Med. Hyg.,* 30, 204, 1981.

58. **Causey, O. R. and Dos Santos, G. V.,** Diurnal mosquitoes in an area of small residual forests in Brazil, *Ann. Entomol. Soc. Am.,* 42, 471, 1949.

59. **Strode, G. K.,** *Yellow Fever,* 1st ed., McGraw-Hill, New York, 1951, 710.

60. **Robinson, D. M., McManus, A. T., Cole, F. E., Jr., and Pedersen, C. E. Jr.,** Inactivated Mayaro vaccine produced in human diploid cell cultures, *Mil. Med.,* 141, 163, 1976.

Chapter 32

MURRAY VALLEY AND KUNJIN ENCEPHALITIS

Ian D. Marshall

TABLE OF CONTENTS

I. HISTORICAL BACKGROUND

A. History of Epidemics

Epidemics of a severe, highly lethal encephalitis were reported in eastern Australia during the summers of 1917 and 1918, the first in Queensland (Q.) and New South Wales (N.S.W.) and the second additionally involving the Murray Valley, including northern Victoria (Vic.). By the end of the second epidemic, 172 cases had been reported in N.S.W., 118 of them fatal, and a further 13 cases in northern Vic. with the outcome unrecorded. The disease was considered to be previously undescribed and was labeled "Australian X disease"; it was thought to be spread by human carriers rather than by insect vectors.[1] In a poorly recorded epidemic in 1922, seemingly centered on Brisbane and nearby Ipswich, one report stated that 49/79 died of Australian X disease. Between January and April 1925, nine cases were reported at Broken Hill in the far west of N.S.W. and one at Tibooburra, 200 miles to the north. Six of these died. In April and May of the same year at least 11 cases occurred in Townsville, Q. All were children under 8 years and ten died.[1]

The complete lack of reports during the next 25 years is the more remarkable because in northern Australia and New Guinea an indeterminate number, certainly in excess of 1 million, of young Japanese, American, and Australian servicemen were present in usually lightly populated enzootic-endemic areas for about 4 years. They were plagued with dengue fever, and smaller outbreaks of epidemic polyarthritis (Ross River virus) were identified against this background; an enormous medical effort was directed against malaria, and even if some deaths with neurological involvement were misdiagnosed as due to "cerebral malaria", it is surprising that an identifiable epidemic of arbovirus encephalitis did not occur.

In the summer of 1950—1951 there were 45 reported cases of severe encephalitis accepted

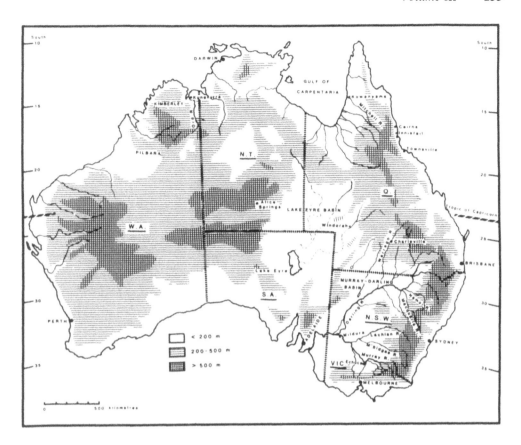

FIGURE 1.   Mainland Australia showing localities of human infection, field research activities, and principal drainage basins.

by Anderson[1] as due to the newly named "Murray Valley encephalitis" (MVE) virus, 19 (42%) of which died. All except two cases were in the broadly defined Murray Valley: the Murray, Lachlan, Murrumbidgee Rivers and their tributaries in southern N.S.W. and northern Vic., the anabranches of the Darling River where they enter the Murray, and the Murray and its tributaries in South Australia (S.A.) (Figure 1). The one fatal case in S.A. was infected in dry farming country to the west of the Murray,[2] and one fatal case occurred at Narrabri in central N.S.W. on the Namoi River, a tributary of the Darling River.[3] The first case occurred in late December 1950, the last in April 1951, and the peak incidence was in February 1951 (27 cases). The age of patients ranged from 7 weeks to 69 years, 62% being less than 15 years, and 69% were male.

Concurrent with this epidemic was the first spectacular epizootic of myxomatosis of rabbits, the virus spreading from experimental sites on the Murray River near Albury westward down the river into S.A. and northward up the Darling through N.S.W. and into southern Q. Enterprising journalists quickly linked the two events, arguing that the mosquito-borne myxoma virus, deliberately introduced to control the rabbit population, must be the same mosquito-borne virus causing encephalitis and death in humans. Public disquiet was such that the responsible minister was pleased to be able to announce to the House of Representatives on March 8, 1951 that three eminent scientists had been inoculated with myxoma virus some weeks earlier, that no local lesions or general signs developed, and no antibody to myxoma virus could be found 3 weeks later![4]

In March 1956, two cases of MVE were serologically confirmed on the Murray River upstream from Mildura. Seroepidemiological investigations suggested patchy distribution of the virus confined to the close vicinity of the river.[5] Similarly, in March and April 1971,

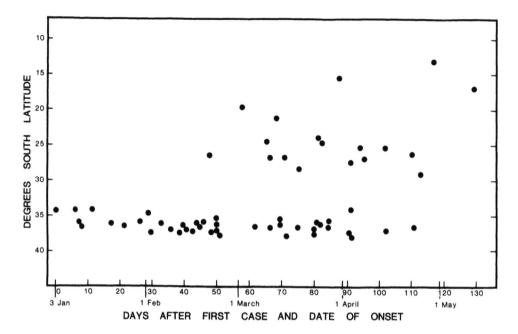

FIGURE 2.   The Murray Valley encephalitis epidemic of 1974, with cases plotted at date of onset and at the latitude where infection was probably acquired.

enhanced MVE virus activity was detected in the Murray-Darling basin with serologically confirmed severe encephalitis in one child at Charleville at the northern limit of the basin and another at Henty in the Murray Valley near Albury. Antibody was detected in domestic fowls at some localities, and in horses with "nervous disease" in the Murray Valley.[6,7]

The next substantial epidemic occurred in 1974, and eventually involved every mainland state. As in 1951, there were almost simultaneous infections over an enormous area of the Murray Valley, and the eight cases with onset in January were scattered from Albury, N.S.W. to Goolwa near the Murray mouth in S.A. There had been 21 cases in the Murray Valley by February 19, 1974, when the first case occurred outside the valley — at Windorah in the "channel country" of southwestern Q. (Figures 1 and 2). The last of 39 cases in the Murray Valley had onset of symptoms on April 15, and the final case of the series of 58 occurred at Kununurra, northern Western Australia (W.A.) on May 9 — the only case to occur in that state. There were ten cases scattered through Q., and the five in the Northern Territory (N.T.) included two at Alice Springs, near the geographic center of the continent, and the ten in S.A. included two in the arid Musgrave Ranges. There were 13 deaths, and unlike the epidemic of 1951, there was a double peak of age prevalence, one under 10 years old and a second over 50 years old.[8] Retrospectively, several of the 1974 cases might have been due to Kunjin (KUN) virus rather than MVE.[9]

In March 1984 an isolated, severe case of encephalomyelitis contracted in northern Vic. was clearly diagnosed as due to KUN infection.[10]

Increased arbovirus activity associated with the development of a large water impoundment and a small irrigation area on the Ord River at Kununurra led to a series of isolated cases and small outbreaks of encephalitis in the Kimberley and Pilbara regions of northern W.A. Eight cases were reported in 1978, two in 1979, eight in 1981, and two in 1984.[11]

B. Discovery of the Agent and the Major Vectors

Anderson[1] has reviewed and referenced the early work on Australian X disease, and as the original source material is not readily accessible, it is not listed here. Investigations, however, are attributed to the original workers.

Breinl, when investigating the Townsville, Q. epidemic of 1917, inoculated monkeys intracerebrally with cerebrospinal fluid and crushed spinal cord and produced an infection which was fatal in 16 to 18 days. Histological sections of the CNS from the infected monkeys resembled those of human cases. Breinl concluded that Australian X disease was an encephalitic form of anterior poliomyelitis.[1]

Cleland and Bradley inoculated pathological material intraperitoneally into monkeys and other animals in 1917 but failed to isolate an agent. During the 1918 epidemic, they used the intracerebral route to inoculate monkeys with brain material from patients who had died 2, 3, and 5 days after onset of symptoms, and recovered virus pathogenic for monkeys from each specimen. One of the isolates was maintained for five passages in monkeys, and by the i.c. route, the virus was pathogenic for a calf, a foal, and a proportion of sheep, and on these and other grounds they disputed Breinl's suggestion that the agent was poliomyelitis virus.[1]

During the outbreak in Broken Hill, N.S.W. in 1925, Kneebone and Cleland directly infected sheep with human brain and maintained the virus through three but not four passages.

None of these pioneering isolates was available when the next epidemic occurred in 1951, so the direct comparison of Australian X disease virus and Murray Valley encephalitis virus could not be made.[1]

During the 1951 epidemic, French[12] isolated virus from the post-mortem brains of 3 patients, but not from 14 other fatal cases, nor from any specimens such as blood or throat washings taken before death. He found the chorioallantoic membrane of developing chick embryos more sensitive to infection than infant or weaned mice, and in 2/3 cases virus was recovered only in this host. It was shown that the chorioallantois is infected even when antibody is present.[12] Miles et al.[2] about the same time isolated virus from the brain of the one fatal South Australian case; the girl had died shortly after onset of the disease and there was no problem in recovering the virus in both infant and weaned mice. Both groups found the virus to be related to, but not identical with, the virus of Japanese encephalitis, prepared complement-fixing antigen from chick embryos, and demonstrated histological and host range similarities between the viruses of MVE and Australian X disease.[2,12]

Little was known of the mosquito fauna and bionomics in the Murray-Darling basin in 1951, but interest was quickly generated by the MVE epidemic and the very much more impressive epizootic of myxomatosis. The nature of MVE virus was not fully appreciated early enough to carry out mosquito collections during the epidemic, and a major effort in the summer of 1951—1952 to find invertebrate and vertebrate hosts of MVE virus was frustrated by the apparently complete disappearance of the virus from the epidemic area; not only was there no clinical or serological evidence of virus activity in the human population, but recent infection could not be demonstrated in young chickens, domestic animals, or nestling birds.[13] When this was established, further processing of mosquitoes for virus isolation was abandoned. However, Reeves et al.[14] were impressed with the taxonomic and biologic similarities of *Culex annulirostris*, the major summer mosquito in the Murray Valley, *Cx. tarsalis*, a major vector of St. Louis and western equine encephalitis in California, and *Cx. tritaeniorrhynchus*, the major vector of Japanese encephalitis in Japan, and postulated that this would be a major vector of MVE, with *Cx. quinquefasciatus* as a possible urban vector, as it is of St. Louis encephalitis in some parts of the U.S.[14]

McLean[15] demonstrated laboratory replication and transmission of MVE by *Cx. annulirostris*, and a number of other culicines, but it was not until 1960 that MVE was finally isolated from pools of wild-caught mosquitoes. Doherty et al.[18] had carefully chosen the Mitchell River Mission (Kowanyama) on the west coast of Cape York, to carry out field investigations in an attempt to validate hypotheses of enzootic foci of MVE in tropical northern Australia.[1,13,17,18] During the first 2 years of investigations at Kowanyama, at Cairns on the east coast of Cape York, and at Normanton near the south coast of the Gulf of

Carpentaria, ten arboviruses new to Australia were isolated from pools of mosquitoes, in addition to four strains of MVE from *Cx. annulirostris* collected at Kowanyama. Among the "new" viruses were four flaviviruses: Kunjin and Kokobera from *Cx. annulirostris* at Kowanyama, Edge Hill from *Aedes vigilax* and *Cx. annulirostris* at Cairns, and Stratford from *Ae. vigilax* at Cairns.[18] Thus, Kunjin virus, the second subject of this chapter, quietly appeared as an incidental isolate from wild-caught mosquitoes. The isolation of these flaviviruses immediately cast doubt on the interpretation of the many serological surveys carried out from 1951 to 1960, when it was thought that dengue viruses were the only flaviviruses complicating the delineation of the geographical distribution of MVE virus.

The hypothesis that *Cx. annulirostris* is an important vector of MVE virus was developed from observations in the Murray Valley 1400 miles south of Kowanyama. That the hypothesis was validated in the tropics was strong evidence of its pivotal role, at least as a primary cycle vector. It remained to assess its role during an epidemic, and it was another 14 years before such an opportunity arose.

During February 1974, at the height of that epidemic, 38 strains of MVE were isolated from pools of *Cx. annulirostris* and 1 from *Cx. australicus* at locations along the Murrumbidgee and Murray Rivers.[19,20] The mosquitoes were collected in relatively undisturbed river redgum (*Eucalyptus camaldulensis*) flood plains, on the fringes of river towns, and on the edge and within permanent and ephemeral swamps. Several sites were close to suspected sources of human infection. During February and March 1974, 1 strain of MVE and 11 of KUN were isolated from *Cx. annulirostris* collected at Charleville, southern Q., on the Warrego River at the northern end of the Murray-Darling basin. Although there were no cases at Charleville, there were several in southern Q.

C. Social and Economic Impact

Despite the rarity of epidemics in the Murray Valley, the economic disruption when they do occur is substantial. In both 1951 and 1974, seasonal itinerant labor on which the fruit- and vegetable-growing industries depend was extremely difficult to recruit, and road transport avoided the valley, making marketing of perishable crops difficult. However, the major losses were in the tourist industry. The dry warmth of the valley has always been a lure to vacationers, particularly in the winter, but in the face of declining agricultural returns, by 1974 this industry had carefully nurtured year-round activity based on water sports, tennis, fishing, paddle steamers, "pioneer villages", duck shooting in season, and a range of cultural pursuits such as sculpture exhibitions, drama festivals, etc. The loss to the hotel-motel industry in Swan Hill alone was estimated by the shire clerk at 3×10^6, and throughout the valley probably $15 to 30×10^6. Many of the camping and caravan parks are situated on the flood plains of the rivers, and if these had not been flooded and largely abandoned in later spring, the disease toll could have been greater. As it was, 10/41 cases contracted in the valley occurred in tourists. The ratio was much lower elsewhere.

Although each epidemic and the sporadic cases provide a tragic legacy of mentally and physically damaged survivors, in Australia the prevalence of such cases compared to those due to other causes is minor.

II. THE VIRUS

A. Antigenic Relationships

MVE and Japanese encephalitis viruses are closely related, with distant relationship to St. Louis encephalitis; West Nile and KUN are closely related, with KUN having some relationship with MVE, and by cross-protection studies, West Nile and Japanese encephalitis have some relationship. These five flaviviruses, and Stratford, Kokobera, and Alfuy, are in the same antigenic subgroup of mosquito-borne flaviviruses.[22-24] Edge Hill also is placed in this subgroup by some,[22,23] and with Uganda S and Banzi in another subgroup by others.[24]

B. Host Range

Both MVE and KUN infect a wide range of laboratory hosts, but unless inoculated i.c. with heroic doses, infection is usually symptomless. Infant mice of susceptible strains are lethally infected by both viruses by any inoculation route and without adaptation. Age-associated resistance in peripherally inoculated mice develops quite rapidly, and is difficult to overcome by passage-induced increases in virulence. Symptomless MVE viremia has been demonstrated in guinea pigs, young chickens, and rabbits. Unlike Japanese encephalitis virus, MVE virus is lethal in peripherally infected hamsters.

C. Strain Variation

Barrett[25] used a battery of serological tests in an attempt to detect antigenic variation in five strains of MVE virus isolated between 1951 and 1969 from fatal cases or pools of mosquitoes and locations ranging from Port Moresby, Papua New Guinea, to the Murray Valley. Four were virtually indistinguishable from each other in all tests. The remaining strain, from a fatal infection contracted in 1969 by a tourist in northwestern Australia, appeared to be different from the other four in some tests, but convincingly only in a plaque-reduction neutralization test, where virus-heterologous antiserum reactions produced relatively low neutralization indexes. Four strains of KUN were not significantly different, although gel diffusion and cross-absorption tests suggested that there were quantitative differences in the expression of some common antigens.[25]

Boyle[26] found that plaque size on VERO cell monolayers was generally stable for a widely varied collection of MVE strains, but was a variable property of KUN, even with any one strain on the same monolayer. The plaques of one MVE strain from Barmah Forest on the Murray River were consistently and significantly smaller than those of other strains, but numerous serological tests with MVE and KUN strains, and virulence tests in newborn and 5-week-old mice failed to reveal further differences.

Unlike Ross River virus, the other principal arboviral disease agent indigenous to Australia, MVE and KUN viruses appear to be remarkably stable in antigenic and biologic properties.

Faragher et al.[27] developed a technique based on the detection of restriction site polymorphisms in cDNA copied from purified Ross River viral genomic RNA, and as in earlier biologic investigations, found substantial differences between strains. There were at least three genetic types and six subtypes with significantly different restriction profiles, and some of these were stable over a number of years in a given geographical location. It was estimated that 1.5 to 5% sequence divergence exists between genetic types of Ross River virus.[28]

Lobigs et al.[29] from the same group adapted the technique for use with flaviviruses. Initially, cDNA was generated from the virion RNA of partially purified extracellular KUN, MVE, and 17D yellow fever viruses and digests prepared with the restriction enzymes Hae III and Taq I. Gel electrophoreses showed that the cDNA fragments had size profiles unique to each of the three flaviviruses. Strains of KUN were then selected from six locations up to 2700 km apart collected over a period of 25 years from three different species of mosquito and the cervical spinal cord of a moribund horse. The maximum sequence divergence found between any pair of these KUN isolates was estimated as less than 1%, i.e., there was no significant difference.[29]

Strains of MVE virus are currently being investigated by the same technique.[30] These are from the same general locations as the KUN strains, except for two from New Guinea, and some of the strains are from human brain and bird serum. Again there is a remarkable homogeneity between all Australian strains. However, the two strains from New Guinea are significantly different from each other and both differ greatly from the Australian strains. One strain was isolated from a pool of mosquitoes collected on the Sepik River near the north coast of New Guinea and the other from the brain of a native who died in Port Moresby hospital on the south coast of the island. The two locations are separated by a mountain range up to 3000 m in elevation. Compared with the 1951 MVE prototype strain, there are 33 changes in 420 nucleotides in the Sepik strain, i.e., 8%, in contrast to less than 1% between any two Australian strains. This finding indicates that it is unlikely that there is frequent interchange of MVE virus between Australia and New Guinea, and suggests that long time intervals are required for detectable evolution in isolation.

The apparent genetic stability of these viruses contrasts with the distinctive geographic "topotypes", and the virulence variants within topotypes, described for St. Louis encephalitis virus.[31] At the molecular level these topotypes and virulence variants were characterized by T_1 oligonucleotide fingerprinting, and the 40 large oligonucleotides selected for virus strain comparison represent approximately 10% of the total RNA genome,[31] whereas the restriction enzyme digests used in the KUN and MVE analyses scan about 4% of the genome.[29] Thus, the apparent stability and variability observed in the two investigations might merely be a reflection of the sensitivities of the two systems used.

D. Methods for Assay

MVE and KUN viruses can both be assayed by i.c. or i.p. inoculation of infant mice, and neutralizing antibodies can be assayed in the same host. Although cell culture types are more restricted and requirements more demanding than with alphaviruses, the assay method of choice is plaquing under agar or equivalent, or cytopathic effect in microculture. Although these systems are rather less sensitive than i.c.-inoculated infant mice, assays and logarithm neutralization indexes are more accurate, reproducible, convenient, and economic than in mice. For MVE and KUN viruses, but not necessarily other Australian flaviviruses, VERO and BHK are the most satisfactory. Varieties of pig kidney cells are less sensitive but produce the most clearly defined plaques and respond to a wider range of flaviviruses.

Virus isolation from pathological and field material can be carried out in the above systems; the large number of specimens collected during the 1974 epidemic were concurrently inoculated i.c. in infant mice and on VERO cells under agar; almost all MVE and KUN isolates caused encephalitis and death in mice and produced plaques without passage. Isolation in two systems avoided the need for reisolation. Most other viruses circulating at that time also infected both systems. If the investigating laboratory can restrict diagnosis and research to MVE and KUN viruses, mosquito cell cultures are probably the present method of choice for processed mosquito pools, but as CPE is not reliable, this is coupled with a system to reveal the presence of these two viruses: fluorescent antibody, plaquing on vertebrate cell monolayers, ELISA, etc. Plaque reduction neutralization tests with thoroughly assayed and authenticated polyvalent or monovalent antisera can be used directly to characterize these two viruses. It might be necessary to reduce the size of pools of some species of mosquito from 50 to 25 because of toxic effects in mosquito cells. Although anachronistic, the chorioallantoic membrane of the developing chick embryo is probably the most sensitive single host for recovering MVE (and presumably KUN) from vertebrate tissue or blood. French[12] in 1951 demonstrated this system to be more sensitive than infant mice and other laboratory animals, and this was confirmed and extended to cell cultures by Lehman et al.[32] during the 1974 epidemic. It has been shown experimentally that the chorioallantoic membrane is infected and pocks are formed in the presence of homologous antibody, a circumstance which neutralizes virus infectivity in other hosts. There has not been an opportunity to compare the sensitivity of mosquito cell cultures or ELISA methods to pathologic material.

III. DISEASE ASSOCIATIONS

A. The Disease in Humans

Robertson and McLorinan[33] presented clinical aspects of 26 patients during the 1951 epidemic, and Bennett[34] those of 22 patients from the 1974 epidemic. Onset was sudden, fever invariable with daily peaks of up to 40.6°C, anorexia and frontal severe headache common, and about half the patients suffered nausea, vomiting, and diarrhea. Several patients told of a vague sensation of dizziness or giddiness as one of the first symptoms. The first indication of disturbed brain function usually appeared 2 to 5 days after onset, with lethargy, drowsiness, irritability, and a dullness deepening into confusion, disorientation, and ataxia. About $1/3$ of the patients had convulsions and fits.

Bennett[34] grouped his series of 22 into mild (11 patients), severe (7), and fatal (4). In the mild group neurological involvement stabilized between the 5th and 10th day of illness, but all had frank encephalitis with varying levels of consciousness. All had neck stiffness and one lapsed into coma responding only to painful stimuli. Only one required a period of artificial respiration. They were discharged from the hospital after periods varying from 2 weeks to 3 months. Reviews up to 16 months later indicated that seven had made a complete recovery, but four had mild degrees of impaired coordination or emotional problems.

The initial disease in the seven suffering the severe form was similar to the mild cases but the neurological signs were progressive and all lapsed into coma, three so deeply that they did not respond to painful stimuli. There was pharyngeal paralysis in two and respiration failure in the other five. Bennett considered that these five would have died without artificial respiration. All seven had severe residual disabilities, including degrees of paraplegia or quadriplegia, various mental defects, and one had features of Parkinson's disease.

The four fatal cases in this series died 6 to 31 days after onset, and all had profound nervous system involvement. Pharyngeal paralysis developed in one and respiratory failure in the other three, for which artificial respiration was given. At the time of death, three had barely detectable CNS activity, but the other, although severely paralyzed and in a respirator, remained conscious but died from pneumonia or fungemia.

The range of signs and symptoms described by Bennett[34] and presented in an abridged form here were originally documented by Robertson and McLorinan[33] in their 1951 series. The Bennett[34] series has been favored here because he quantifies that most important segment of patients: those who survive with major sequelae. Of the 7/22 (32%) with severe sequelae, Bennett states categorically that five would have died without the use of artificial respiration. In the 1951 series, respiratory difficulties were present at a early stage in fatal cases, but artificial respiration was resorted to far less frequently and seemingly less effectively than in 1974. In the 1951 epidemic 19/45 (42%) reported cases were fatal,[1] compared with 13/58 (22%) in 1974.[8] This includes the 4/22 (18%) in Bennett's series. If the five who survived because of artificial respiration are added to this series, the case fatality rate would be 41%, virtually the same as 1951. Of course, the five "saved" by this treatment all have severe residual disabilities, and that is a dilemma as old as medicine.

All investigators stressed that clinical[1,33,34] and pathological[35] features of the disease were essentially similar to the original descriptions of Australian X disease.

Clinical expression of the disease in 1951 and 1974, however mild, was invariably encephalitis or encephalomyelitis.[1] As with other arboviral encephalitides, the morbidity rate is very low, perhaps one clinical case for every 800 to 1000 infections.[1] There is little clinical information relating to the 21 cases reported from northern W.A. from 1974 to 1984.[11] There have been no fatal cases in this series, although in 1969 a girl who died in Toowoomba, Q. was thought to have been infected with MVE while traveling in northern W.A.[36] Some of these cases have had profound neurological involvement, with sequelae reported in at least two cases, but many, possibly most, have had no significant encephalitic signs or symptoms. This has led to the suggestion that the "endemic" strains of MVE virus active in these areas are attenuated, although supporting laboratory evidence is lacking.

One of the eight cases reported from the Kimberley area of W.A. in 1978 had a monospecific HI and neutralizing antibody response to KUN virus.[11] The case was reported as frank encephalitis, but details have not been published. Prior to this, two accidental laboratory infections with KUN had been reported.[37] Both illnesses were mild, in one case with rash and lymphadenopathy; KUN virus was isolated from the serum of this patient 1 day before and 1 day after onset. Doherty et al.[9] reviewed the serology of 45 patients accepted as MVE encephalitis cases during the 1974 epidemic, and in five of these, HI and neutralization tests indicated that KUN was the more likely causative agent.[9]

In 1984, during a summer of enhanced arbovirus activity in the Murray Valley, a patient with undifferentiated encephalitis was eventually serologically diagnosed as a KUN infection.[10] The encephalomyelitis was severe, and bulbar and truncal motor involvement was suggestive of damage to cranial and anterior horn cells, as frequently described in severe cases of MVE. Respiratory support was required and the slowly resolving neurological involvement left wasting of major upper limb muscle groups. This is the first published account of serologically confirmed KUN virus encephalomyelitis.[10]

B. Domestic Animals

Gard noted increased incidence of nervous disease in horses in the summers of 1971 and 1974, which coincided with periods of MVE/KUN virus activity in the Murray-Darling basin.[7] Attempts to recover virus from usually inadequate pathological specimens failed, although two showed histological evidence of encephalitis. Of 52 cases, 13 had serological evidence of recent infection with MVE. Attempts to infect two horses with a low dose of MVE failed.

In 1984, KUN virus was isolated from the spinal cord of a horse in the Murray Valley which was moribund with advanced signs of encephalomyelitis.[38]

During epidemics of MVE there are frequent reports of dogs dying, but although prevalence of antibody is high, virus has not been recovered from the few specimens examined. Dogs are a favored blood source for *Cx. annulirostris*, and it is likely that death is due to constant worry and blood loss.

Pigs, cattle, and sheep are infected, but no disease association has been reported.

C. Wildlife

Although viremia is common, no clinical signs have been reported in wildlife either in nature or in the laboratory. This phenomenon is frequently regarded as an example of host and parasite evolving together to a benign relationship, but a morbidity rate such as that seen in man, for example, would pass unnoticed in wildlife. It seems unlikely that the low morbidity rate in man is genetically derived through long and close relationship with the virus.

D. Applicable Diagnostic Procedures

During an epidemic tentative diagnosis is made on signs and symptoms, and whether the patient was in a region of known MVE/KUN virus activity in the previous 4 weeks. Serum is checked for rising titers of HI antibody and the appearance of CF antibody; the presence of CF antibody is a reasonably reliable indication of recent infection but is not always detectable. Plaque-reduction or infant mouse protection neutralization tests are carried out on HI positive sera at least against MVE and KUN, but eventually against other Australian flaviviruses. In recent years ELISA tests have been developed by some diagnostic laboratories and are being used in place of HI. Wiemers and Stallman,[39] during the 1974 epidemic, found IgM detected by HI a diagnostically useful test when applied to suspected cases occurring in Q, particularly when antibody was present in the acute sample, when the HI titer was static, and when CF antibody was not detectable. Twelve cases were confirmed or solely diagnosed on the presence of IgM. Doherty et al.[9] reviewed the antibody history of 53 patients diagnosed as MVE infections in 1974. As with Japanese and St. Louis encephalitis, there was no clear pattern of HI antibody response with regard to timing and magnitude of the rise and fall of titer. Late rises in antibody titer, including IgM, were suggestive of recrudescent, persistent infection in two cases. In one of these, IgM was present in 15 serial samples taken from day 7 to day 190; in general, IgM persisted for at least 40 days and in some cases to over 100 days.

The diagnosis of MVE or KUN infection by demonstrable IgM should be used with discretion. In the uncomplicated situation of a patient with clinical encephalitis during an epidemic, rapid diagnosis by detection of IgM is sufficient, although a negative result should obviously not exclude MVE or KUN as the causative agent. Because of the rather long persistence of IgM antibodies, diagnosis of a sporadic case in this way should be accepted only if clinical criteria are present, and for the definition of a syndrome other than encephalitis as being due to MVE or KUN infection, positive IgM should be supported by classical serological criteria. The indicator case of an epidemic in the temperate areas of Australia should also be confirmed, as quickly as possible, by as many serological tests as are available on sequential serum samples before publicity and expensive control measures are launched.

There have been no reports of adverse effects of the virus on the vector.

IV. EPIDEMIOLOGY

A. Geographic Distribution

In the tropics, isolates of MVE virus have been made from material collected at about 4°S latitude on the Sepik River near the north coast of New Guinea, from the Brown River area near the coast of the Gulf of Papua, from the west coast of Cape York Peninsula, from the Darwin area of the N.T., and from northern W.A. South of Capricorn, isolates have been made only in the Murray-Darling basin, from Charleville on the Warrego River in southern Q. and Narrabri on the Namoi River, northern N.S.W. to the Barmah Forest, Vic., and on the Murray River at about 36°S. Serologically confirmed cases and seroepidemiological studies fill in the gaps between these isolates except for the highlands of New Guinea. There is no substantial evidence of occurrence on the temperate coastal areas south of Brisbane on the east coast, in W.A. south of the Tropic of Capricorn, or west of the S.A. border. There has been no substantial evidence that the virus has been active on the east coast north of Brisbane since 1925.

Although there have been reports of a low prevalence of MVE antibody in human sera in Southeast Asia and particularly Indonesia, Japanese encephalitis virus is known to be active in this region, and until MVE virus is also isolated the identity of these antibodies remains enigmatic.[40] It is generally accepted, without much evidence, that MVE occurs to the east and south of an extension of the Wallacea zone as an arc to the south of the Philippines and north of New Guinea.

KUN virus has been isolated mainly from mosquitoes and from all the same general areas of mainland Australia as MVE virus. It has not yet been found in Papua New Guinea, but unlike MVE, KUN has strayed into the Oriental zoogeographic zone; there have been three KUN isolates from pools of *Cx. pseudovishnui* collected in Sarawak, Borneo.[41]

B. Incidence

Epidemics and sporadic cases are too rare to allow a meaningful calculation of disease incidence, and appropriate longitudinal serological surveys have not been carried out in Australia. Hawkes[42] tested sera from 793 native New Guinea children four times over a period of 14 months. The children were aged 6 years and under. Villages were divided into ecologically distinct zones from the Sepik River to the foothills above Maprik. HI and neutralizing antibody patterns to flavivirus antigens were complex, but in basic HI tests 12.3% of children in the river zone converted from negative to positive in the 14 months, 5.8% in the Sepik Plains zone, 5.2% in the Peripheral Plains zone, and 2.7% in the foothills zone.

C. Seasonal Distribution

There have been no serologically confirmed cases of MVE or KUN infection in humans, nor isolation of MVE or KUN viruses from any source from anywhere in Australia between June and October, inclusive. This is the southern winter and the tropical dry season. Most of the detectable activity occurs during the first 3 to 4 months of the year, during and particularly at the end of the tropical wet season, and the southern dry summer.[43]

D. Risk Factors

From 1951 to 1984 there were 104 cases of encephalitis in the Murray-Darling basin of southeastern Australia acceptably confirmed as due to infection with MVE or KUN virus.[1,8,10] As KUN virus was first isolated in 1960 and there is still some reservation about differential serological diagnosis, all cases are here regarded as caused by MVE or KUN.

Of the 104 cases, 52 (50%) occurred in children up to 10 years old, including 13 in the 1st year of life. Twenty-four (23%) were over 50, six of these cases being 70 to 76 years old. Only two cases (2%) were between 31 and 40, and there were only nine (8.5%) in the age group of 21 to 40. The twin peaks of morbidity in the young and the old were not apparent in the 1951 epidemic; 19 of the 24 cases in the over-50 group were infected during the 1974 epidemic.

There was a high preponderance of male cases in both 1951 and 1974; overall, 64% male, 36% female. This does not seem to be related to occupation, as 60% of the under 10-year-old cases were male.

The case fatality rate was 29%. As with morbidity, the case fatality rate was considerably higher in males: 34% compared to 22% in females.

None of the available analyses of epidemics in the Murray Valley have listed race. The inference is that all reported cases were Caucasian.

With only 125 to 130 recognized cases in Australia and New Guinea over the past 34 years, the identification of risk factors is rather academic. Despite this, and ignoring the enigmatic cases from the Kimberleys in W.A. and the lone New Guinea native, several at least tentative factors emerge: the very young and moderately old male caucasian visiting or living in the Murray-Darling basin during the course of a rare epidemic (December to April), are those principally at risk of clinical infection and death. In some of the clinical cases retrospectively studied after the 1974 epidemic, there was a high KUN antibody level in the first serum sample, consistent with the concept of an "original antigenic sin" response.[9] The 3 best documented in this group of 4 were aged 60, 71, and 74, and apart from possibly misdiagnosing the causative agent of the current illness, the more important implication is that waning immunity to KUN or MVE might enhance or provoke a clinical response to heterologous infection. This possible risk factor should be addressed in future epidemics.

E. Serologic Epidemiology

McLean and Stevenson[44] added serological evidence to the many features identifying MVE with Australian X disease. Broken Hill, a mining town in far western N.S.W. recorded epidemics of Australian X disease in 1917, 1918, a small outbreak in 1925, but no cases in 1951. When bled in 1952, a higher proportion of residents born before 1918 (19/92) carried neutralizing antibody to MVE virus than those born since 1928 (2/69).

The geographic distribution of MVE and hypotheses of epidemic occurrence have been developed from studies of antibody prevalence. The 1951 epidemic was scarcely spent before it was shown that virus activity was confined to areas north and west of the Great Dividing Range,[45] and as the widely accepted concept of arbovirus ecology at that time was of virus survival in tropical maintenance cycles with periodic epidemic excursions into temperate areas, investigators rapidly moved north. Dengue and MVE were the only antigenic group B agents known to be in Australia, but the fact that there are at least four other viruses more closely related to MVE than is dengue probably did not materially affect the general delineation of geographic occurrence, with the possible exception of east coastal areas of Q. Here, evidence for the presence of MVE virus since the clinical cases in Townsville in 1925 has been solely serological.

Serological surveys of aborigines and Caucasians following the 1951 epidemic provided abundant evidence that flaviviruses were enzootic-endemic in northern Australia and the lowlands of Papua New Guinea.[1] In more detailed later surveys, it was shown that flavivirus antibody prevalence in northern Q. west of the Great Dividing Range and in the N.T. reaches 100% by the age of 10 in some aboriginal settlements on the Gulf of Carpentaria.[46] KUN seems more prevalent than MVE in most communities, but with up to five flaviviruses active, the specificity of even neutralization tests is suspect. In a prospective study of a small group of aborigines at Kowanyama, increases in antibody titers following presumed reinfection with flaviviruses occurred in antibody positive subjects as frequently as conversions.[47]

In New Guinea, Anderson et al.[48] found a high prevalence of MVE neutralizing antibodies in the lowlands but not the highlands of Papua New Guinea, with highest activity in the swamps of the Aramia River near the coast of the Gulf of Papua. Wisseman et al.[49] confirmed that only the highland plateau was free of flavivirus activity, with the inhabitants of some coastal lowlands recording 95 to 100% positive, and communities in the low hill country and deep mountain valleys at intermediate rates. They used seven flavivirus hemagglutinins and, as with Hawkes[42] (see Section IV.B), they suspected that more than one flavivirus was contributing to a complex reaction pattern. In Hawkes' longitudinal survey of native children in the Sepik District, transmission of KUN and Kokobera occurred throughout the year, and MVE in two of the three time intervals between samplings.[42] "Wet" and "dry" seasons are not as pronounced in the Sepik District as in northern Australia, where transmission is interrupted in the "dry" season. MVE and Kokobera viruses were isolated from mosquitoes collected in the Sepik district, but despite medical surveillance of the region, no encephalitis attributable to arbovirus infection was identified over a period of about 5 years. Similarly, at Kowanyama, where close surveillance extended over a period of about 13 years, no cases of arbovirus encephalitis were recognized.

In W.A., sera from aboriginal and white communities in the Kimberley District of the tropical north of the state were collected in 1958—1960, and 92% and 60%, respectively, had antibody to MVE hemagglutinin.[50,51] In neutralization tests a high proportion appeared to be due to KUN rather than MVE. These sera were collected before the first component of the Ord River irrigation project was constructed. Aborigines in the central desert of W.A. were negative for flavivirus antibodies, but surprisingly, all had alphavirus antibodies. In the populous southwest of the state, the prevalence of flavivirus antibody was insignificant and probably not locally acquired.

Unlike the Sepik District and the Gulf of Carpentaria aboriginal communities, clinical MVE has been documented in young aborigines and young adult whites in the Kimberleys.

It is apparent that there is a high prevalence of MVE and KUN infection in ecologically appropriate regions between the equator and the Tropic of Capricorn. More recently, Hawkes et al.[52] commenced an important and ambitious prospective survey of arboviral infections in N.S.W. using regular blood donors and serial bleeds of patients through pathology laboratories. To date, only base-line results have been published on the sera from 16,842 long-term residents of all districts of N.S.W. Flavivirus antibody prevalence rates were low on the coast and tablelands (2 to 8%), moderate on the western slopes of the Great Dividing Range (6 to 11%) and high (26 to 42%) on the western plains, i.e., the Murray-Darling basin. Much of the latter region is arid to semiarid, and most of the population is concentrated in small- to moderate-sized towns on the rivers. Some of these were pockets of apparently very high virus activity. Bourke and Brewarrina on the northern reaches of the Darling River had rates of 78% and 55% in a region with an overall rate of 42%, and about one third of the regional population developed antibody by the age of 20. At Bourke the rate was the same in aborigines and Caucasians (about 90 of each). These rates are as high as in Caucasians living in tropical W.A. Towns in the Murray Valley ranged from 19% to 30%. There was a slight preponderance of male reactors throughout the state. Antibody prevalence rates increased with age in a stepwise pattern suggestive of continuous rather than epidemic exposure.[52] Although there is always uncertainty about the identity of the eliciting flavivirus, HI reaction patterns indicate that KUN and MVE are responsible for at least a proportion of the infections, particularly in inland areas.

V. TRANSMISSION CYCLES

A. Evidence from Field Studies
1. Vectors

The first isolation of MVE virus from a nonhuman source was from *Cx. annulirostris* collected at Kowanyama (Mitchell River Mission) during March and April 1960, and the prototype isolation of KUN virus was also from *Cx. annulirostris* collected during the same period.[18] Since then there have been 108 isolates of MVE and 161 isolates of KUN from mosquito pools collected from the Sepik River at latitude 5°S through to the Murray Valley at about 35°S (Table 1). Although an overwhelmingly high proportion of these isolates has been from *Cx. annulirostris* (100 or 93.5% MVE, 158 or 94.6% KUN), year-round transmission by this mosquito has not been demonstrated even at those study sites (Kowanyama, Q. and Kimberley district, W.A.) where adults of this species are active throughout the year.[43,47,53]

The first opportunity to incriminate an epidemic vector was in 1974. Between February 4 and 12 at the height of an epidemic in the Murray Valley, a total of 136,077 *Cx. annulirostris* yielded 238 isolates of 11 antigenically distinct viruses including 38 strains of MVE and 111 strains of KUN; 180 *Cx. australicus* yielded 1 isolate of MVE and there were no isolates of any viruses from 917 culicine mosquitoes of other species nor from 1184 *Anopheles annulipes*.[19,20] Many of the isolates were from trapping sites close to places of human infection.

At Charleville, southwest Q. and near the northern limit of the Murray-Darling basin in February and March 1974, 22,776 *Cx. annulirostris* yielded 33 isolates of 9 viruses including 1 of MVE and 11 of KUN. No other species yielded virus, and, although clinical cases were not reported at Charleville, several were confirmed elsewhere in southern Q.[21]

There seems no doubt that *Cx. annulirostris* is the major epidemic vector for both MVE and KUN, and at least in the summer or wet season, must also play an important role in maintenance cycles.

2. Vertebrate Hosts

MVE virus has been recovered from the blood of a white-faced heron (*Ardea novaehollandiae*) in the Murray Valley,[20] and Alfuy virus, a very close relative of MVE, from the blood and organs of a swamp pheasant (*Centropus phasianinus*) collected on April 1, 1966 at Kowanyama.[54] MVE virus has also been recovered from the blood of three sentinel chickens, two in the Murray Valley,[55] and one at Charleville.[21] KUN virus has been recovered from the heart muscle of a yellow oriole (*Oriolus flavocinctus*) at Kowanyama, and from the CNS of a horse in northern Vic.[38] At Kowanyama no viruses were recovered from birds collected in dry seasons, and in the Murray Valley the isolations were made in summer and early autumn.

Anderson[1] and Miles and Howes[17] found the horse useful in delineating geographic occurrence of the disease. They found 51/91 sera collected from horses in the Murray Valley

Table 1
SOURCES OF MURRAY VALLEY ENCEPHALITIS AND KUNJIN VIRUS ISOLATES, OTHER THAN FROM *HOMO SAPIENS*

		Latitude									
	5°S (New Guinea)	15°S (North Q.)		16°S (Kimberley, W.A.)		26°S (Charleville, Q.)		35°C (Murray Valley)		Total	
	MVE	MVE	KUN	MVE	KUN	MVE	KUN	MVE	KUN	MVE	KUN
Culex annulirostris		6	13	55	21	1	11	38	113	100	158
Cx. australicus		1							1	1	1
Cx. bitaeniorrhynchus				1						1	0
Cx. quinquefasciatus			1							1	1
Cx. squamosus			2							0	2
Cx. pullus			2							0	2
Aedes normanensis		1		2						3	0
Ae. tremulus					1					0	1
Culicines, mixed pool	1									1	0
Anopheles bancroftii			1							0	1
An. farauti			1							0	1
Ardea novaehollandiae (white faced heron)								1		1	0
Equus caballus (horse)									1	0	1
Sentinel chickens							1	2		3	0

in 1951 had antibodies to MVE,[1,17] compared with 4/191 from southern Vic. Gard et al.[7] noted increased numbers of cases of "nervous disease" in horses reported during 1971 and 1974, both years of MVE and Ross River virus activity in the Murray-Darling basin. Several of these cases showed rising titers of antibody to MVE, KUN, or Ross River virus. Strangely, fatal cases seemed to be associated with Ross River virus infection rather than the flaviviruses. Neither disease nor viremia has yet been successfully induced by experimental inoculation of horses, but their potential as an amplifying host of these viruses was demonstrated by the isolation of KUN virus from a moribund horse, as previously mentioned.[38]

Feral pigs are widely distributed in N.S.W. and Q. where some of the major colonies have been well established for over 110 years. The highest densities of pigs in N.S.W. are in the major marsh systems of the Macquarie and Lachlan Rivers and in other parts of the extensive flood plains associated with the Murray-Darling basin. Gard et al.[56] examined sera from pigs collected between January 1971 and February 1976 from three widely separated areas of northern N.S.W.: Yantabulla on the Cuttaburra watercourse, Moree on the Gwyder River, and the Macquarie Marshes downstream from Warren. MVE HI antibody was found in 69% of 411 sera from Yantabulla, 64% of 55 sera from Macquarie Marshes, and 25% of 151 sera from Moree. Serum dilution plaque reduction tests were interpreted as indicating that MVE virus infected Yantabulla pigs in the summers of 1971—1972 and 1972—1973, that more than one flavivirus had infected pigs shot in March 1974, and those sampled in January 1976 had been infected with KUN virus. Yantabulla had experienced a series of annual sustained floods in the frequently dry watercourse between 1968 and 1974, and again in 1976. Such conditions not only rapidly increase the feral pig population and mosquito activity, but large numbers of waterfowl and waders are attracted to the area, and if conditions are sustained, they will remain and breed.[56]

Birds, particularly waterbirds, were an attractive hypothetical vertebrate host virtually from the time when the virus was first characterized in 1951. Human infection was confined to the flood plain of the river system; the causative agent was related to Japanese and St. Louis encephalitis viruses, both of which appeared to involve birds in the primary cycles. Suspicion was falling on *Cx. annulirostris* as a principal vector, again partly due to its relationship to *Cx. tritaeniorrhynchus* and *Cx. tarsalis*, known vectors of Japanese and St. Louis encephalitis, and in the epidemic area, *Cx. annulirostris* is a summer mosquito largely confined to the variable breeding waters of the flood plain.

Anderson[57] collected bird sera from rookeries in the Mildura area between July and November 1951; it was a high flood year and a season of intensive breeding. Of the 99 sera from water birds, 40 had antibody to MVE virus, as did 11/60 land birds. Numbers of individual species were small, but grouping them to orders, Ciconiiformes (73%), Pelecaniformes (64%), and Anseriformes (45%) were the most commonly infected.

At Kowanyama birds were collected during periods in the dry and wet seasons from 1964 to 1967.[54] Over 40% of 74 sera and 20% of 332 blood samples reacted with MVE hemagglutinin. The species with the highest proportion of reactors were the brolga (*Grus rubicundus*) and the nankeen night heron (*Nycticorax caledonicus*) with 25/27 and 7/10 positive, respectively. HI, mouse neutralization, and plaque inhibition tests all gave results suggesting that antibody was to members of the MVE-KUN-Alfuy subgroups, and many samples reacted to equal titers to the three viruses.

At the time of the 1974 epidemic, the only hypothesis explaining the periodic epidemics in the Murray Valley required the movement of viremic waterbirds into the area from northern regions of endemicity, so the serological survey concentrated on these species.[59] There was a preponderance of reactors in the orders Ciconiiformes (35/64, 55%) and Pelecaniformes (33/79, 41%). A large sample of ducks was taken because banding had produced evidence of rapid flight from the Darwin area to the Murray Valley. However, only 10/222 (4.5%) reacted. Within Ciconiiformes, the highest prevalence was in nankeen night herons with

22/25 (88%) positive. Three other species of heron and the sacred ibis ranged between 30% and 40%, and in Pelecaniformes, darters and three species of cormorant were all above 40% positive. The sera were also tested against Sindbis antigen, and although highest prevalence was again in herons and cormorants (46% and 56% positive), reactors were spread more evenly across other orders. For instance, 70/222 (32%) ducks had antibodies to Sindbis. The principal apparent vector of Sindbis virus in the Murray Valley is also *Cx. annulirostris*. This not only suggests that herons and cormorants are more attractive or accessible than other species to blood-seeking *Cx. annulirostris*, but also that they are more susceptible than other species to infection with MVE/KUN.

There were record levels of spring flood in the 2 years following the epidemic; these rapidly receded in late December—January each year so that while birds and other wildlife continued exuberant breeding,[60] summer mosquito populations were less than 1/10 those of 1974.

In the spring, summer, and autumn of 1974—1975, the collecting sites trapped during the epidemic were revisited; 76/174 (44%) juvenile nankeen night herons had flavivirus antibody, 11 of them with HI titers of 1/320 or more; 75/78 (96%) adults of the same species had antibody, 24 with high titers. Other Ciconiiformes had increased antibody prevalence over the epidemic year: 80/178 (45%), 11 with high titers, but cormorants were already losing evidence of involvement: 60/274 (22%), five with high titers. Documentation of the extraordinary surge in waterbird breeding includes flocks of mostly juvenile nankeen night herons following and feeding on a mouse plague in dry wheat country, and the precocious breeding of over 2000 pairs of juvenile nankeen night herons (<2 years) in clumsily built nests close to the water, with their probable parents in nuptial plumage nesting more traditionally on the treetops. In this tiny 32-ha swamp near the Lachlan River there were also breeding colonies of 11 other species of herons, cormorants, ibis, and spoonbills.[60] Similar profligate activity was occurring in suitable and unsuitable habitats throughout the Murray Valley and probably the entire Murray-Darling basin.

The inevitable population crash occurred in the spring and summer of 1976—1977, drought conditions having replaced three successive high spring floods. Sporadic breeding returned to the more permanent swamps.

In most species antibody prevalence started decreasing during the 1st or 2nd year after the epidemic, but percentage incidence and antibody titers in nankeen night herons continued to be relatively high. Prevalence in most colonies throughout the Murray-Darling basin is over 40%, and in adults at Barmah Forest usually 90 to 100%. In the most recent sample at Barmah Forest in 1980, 17/17 adults had MVE/KUN antibodies, six with titers of 1/320 or more. Plaque reduction tests indicated that MVE and KUN viruses were about equally involved. At Barrenbox Swamp near Griffith during the same week, 15/19 nankeen night herons were positive, and the five high titers included one at 1/960 and one >1/1280.[61]

At the Ord River study site in northern W.A., Liehne et al.[58] found MVE antibody prevalence to be high in both waterbirds and others; indeed, reactors were spread fairly evenly over all species tested. All avian sera were collected in the vicinity of the town of Kununurra and the irrigation area, and here the distinction between aquatic and land habitats might become rather blurred. For example the little corella (*Kakatoë sanguinea*), a psitticine, feeds on irrigated sorghum and other grain crops during the day, and at night roosts over water in the dead limbs of drowned trees. They and many others are probably as closely associated with *Cx. annulirostris* as the conventional waterbirds.

Birds appear to be involved in maintenance cycles of MVE and KUN, but not necessarily to the exclusion of other native and feral vertebrates. Both in 1950—1951 and 1973—1974, unusually exuberant water-bird breeding in the Murray Valley before and during the epidemic might also have provided the large number of susceptible vertebrates required to amplify the maintenance cycle to epidemic levels of virus activity. For an epidemic to occur, this circumstance must be combined with a grossly abnormal population density of *Cx. annulirostris* persisting through the summer, as witnessed by the fact that vertebrate but not mosquito populations increased to even higher levels in the two MVE/KUN virus-free seasons following the epidemic, albeit that immune rates in possibly critical hosts were also high. There appears to be some peculiar but not completely defined relationship between MVE, and to a lesser extent, KUN and the nankeen night heron. The greatly enhanced breeding activity of this bird in the Murray Valley might well presage an epidemic. Antibody to KUN virus is more widely distributed across bird species than antibody to MVE, KUN virus was more commonly isolated from mosquitoes than MVE during the epidemic, and KUN but not MVE activity has been detected on several occasions in the Murray Valley since 1974. The most pronounced KUN activity in 1979 and 1984 was preceded on each occasion by a surge of waterbird breeding that did not include heron species.[61]

B. Evidence from Experimental Infection Studies
1. Vectors

MVE and KUN viruses have been isolated from ten species of Culicine mosquitoes, but only two of these have been examined for vector competence in the laboratory. In one other, MVE replication has been demonstrated but transmission not attempted. On the other hand, a further 14 species have been used successfully with MVE in laboratory investigations including species exotic to Australia such as *Ae. polynesiensis, Ae. pseudoscutellaris, Cx. tarsalis,*[62] and *Cx. tritaeniorrhynchus.*[63] Obviously, MVE is capable of replicating in and being transmitted by a wide range of Culicine mosquitoes, and it is perhaps surprising that field isolations have been made only from a small proportion of the potential vectors available.

McLean[15] demonstrated replication of MVE virus in 11 species of Culicine mosquitoes after feeding on blood-and-virus-soaked cotton pledgets and/or viremic chicks. These included *Cx. annulirostris* and *Cx. quinquefasciatus.* For both he was able to construct classic growth curves after their feeding on small doses of virus. Both transmitted virus after but not during the eclipse phase, and *Cx. annulirostris* had higher transmission rates than *Cx. quinquefasciatus.*[15]

Cx. quinquefasciatus is a recognized vector of St. Louis encephalitis in the U.S., and as it is domesticated throughout most of Australia, it has been considered as a potential epidemic MVE vector, particularly in the river towns and particularly because it feeds on chickens and man. Altman,[63] using a colony of *Cx. quinquefasciatus* originating in Malaya, found replication and transmission of MVE effective but rather erratic, and generally it was a far less competent laboratory vector than colonized *Cx. tritaeniorrhynchus* tested concurrently. Carley et al.[64] used a colony of *Cx. quinquefasciatus* established from Brisbane stock, and feeding very large doses of virus achieved replication of both MVE and KUN, although 0/10 other arboviruses multiplied in feeding experiments. Transmission was not attempted.

Kay et al.[65] collected and briefly colonized *Cx. quinquefasciatus* from Brisbane, Charleville, Cairns, and Kowanyama in Q., from Mildura in Vic., from Darwin in the N.T., and Port Hedland in W.A. Although the Brisbane colony had been established 4 years before the others, it appears to be from a more recent stock than that used by Carley et al.[64] All of these colonies were at best poorly susceptible to MVE virus, with ID_{50} of over 10^5 mouse LD_{50}, and were almost completely refractory to infection with KUN. Transmission of MVE to chickens was insignificant, and negative with KUN. It is difficult to reconcile these findings with earlier reports. MVE virus was isolated from *Cx. quinquefasciatus* once in the Ord Valley, W.A.[53] and KUN once at Kowanyama, Q.[66] This species has not been actively sought during a temperate-region epidemic, and although it seems unlikely to be a major vector, this should be done.

Wild-caught *Cx. annulirostris* collected near Brisbane had an ID_{50} infection threshold of $10^{2.9}$ mouse LD_{50} of MVE, and after infection with large virus doses, transmission rates to chickens reached 85% 10 to 12 days after infection.[67] KUN virus was not evaluated.

In an attempt to assess genetically based variation in vector competence with MVE and KUN viruses, *Cx. annulirostris* were collected from most of the geographically widely dispersed locations as were the *Cx. quinquefasciatus* described above. The MVE ID_{50} in the ten colonies ranged from $10^{1.7}$ to $10^{3.9}$ mouse LD_{50}, and KUN in eight colonies from $10^{2.7}$ to $10^{4.8}$ mouse LD_{50} per mosquito. Transmission rates after high doses ranged from 50 to 89% for MVE and 0 to 55% for KUN.[68] These variations appeared to be random, and far more intensive investigations would be required to determine whether such variations are genetically based. It would also be informative to assess vector competence in colonies derived from different habitats over a relatively small area, as well as the very widely dispersed colonies examined here. In general, laboratory vector competence with MVE was lower than field evidence would suggest, and was so poor with KUN that without the abundant field evidence to the contrary, *Cx. annulirostris* would have to be rejected as a possible vector.[68]

2. Vertebrate Hosts

McLean[69] demonstrated viremia in young chickens, infected mosquitoes by feeding them on this source, and transmitted the virus via infected mosquitoes to normal chickens. There was no illness or death. Levels of viremia were less in 28-day-old chickens than in 4-day-old chickens, but even at 6 months a low-level viremia developed. Maguire and Miles[70] infected day-old chicks with MVE virus and detected virus in cultured spleen up to 21 days after cessation of viremia, but cortisone and various stressing agents did not prolong the primary viremia nor induce a second viremia.

The paradox of the high prevalence and frequent high titer flavivirus antibody in nankeen night herons was investigated by Boyle et al.[71,72] Nankeen night herons, plumed egrets (*Egretta intermedia*), little egrets (*Egretta garzetta*), and white necked herons (*Ardea pacifica*) were inoculated intradermally/subcutaneously with 0.1 or 0.2 mℓ of diluted virus. Nankeen night herons were aged approximately 1.5 to 37 months when inoculated with small doses of low passage strains of MVE or KUN virus isolated in the Murray Valley during the 1974 epidemic. Maximum viremias tended to be higher in birds up to 2.5 months old (10^4 to 10^5 $LD_{50}/m\ell$) than in birds more than 8 months old (10^3 to 10^4 $LD_{50}/m\ell$). Although sensitive to very small doses of MVE (<1 VERO pfu), the dose influenced the time of onset and sometimes the level of viremia. Viremia usually lasted 3 to 5 days. There was no difference in response to MVE or KUN virus.

Only high mouse brain passage Nakayama strain of Japanese encephalitis was available, and this produced viremia in 6/7 nankeen night herons up to 3 months old inoculated with $10^{2.3}$ to $10^{4.3}$ mouse LD_{50} per bird. The one bird without viremia developed HI antibody. Titers of viremias were not always related to size of inoculum and were lower and of shorter duration than in birds inoculated with similar doses of MVE and KUN viruses.

Although there were minor differences in time of onset and decline in viremia, little egrets and white necked herons responded in much the same way as nankeen night herons, but nankeen night herons were more susceptible than plumed egrets to infection with low doses of KUN virus.

The levels of viremia reached in birds infected with MVE and KUN viruses were not exceptionally high nor long lasting, but were similar to those found by McLean[15,69] in young chickens, and in his series almost all *Cx. annulirostris* became infected after feeding during peak viremia. Although Japanese encephalitis viremias were considerably lower, in 2/3 nankeen night herons they would have been sufficient to infect at least 50% of the *Cx. tritaeniorhynchus* used by Gresser et al.[73]

Antibody responses were closely monitored and analyzed.[72] Epidemiologically useful observations can be summarized.

HI antibodies were first detected 5 to 6 days after infection and reached maximum titers of 1/320 to 1/2560 by 10 to 20 days. Between 60 and 120 days, titers had declined to 1/20 to 1/320, and then tended to remain stationary for at least 2.5 years after primary infection. Titers were two- to eightfold higher to infecting virus than heterologous virus (MVE and KUN). In field-collected sera a titer of 1/320 to 1/640 would indicate infection within the previous 6 months. When reinfected with homologous or heterologous virus, higher antibody titers are reached than after primary infections, but titers again decline to 1/ 320 to 1/640 within 6 months of the secondary infection. Neutralizing antibody responses paralleled HI antibody response, but were more specific and levels remained high for longer periods.

IgM (19S) antibody, after MVE infection, was detectable from day 6 or 7 to about day 20, but after KUN virus infection persisted for over 27 days. From characteristic cross-reaction patterns it was possible to determine whether MVE, KUN, JE, or WN virus had induced the antibody.[72]

Seven nankeen night herons were secondarily challenged with homologous or heterologous virus. None of those challenged with homologous virus had a second viremia when inoculated up to 152 days after primary infection. Responses to heterologous challenge were variable; 1/4 birds had KUN viremia of up to $10^{2.4}$ pfu/mℓ for 4 days when challenged 152 days after primary MVE infection, and another bird had a primary KUN viremia and low-titered secondary viremia when challenged with MVE 22 days later. The second viremia was serologically confirmed as MVE virus. Six of the seven birds had rapidly rising secondary HI responses, often reaching very high (1/2560, 1/5120) titers.[72]

Nankeen night herons are long-lived birds, and secondarily infected birds might explain the high HI antibody titers often found in field specimens. Although these investigations would have been more convincing if *Cx. annulirostris* had been used to infect the birds and to sample viremia, the infecting doses used were usually very small, and it seems highly likely that individual nankeen night herons in nature could contribute to the maintenance of both viruses, and with a longer interval between infections than was practical here, perhaps also recirculate homologous virus after reinfection.

The use of vector mosquito species as virus donors and recipients is theoretically the optimal method for the elucidation of potential transmission cycles, although in practice there is often frustrating lack of correlation between viremia assays and mosquito infection rates. Despite several enlightened earlier attempts, it was not until 1975 that *Cx. annulirostris* were successfully colonized.[74] Kay et al.[75,76] were the first to use this Mildura colony and others they had established from Brisbane stocks to investigate potential transmission cycles of MVE virus in a rather random selection of domestic, feral, and native vertebrates.

It was confirmed that cattle and sheep are unlikely to contribute to maintenance or amplification of MVE, and that their relatively transient antibody response renders them unsuitable for serological survey. Dogs are a favored blood source of *Cx. annulirostris*, but if the 11 crossbreeds tested are representative, they are also minor hosts to MVE virus.[75] Although feral mice showed transient viremias, they too are unlikely to play a significant role.[76]

Feral pigs frequently have high MVE/KUN antibody prevalence in serological surveys, but their response to laboratory infection was erratic. Three of nine young domestic pigs had significant levels of viremia 1 and 2 days after intravenous inoculation of high doses of virus, and 5/6 a trace response for 1 to 3 days. Of 15 feral pigs, 5 had trace viremias for 1 to 4 days, and this included 2/4 which received i.v. doses of $10^{9.6}$ mouse LD$_{50}$. None of the pigs fed upon by infected *Cx. annulirostris* or *Ae. aegypti* had more than trace viremias. None of 22 *Cx. annulirostris* which fed on a pig with trace viremia was infected, although all 24 pigs developed detectable HI antibody. This response also was variable in level and persistence. Boyle[26] also found young pigs disappointingly inconsistent in viremia response to i.d.-s.c. inoculation of small doses of MVE and KUN, with 4/12 giving moderate viremias and 2 others a trace. The role of the pig in the ecology of Japanese encephalitis in Asia is of supreme importance, but this might not be so with MVE/KUN. However, feral pig densities are extremely high in some areas, so that even if only a relatively small proportion respond with viremia, there might be a significant contribution to maintenance or amplification.[75]

Two species of macropod were investigated with widely divergent results.[75] The response of the gray kangaroo (*Macropus giganteus*) was erratic, but all of 14 had at least trace levels of viremia and 8 had levels ranging from $10^{2.3}$ to $10^{5.3}$ mouse $LD_{50}/m\ell$ persisting up to 5 days. Antibody response was high and persistent. Infection of *Cx. annulirostris* feeding on gray kangaroos was also rather erratic, but the observed overall rate of 9% was close to the statistically expected rate for the levels of viremia at the time of feeding. Nevertheless, it was not a very impressive result for an otherwise promising vertebrate host. The red and the gray kangaroo are the common large species of the dry inland plains, and MVE/KUN antibody prevalence rates in the Murray-Darling basin are generally high and parallel those of feral pigs in the same localities.[61] The agile wallaby (*Macropus agilis*) is probably the most common macropod in tropical Australia, and prevalence of flavivirus antibody is reported to be high, but apparently not to MVE/KUN.[77] Following exposure to MVE-infected mosquitoes, 1/10 had detectable viremia and the HI antibody response was generally low and transitory. The rather frequent frustrations of this laborious technique are well illustrated by the observation that 5/43 mosquitoes were infected after feeding on a gray kangaroo with a viremia of $10^{3.9}$ mouse $LD_{50}/m\ell$ — the highest of those fed to mosquitoes. Yet 3/40 mosquitoes were infected after feeding on an agile wallaby in which no virus could be detected by mouse inoculation.

The least expected results of this series were provided by domestic and feral rabbits (*Oryctolagus cuniculus*), and, in particular, by 3/16 of the feral rabbits. These had long-sustained, very high titered viremias persisting even in the presence of high titered antibody, and are best regarded as aberrant responders, or the consequence of an assay problem, until the results can be repeated. All other rabbits produced viremias of about $10^{2.8}$ to $10^{4.0}$ for 2 to 4 days. Only 1 feral rabbit was exposed to 15 mosquitoes on day 4 when viremia was at $10^{5.6}$ $LD_{50}/m\ell$; 7 of these were infected. Three domestic rabbits infected i.v. were exposed to *Cx. annulirostris*, 28/48 were infected 1 day after inoculation, 5/31 on day 2, and 2/28 on day 3, by which time there was only a trace of viremia. All rabbits produced high titered and sustained antibody.[75] Although rabbits have been used on many occasions for antibody production, and in early investigations have been inoculated intracerebrally with massive doses of MVE in unsuccessful attempts to produce encephalitis, there seems to be no previous report of viremia. Similarly, there have been no systematic serological surveys of feral rabbits for arboviruses. In the Murray-Darling basin, 106 rabbit sera were incidentally collected between 1974 and 1980, but only two of these had HI antibody to MVE and KUN, both at high titers. Although rabbits were collected in relatively small numbers from many localities ranging from the Murray Valley to the Q. border, the two reactors were from the same grazing property on the Cuttaburra waterway and 2 years apart. This was also the source of one of the feral pig colonies examined by Gard et al.[56]

Birds tested by Kay et al.[76] comprised black ducks (*Anas superciliosis*) and three psittacines: the galah (*Eleophus roseicapillus*), little corella (*Cacatua sanguinea*), and the sulfur crested cockatoo (*Cacatua galerita*). All species responded to infection, mostly via infected *Cx. annulirostris* and generally to levels of viremia comparable to most of the mammals tested. Antibodies to MVE/KUN have been found in all these species.[58,61,75]

C. Maintenance, Overwintering, and Interepidemic Survival

The apparent epidemic and enzootic vector is *Cx. annulirostris*, and yet nowhere within its range does it appear capable of maintaining the insect-vertebrate cycle throughout the year. In temperate areas with continental climate, *Cx. annulirostris* activity declines in autumn and ceases in May. Adults can be found at resting sites throughout the winter, but there is virtually no activity until October.[43,78-80] The hypothetical capacity for overwintering adults to carry arboviruses through to the following spring has been examined far more closely elsewhere with other arboviruses, and it seems unlikely to be a reliable survival strategy.[81] In the spring and summer following both the 1951 and 1974 epidemics, no arboviruses were found in *Cx. annulirostris* collected in the Murray Valley; this despite the fact that in February 1974, 238 strains of 11 distinct arboviruses were isolated from pools of *Cx. annulirostris*.[13,20,43] Although MVE virus has not been isolated from any source in the Murray Valley since 1974, other arboviruses, including KUN, have been found frequently enough to suspect that many persist locally, or at least in the Murray-Darling basin, but their survival strategies are conjectural.[43]

A more intriguing circumstance of the apparent inability of *Cx. annulirostris* to maintain a continuous cycle of MVE or KUN is that observed in the monsoonal tropical areas of northern Australia. Here *Cx. annulirostris* is active throughout the year, although in the dry season their activity is limited to areas of suitable permanent or persistent breeding water, as are many of the vertebrates on which they feed. Intensive investigations at Kowanyama over 17 years introduced a bewildering array of previously unsuspected arboviruses, and added considerably to the basic knowledge of arbovirus ecology, but produced no evidence that mosquito-vertebrate transmission of flaviviruses continues through the dry season, and suggested that the extent of transmission in the wet season varies widely between years.[47] Investigations at Kununurra in the Kimberley district of northern W.A. have more consistently yielded isolates of MVE and KUN from mosquitoes, but as at Kowanyama, these are mainly toward the end of the wet season (March to May). On one occasion, three isolates of MVE were obtained from *Cx. annulirostris* during the very wet end of the 1974 dry season (November—December).[53] There seems to be no reason for this interruption to primary cycle transmission in tropical areas; antibody prevalence in host vertebrates is high, but not so high that transmission would be interrupted. *Cx. annulirostris* was shown not to be a particularly efficient laboratory vector, so perhaps transmission effectively ceases when the population falls below a critical threshold density. However, wherever arboviruses have been studied in eastern Australia, activity is overwhelmingly concentrated in the first 3 months of the year, and no indicator of MVE/KUN activity has ever been clearly positive from May to November, inclusive.[43] The W.A. mosquito isolations are recorded in time periods of up to 3 months, but no MVE/KUN activity is apparent from June to October at least. It is quite possible that transmission is simply below detectable levels, and this could apply to interepidemic survival of MVE and overwintering of KUN in temperate areas as well. Whatever the survival strategy in tropical areas, this period of occult persistence includes those months when, by one strongly held hypothesis of the generation of epidemics, infected birds should be taking off on their southward flight to seed the Murray Valley with MVE and KUN viruses.[8,17] In fact, no significant aberrations of mosquito, bird, or virus activity were observed at Kowanyama during the second half of 1970, nor at Kununurra or Kowanyama during the second half of 1973, which would have provided an intimation that virus activity was about to quicken in the Murray Valley.

In the laboratory, MVE virus has been shown to be transovarially transmitted in the eggs of *Ae. aegypti*,[82] and KUN in *Ae. albopictus*.[83] Field observations indicate that transovarial transmission probably does not occur in *Cx. annulirostris*, but in other species such as *Cx. australicus*, it is a completely unexplored possibility. In *Aedes* spp. that have dessication-resistant eggs, highly speculative but plausible scenarios can be developed. In the temperate regions, the usually dry summer months are inimical to the activity of *Aedes* spp., and eggs are often hatched by the autumn rains. Several species are active through the winter, but the main surge of activity is in the spring, and depending on the local aquatic conditions, this activity can occasionally extend into summer. Probably in 1950 and certainly in 1973, the spring was hot and the Murray Valley flooded, and this produced early massive breeding of *Cx. annulirostris* coinciding with persisting massive populations of day-feeding *Aedes* spp. This intense activity overlapped for 8 to 10 weeks before the *Aedes* population dwindled at the end of January. If MVE and KUN are transovarially transmitted in one or more of these *Aedes* spp., in nonepidemic years this could be a small-scale event of limited transmission to vertebrates in winter or spring, but if temperature, surface water, and wildlife breeding are all hospitable to concurrent *Aedes* and *Cx. annulirostris* build-up, the latter could initially be infected by *Aedes* via an infected vertebrate, amplify the virus activity, and spill over to the human population. In Charleville, southern Q., as in the Murray Valley, *Aedes* populations are as erratic in size and composition as the weather, and there, too, maximum arbovirus activity appears to occur when large *Aedes* and *Cx. annulirostris* populations overlap.[84] At Kowanyama it is the dry season (southern winter) that is inimical to *Aedes* spp. activity. Transmission seems not to be maintained by the dry-season *Cx. annulirostris* population, and perhaps MVE and KUN virus survive this period in the eggs of *Aedes*, and it is they that start the wet-season virus activity and introduce MVE and KUN into the burgeoning wet-season *Cx. annulirostris* population. As elsewhere, *Aedes* populations are difficult to predict, and precise conditions for the population explosions by one or other species are not known. Although MVE and KUN are more frequently detectable at Kowanyama than in the Murray Valley, nevertheless there might be a predisposing confluence of *Aedes* spp. and *Cx. annulirostris* before MVE and KUN reaches a threshold of detectability. At Kununurra, where MVE and KUN can be detected in most years, the critical investigations are based on a rather small area of irrigated farmland and associated waterworks, and an enormous bird population. The irrigation system might stabilize the relationships between *Aedes* spp. and *Cx. annulirostris* so that the detection of MVE and KUN in the wet season becomes more predictable.

It took about 75 years for the idea of persistence of arboviruses by transovarial transmission in mosquitoes to be demonstrated as valid after numerous failed experiments and unsubstantiated observations. Perhaps the long-canvassed idea of the recirculation of virus in the bloodstream of a persistently infected vertebrate at a level sufficient to infect a few mosquitoes in spring will also eventually be validated.[81] That simple scenario for the flavivirus encephalitides would make all the many other hypothetical survival strategies immediately redundant.

VI. ECOLOGICAL DYNAMICS

A. Environment
MVE and KUN are viruses of the plains: tropical coastal plains and hot, dry, inland plains. Well over 90% of cases and virus isolations have been at elevations below 200 m and the rest lower than 500 m. Although there has been no strong evidence of MVE/KUN activity on the eastern Q. coast since 1925, several cases occurred in 1974 on the low eastern slopes of the Great Dividing Range at about the Tropic of Capricorn (Figure 1).

The coastal plains on the Gulf of Carpentaria consist of tropical tussock grassland and large areas of open, low eucalyptus forest intersected with more densely timbered areas along watercourses. There are no major river systems, drainage from the low tablelands being by individual streams with a few ill-defined tributaries.[85] In the wet season (January to March) these streams are in full flood, much of the intervening country is submerged, and there are many transient ground pools. During April and June the streams recede to within their banks and flow is reduced, transient pools disappear, and permanent waters are reduced in area. In the dry season (July to October) the streams stop flowing and are reduced to scattered waterholes. The size of permanent swamps are further reduced. The unpredictable rains of November—December enlarge the residual pools in the streambeds, but flow occurs only rarely. Transient pools sometimes appear late in December.[85] In northern W.A. the low tablelands are closer to the coast and the north-flowing Ord River has cut a valley flanked by low escarpments which broadens out to a coastal plain before discharging into the Timor Sea. The billabongs and marshes of the lower Ord are difficult of access, but would have been the main bird-mosquito ecosystem before impoundment of the river about 100 km upstream at Lake Argyle and the Diversion Dam. The latter is used to direct water into the irrigation areas, and has created permanent swamps through breached natural levee banks. The effect has been a consolidation and extension of an aquatic ecosystem which previously was an annually transient system as at Kowanyama. The same monsoonal wet-dry seasons apply, but now the aquatic ecosystem in the Kununurra area is permanent.[86] MVE/KUN activity is largely confined to the northerly Kimberley district, but more recently activity has been detected in the coastal areas of the Pilbara district down to about the Tropic of Capricorn.

In eastern Australia, MVE/KUN activity seems to be confined to the drainage basins west and north of the Great Dividing Range. The Carpentaria basin is separated from the Lake Eyre basin by the Barkly and other tablelands. This enormous basin drains to Lake Eyre, below sea level in the deserts of south-central Australia. Although the rivers and ''channel country'' of western Q. are under the influence of the monsoonal rainfall pattern of the north coast and receive precipitation in most years, this is not as reliable as in coastal regions, and water flow through to Lake Eyre is a very rare phenomenon; 1973—1974 was such a season. There are some notable waterbird breeding areas, such as those of the pelican on Coopers Creek. The Grey and Barrier ranges separate the Lake Eyre basin from the Murray-Darling basin which contains the major river system of the continent. The most northerly headwaters of the tributaries of the Darling River are those of the Warrego, rising at about 25°S, and the most southerly at about 37°S; the system drains water from subtropical areas influenced by the monsoonal system to snowpack from the southern Alps. There are natural grassland downs in the north, ranging to red ''mallee'' sand in the south, but mostly the basin consists of sparsely forested, dry inland plains, much of it semidesert saltbush savannah or drought- and fire-adapted low-growing mallee eucalyptus. Generally harsh, uncompromising country which harbors the largest of the still-extant macropods, the red and the gray kangaroo; the largest indigenous bird, the flightless emu; a diversity of smaller marsupials, and reptiles such as frill neck, bluetongue lizards, and goannas. These now compete with feral pigs, goats, and south of Capricorn, rabbits, as well as introduced domestic cattle and sheep. Many of the watercourses flow only after flooding rains, and some of these ephemeral streams in the north of the basin form vast lakes on the heavy clay soils which might persist for up to 5 years, while in the south they sometimes disappear into sand or shallow salt lakes. Imposed on this harsh landscape is the vegetation associated with the main river system. From Bourke on the Darling and Echuca on the Murray, the fall to the Murray mouth at sea level is about 100 m, so the anastomosing rivers winding through the semidesert plains develop lakes, anabranches, and billabongs, and flow into and out of extensive swamps and marshes. In the tropics and the temperate inland regions of Australia, the stream systems are lined with a band of relatively dense riverine forest which defines the regular flood plain. In places, this forest will be a ribbon of about 50 m on both banks of the river, but will range out to 50 km or more where there are extensive semipermanent swamps such as the Macquarie Marshes. This riverine forest and swampland forms a continuous uniform ecological network superimposed on the generally arid basin, and a biological conduit from southern Q. to the coast of S.A. Paradoxically, this is also one of the major waterbird breeding areas of the continent. Rather than regard the Murray Valley as a separate ecosystem when considering the epidemiology of MVE/KUN, the fluvial system of the whole of the Murray-Darling basin should be regarded as a single ecological entity.

Irrigation was being practiced in the Murray Valley before the first cases of Australian

X disease occurred, but the early epidemics were not in, or even near, the irrigated area. Since then, irrigated farming has become very extensive and diversified throughout the broadly defined Murray Valley, and in more recent years, in the central and northern areas of the Murray-Darling basin, where cotton and coarse grain crops have become major industries. The Murray Valley irrigation areas have provided a large, stable summer population of *Cx. annulirostris*, and other mosquitoes in season, but this in itself has not increased the frequency of MVE/KUN epidemics, although arguably it might have influenced the incidence of Ross River virus infection. No doubt once an epidemic is triggered, *Cx. annulirostris* breeding in irrigation waters will contribute to the maintenance and spread of the viruses, but the generation of an epidemic seems to require initially a major pluvial aberration which almost seems to be anticipated by wildlife, and particularly bird life, so they reactivate long-neglected breeding areas, and opportunistically set up breeding colonies much closer than usual to human habitation. What might reduce the incidence of epidemics is ever-increasing water control throughout the basin, and particularly flood mitigation, which is reducing the area of wetlands in the Murray-Darling basin. Such draining of swamps, sometimes inadvertent, will be particularly damaging to those species which seem to be primarily involved in MVE/KUN cycles.

B. Climate and Weather

Following the epidemic in the Murray Valley in 1951, and the absence of MVE virus activity in 1952, Anderson and Eagle[1,13] and Miles and Howes[17] developed hypotheses of rainfall patterns predictive of epidemics in the Murray Valley. They reached essentially the same general conclusions: the three epidemics of MVE that had occurred in the Murray Valley followed abnormally high rainfall in spring and early summer in northern rainfall districts, and 2 other years of excess rainfall in some but not all the rainfall districts had resulted in epidemics in Q. only or Q. and Broken Hill, N.S.W. Miles and Howes[17] favored rainfall greater than 110% of normal during spring and early summer in rainfall districts over a wide area of N.T. northern S.A. and the whole of or at least inland Q. Anderson and Eagle[13] placed greater emphasis on the northern regions of the Murray-Darling basin where, in the 5 wet years, rainfall was more than 200% of normal. Both groups considered that the excess rainfall provoked movement of birds and virus down to the Murray Valley in spring.

Anderson et al.[5] became somewhat disenchanted with the hypothesis when the 2 cases which occurred in the Murray Valley in 1956 were not preceded by abnormal rainfall in the Darling watershed nor the north Q. watersheds, and when all the indications were right for an epidemic in 1957, nothing happened.

Although much higher than average rainfall was experienced over almost all of Australia in 1973, the fluvial predictions as defined by the above groups were again not met, and the 1974 epidemic was the most widespread of any so far. Objectivity and availability made rainfall records a highly desirable prediction method, and Forbes[8] set about fine-tuning the records. He found that the epidemics of 1918, 1951, and 1974 were all preceded by rainfall in excess of decile 7 in one or both of the spring and summer quarters of the preceding year and in the spring quarter immediately preceding the outbreak in all four defined "main" watersheds: those fringing the Carpentaria basin, the headwaters of the mainly ephemeral streams draining to Lake Eyre, the Darling River headwaters, and the headwaters of the Lachlan, Murrumbidgee and Murray Rivers. The minor intrusions of 1956 and 1971 are accounted for by failure to reach decile 7 in all major catchment areas or reaching them in the wrong months, and similarly, outbreaks of Australian X disease only in Q. and/or N.S.W. were related to other patterns.

With the earlier hypotheses, there was scarcely enough time to obtain and analyze the rainfall records from remote areas before the first case could be expected, but with Forbes' hypothesis, the high rainfall in the spring or summer quarter of the previous year gives warning that decile 7 rainfall in the next spring quarter will result in an epidemic. Forbes[8] also favors movement of infected birds from tropical areas.

Many of the floods in the Murray Valley result from rainfall in relatively distant watersheds rather than in local rainfall, but local weather is important, particularly with regard to mosquitoes. A cold spring will delay the build-up of *Cx. annulirostris* populations, no matter how much otherwise suitable breeding water is available.[78] The sudden drying out of the flood plain for whatever reason, or unseasonably cold or windy weather, will probably check an epidemic. Russell[79] found that *Cx. annulirostris* populations at a site on the Murray River near Echuca had a great capacity to rebound from very low levels after drought years, or after a setback early in the season such as prolonged cold weather in spring, but this was less likely to occur when temporarily adverse conditions checked or reduced populations during the summer.

C. Vector Oviposition

Cx. annulirostris oviposition activity in the Murray Valley is stimulated by rainfall creating and extending shallow grassy pools. The typical breeding habitat is the permanent and semipermanent swamp with abundant emergent vegetation, but this extends to shallow, still water with emergent grass, irrigation drains, pools in roadside drains, and even in towns. In some areas under favorable conditions of vegetation it has become adapted to the polluted water of sewage ponds.

D. Vector Density, Fecundity, and Longevity

Mosquitoes have only once been trapped during and in the immediate vicinity of an epidemic, and this was during the 1974 epidemic in the Murray Valley.[19,20] Mosquito activity between February 4 and 13, at the height of the epidemic, was almost entirely crepuscular and nocturnal, although there was anecdotal evidence of intense activity of day-biting *Aedes* spp. during the first 4 to 5 weeks of the epidemic, which is much later into the summer than usual. The overall capture rate for 51 trap nights at seven widely dispersed sites in the epidemic region was 2675. At Barmah Forest, about 20 km upstream from Echuca, and still largely flooded in February, ten trap nights yielded an average of 9103 mosquitoes with a maximum of 26,293 in one trap night. The latter yielded 32 viruses including seven strains of MVE and 20 KUN.[19,20]

It is difficult to reconcile an epidemic of this magnitude with a crude MVE isolation rate from *Cx. annulirostris* of only 1/3581, or a combined MVE/KUN isolation rate of 1/913. Yields were no higher at Charleville, southern Q.[21] This low rate was presumably compensated for by absolute numbers, density, and attack rate. Some indication of the density of mosquitoes and viruses is provided by the yield from the most successful trap mentioned above.

Observations on the density of mosquitoes required for detection of MVE/KUN activity is contradictory. The tremendous abundance of *Cx. annulirostris* observed during the 1974 epidemic has not been seen since, and a threshold density calculated on this occurrence would be very high. At Barmah Forest in February 1979, the 3rd successive drought year, the mosquito population was at a very low ebb, and the catch per trap night was only 25. Two strains of KUN were recovered from 485 trapped and aspirated *Cx. annulirostris* and one from a total of two *Cx. australicus*;[43] for KUN a crude isolation rate of 1/243 *Cx. annulirostris*, a very much higher rate than during the epidemic. The total catch of all species from five locations in the Murray Valley in February 1979 was a meager 5218, and these yielded 31 strains of five arboviruses (Sindbis, Kunjin, Edge Hill, and Umbre), an overall isolation rate of 1/168.[43,61] If that and one or two other seasons are regarded as in some way aberrant, it can be said that the higher the density of mosquitoes at Kowanyama, Charleville[66,96] and the Murray-Darling basin, the greater the arbovirus activity, but a threshold density for KUN or MVE viruses cannot be sensibly computed.

Kay[88] found that the age composition of populations of *Cx. annulirostris* at Charleville (mean daily survival rate 74%) and Kowanyama (72%) were similar despite dissimilar climates. Only 12% and 18% of females from Charleville and Kowanyama had taken three or more blood meals and were thus old enough potentially to transmit MVE virus. These estimates were based on the extrinsic incubation period of 7 to 10 days at 27°C, and gonotrophic cycle of 3 to 4 days at both locations. Gonoactive pars and multipars were collected throughout the year at both locations, despite low winter temperatures at Charleville. Russell[89] collected *Cx. annulirostris* at 2-week intervals over 2 years at two locations near Darwin, N.T. and scored them for age composition. The species was recorded throughout the year at both sites, although greatest abundance occurred following the end of the wet season (May—June). The oldest females were 5-parous and the mean proportion parous was the same at both sites (0.45 and 0.44), as were the mean daily survival rates estimated from the proportion parous (83% and 82%). Females old enough to transmit MVE were recorded more frequently at one site than the other, and averaged 13.5 and 8% of the total collections. They were present most months of the year.

Russell[90] also examined a population of *Cx. annulirostris* at Appin, 73 km southwest of Sydney, over a period of 2 years. Here there was no activity between May and October, and the overwintering adult female population was predominantly 1 to 3 parous and in a state of "quiescence" rather than diapause. The duration of the gonotrophic cycle ranged from 4 to 10 days depending on temperature, and females old enough (2-parous) to transmit arboviruses were present from October to May.

From these considerations, the longevity of *Cx. annulirostris* in tropical and temperate areas is rated as only "moderate", and its vector status seems to be largely attributable to abundance.[87]

E. Biting Activity and Host Preference

Biting activity of *Cx. annulirostris* commences quite abruptly and fiercely about 30 min after sunset, eases somewhat after about 2 hr and then continues at a reduced level through the night, with a second, smaller peak before dawn. Attack usually ceases at full sunrise, but on heavily overcast, warm, and humid mornings it will continue to 10 or 11 a.m.

Cx. annulirostris comes freely to CO_2-baited traps, perhaps indicating a catholic appetite. Lee et al.[91] precipitin-tested blood meals of mosquitoes collected from a range of sites, but mostly the restricted animal habitats in farmyards. However, even when collected from resting sites on river flats, *Cx. annulirostris* showed predilection for mammals, particularly rabbits, rather than birds. Man was attacked but not preferentially. They suggested that MVE virus would be more likely to be spread from chickens to man by *Cx. quinquefasciatus* rather than directly by *Cx. annulirostris*. Many reports have commented on the introduced rabbit as a preferred source of blood by *Cx. annulirostris*, and with *Anopheles annulipes*, it has been jointly responsible for vectoring myxoma virus. Feral rabbits do not extend to the tropical north, but Kay et al.[92] using stable traps in a carefully planned spatial study, also concluded that *Cx. annulirostris* at Kowanyama preferred mammals. Of the vertebrates offered, calf, pig, and dog were equally sought, followed by kangaroos and man, with chickens least attractive. Blood meals identified by precipitin tests from mosquitoes caught in the village gave the same general results, with dogs the most popular source.[92] In these investigations *Cx. quinquefasciatus* and *Cx. squamosus* were the only mosquitoes to favor birds, but there have been very few isolations of arboviruses from these species.

Russell[93] modified EVS traps to carry 500 g chickens, rabbits, or guinea pigs, and compared catches with dry-ice baited traps, CDC light traps, and unbaited EVS traps at two study sites, one on the Murray River flood plain, and one in partly cleared sclerophyll forest at Appin, 70 km southwest of Sydney. The dry-ice baited EVS traps attracted most mosquitos at both sites and the unbaited EVS traps scarcely any. However, unlike the Kowanyama trials, and the earlier observations of Lee et al.[91] in the Murray Valley, the chicken attracted significantly more *Cx. annulirostris* than either the rabbit or the guinea pig at both locations.

The behavior of a vertebrate and its tolerance to mosquito attack have been shown to be at least as important as direct attraction or repulsion. Edman et al.[94] found the motionless hunting stance, remarkable tolerance to mosquito attack, and nocturnal feeding habit of a night heron allowed far more numerous successful blood meals to be taken than from waterbirds with more restless habits. In addition, the nesting night heron leaves the nestlings exposed to nocturnal mosquito attack. No attempt has been made to find the attractiveness of the nankeen night heron, cormorants, etc., to *Cx. annulirostris*.

F. Vertebrate Host Density and Immunological Background

Breeding of many species of wildlife in Australia, both birds and mammals, is triggered by water and its effects on pasture germination and growth, colonization of transient and semipermanent pools by waterweeds, insects, fish, and crustaceans, the increased availability of small prey animals, and the production of seeds, both wild and agricultural. This quickening of life extends to mosquitoes and viruses. Virtually every substantiated case history of epidemic MVE/KUN encephalitis begins with a close association, either residential or fleeting, with the Murray-Darling river system and its river-redgum-forested flood plains. In both observed epidemics, waterbirds have exploited the enormously increased wetlands habitat which has persisted for one or two seasons after the epidemic. Mammals tend to be driven out of the riverine habitat by rising flood water and the availability of new pasture and water on the adjoining plains. This would not completely explain the apparent lack of involvement of the rabbit in MVE maintenance and epidemics, for although their warrens are obviously vulnerable to flood, sandhills and old levee banks are usually close by.

As commented upon in Section V.A.2, the wildlife, and particularly waterbird populations, underwent extraordinary expansion over three successive highly favorable years, but MVE/KUN activity was detected only in the first of those years, the summer of 1973—1974. Although details are lacking, it is likely that a similar phenomenon occurred during at least one season following the 1950—1951 epidemic. Obviously there have not been sufficient opportunities to attempt to define threshold levels of vertebrate and concurrent mosquito populations required to generate an epidemic. The prevalence of MVE/KUN antibodies in the bird population was already high before the end of the epidemic, but the extensive breeding throughout the Murray Valley must have provided large actual numbers of susceptible juveniles during the following two seasons. The lack of virus activity was probably due as much to the very much smaller *Cx. annulirostris* populations as to immunity in the vertebrate hosts.[43]

G. Vector Competence

Detailed laboratory estimates of vector competence of *Cx. annulirostris* can be rated as only moderate for MVE and very poor for KUN, although some of the less precise, earlier investigations with MVE suggested greater efficiency (Section V.B.1).[15,16,67,68] These estimates, particularly with KUN virus, must surely be artefactual. Even if the perceived incompetence is compensated by the effect of astronomical mosquito numbers, the fact that at Kowanyama, Charleville and in the Murray Valley the isolation rate of KUN:MVE from *Cx. annulirostris* is at least 3:1 is the opposite of what would be expected. Nevertheless, isolation rates from *Cx. annulirostris* during the epidemic were low, and in the smaller mosquito populations of nonepidemic years, MVE and KUN viruses rarely reach detectable levels, except in the artificially stimulated ecosystem of the Ord River region of northern W.A.

Both MVE and KUN have been recovered from *Cx. australicus*, a sylvan breeding member of the *Cx. pipiens* complex which feeds mainly on birds and rabbits. The MVE isolate was from Lake Poon Boon in the Murray Valley during the 1974 epidemic, and the KUN was from the Barmah Forest in 1979, a nonepidemic year.[20,43] It does not feed on man and does not find CO_2 bait very attractive. It is active in fluctuating numbers in all seasons in the Murray Valley and is an obvious candidate as a primary cycle vector.[95]

H. Movements and Migrations of Vectors and Hosts

Myxoma virus "escaped" from an experimental site near Corowa on the Murray River in November—December 1950, and by March it had spread 1100 mi west down the Murray River and out into the semidesert of S.A. It had spread up the Murrumbidgee and Lachlan Rivers and 1000 mi up the Darling River, and then spilled out over the flooded northern end of the Murray-Darling basin straddling the N.S.W.-Q. border. Outliers of diseased rabbits appeared in the channel country north of and contiguous with Lake Eyre, and on the S.A. coastline. There might have been some human intervention associated with the spread, but the relentless onward movement on several fronts suggested that deliberate capture and release efforts were incidental to the mainstream of movement.[4] In Australia, myxoma virus infects only rabbits (*Oryctolagus cuniculus*) and very occasionally hares (*Lepus europaeus*), and it is spread mechanically by biting insects, contaminated thorns, etc. The initial spread was principally by *Cx. annulirostris*, and in general the disease clung closely to the river-redgum flood plains of the watercourses in the Murray-Darling system, the haunt of *Cx. annulirostris*, which, at the time, was also vectoring MVE virus. In subsequent years, *Anopheles annulipes* became the major vector as the disease spread into the drier country between the rivers. Somehow, *Cx. annulirostris* and the briefly but lethally infected rabbit managed to move the virus at a speed of 9 to 10 mi/day.[4] It is of interest that the one fatal case of MVE in S.A. in 1951 occurred in dry farmland country west of the Murray River in the general area of several isolated myxomatosis outbreaks.

The northerly movement up the Darling River makes it a little less likely that MVE virus was carried south by *Cx. annulirostris* at the same time, at least by movement close to the ground. Long-distance, high-altitude, warm-wind movement has been postulated to move viruses from the tropics to temperate zones, and Sellers[96] has reviewed and extended this difficult and intriguing hypothesis. MVE virus is nominated in the generality as being transported from the tropics to the Murray Valley via infected mosquitoes carried on warm windstreams. As with other theories of pre-epidemic movement of virus from the tropics, this one more than ever is faced with the problem that mosquitoes infected with MVE virus have not been found at tropical field stations at the time they should be joining the airstream south.

Russell[97] measured dispersal of *Cx. annulirostis* from an isolated larval habitat on the Murray River flood plain near Echuca during a period of drought, using dry-ice EVS traps at 1 km intervals for 7 km along four transects. Relative abundance at 7 km was less than 10% of that at 1 km, but on all transects maximum flight range could be assumed to be greater than 7 km. The proportion of parous mosquitoes decreased with distance from the breeding site, although parous females were collected throughout the sampling area.

Migrating birds pay scant attention to Wallacea, and superficially at least, there seems no reason why Japanese encephalitis should not be introduced every spring. Among the many migrants is the Japanese or Australian snipe (*Gallinago hardwickii*) which arrives from Japan mainly to eastern and southern Australian river flats, swamps, and marshes in September, and departs in March, which would be excellent timing if they carried Japanese encephalitis. In fact, they are also patchily distributed in the Kimberleys and along the northern coastline, but not in southwestern Australia, which is roughly coincident with the known distribution of MVE and KUN viruses.

The water birds of the major breeding areas of inland Australia and the northern coastal areas are mostly nomadic rather than migratory, dispersing in all directions from every breeding site, restlessly crisscrossing the continent so that although population numbers in a particular swamp appear static, there might be a mixture of sedentary birds, and birds from other regions replacing those that have left. Ducks, cormorants, ibis, etc., show this behavior, but too few bands are returned from herons to obtain satisfactory movement records. Most returns from nankeen night herons have been from the vicinity of the original banding, although up to several years later. The oldest recovery was 15 years after banding, and the most spectacular was one banded as a fledgling at Barham on the Murray River and shot 9 months later in Papua near the Indonesian border. The nankeen night heron and the black crowned night heron *(Nycticorax nycticorax)* seem largely to respect Wallacea; vagrant black crowns are sometimes seen in the N.T. and New Guinea, and nankeens are either vagrant or at least much rarer than black crowns in Indonesia. Both occur in the Philippines. Wherever the black crown occurs in this region, the virus is Japanese encephalitis, and MVE virus occurs only where the nankeen night heron holds exclusive territorial rights.

I. Human Element in Disease Ecology

Modification to the fluvial systems, irrigation farming, flood mitigation, wetland drainage schemes, sewage disposal, and mosquito abatement measures are all practiced in the tropical and temperate habitats of MVE and KUN viruses, and all modify disease ecology, sometimes attenuating virus cycles and sometimes exacerbating them.

In the Murray Valley the towns are mostly sited on the river redgum flood plains, often with natural and artificial levee banks around the periphery. Social and recreational activities are concentrated on the rivers, lakes, and reservoirs, and on the forested banks. Camping and caravan parks are almost always in the forested flood plain on the rivers in the tourist areas of the Murray Valley, as are *Cx. annulirostris* and waterbirds. The closely settled irrigation areas create their own mosquito problems, and itinerant fruit pickers and farm laborers are frequently exposed to Ross River virus infection and refuse to enter the area during the rare encephalitis outbreaks. The excellent climate and diminishing farm returns have led to the development of a major tourist industry, and visiting city dwellers were among the early cases in both 1951 and 1974.

Although the Kimberleys in W.A. and northern tropical areas also have expanding tourist industries, access to arboviral habitats is only practical in the relatively safe dry season. The new towns associated with major mining developments are planned with mosquito abatement in mind, although the execution does not always meet the planned standards. Mining companies have been encouraged to actively control mosquitoes in the town and at the work site.

VII. SURVEILLANCE

A. Clinical Hosts

Despite long intervals between epidemics, medical practitioners throughout the endemic and epidemic areas are probably now more alert to the possibility that a case of encephalomyelitis might be due to MVE or KUN than ever before. Arbovirus diagnostic services are still unsatisfactory in some states, but this is gradually being remedied. The prospective seroepidemiological studies of Hawkes et al.[52] (Section IV.E), involving close contact with pathological services throughout N.S.W., should provide the first serologically confirmed clinical indicator case at the outset of an epidemic.

B. Wild Vertebrates

Serological and virological surveys in the Murray-Darling basin were carried out for 10 years from 1973. Tests were for MVE, KUN, Ross River, Sindbis, and occasionally Barmah Forest virus. Although essential to research, such surveys are of limited value in surveillance.

The build-up of breeding colonies of water birds is monitored, but not yet systematically. There is exchange of information between officers of the Victorian Health Department and National Parks and Wildlife Service at an informal level, and this probably will be sufficient to alert appropriate authorities.

C. Vectors

Abundance is monitored by Health Departments of N.S.W., Vic. (both with EVS CO_2 traps), S.A. (240 V light traps with each day's catch automatically sorted into separate alcohol jars), and W.A. (EVS CO_2 and sometimes bait traps). Vic. processes a proportion for virus isolation, but immediate surveillance and epidemic prediction can only be on abundance of *Cx. annulirostris* persisting into January and February and on sentinel chicken conversions. Research-oriented mosquito observations and virus isolation from mosquitoes was carried out for 12 years from 1973 in the Murray-Darling basin.[61]

D. Sentinels

N.S.W. and Vic. Health Departments bleed and promptly test sentinel chickens strategically sited along the Murray River and throughout N.S.W. The chickens are bled on a weekly or fortnightly basis from November through April. Most years, many chickens convert to Sindbis positive almost immediately. Significant numbers have converted to KUN positive in 3 out of 10 years of surveillance. Conversions have been late in the season, and extra mosquito control efforts were not mounted.

MVE virus and antibody were recovered from sentinel chickens in the Murray Valley in 1974, but only after human cases had been diagnosed.[55] Chickens are probably better sited now. Virus and antibody were also detected in sentinel chickens at Charleville before human cases occurred in Q., but this was after the epidemic had started in the Murray Valley.[21]

Veterinary virus diagnostic services are well organized in Vic., N.S.W., and N.T., and horses incidentally act as sentinels in Ross River and Kunjin outbreaks.

E. Rainfall Patterns

Rainfall records are scrutinized as soon as available to detect predisposing rainfall patterns as defined by Forbes[8] (Section VI.B).

VIII. INVESTIGATIONS OF EPIDEMICS

The ad hoc investigations of 1974 were the first to have been conducted during the course of an MVE epidemic. Little is known about the ecological dynamics relating to the generation of an epidemic, and it is hoped that present levels of surveillance will continue at least in the Murray Valley, but preferably in the Murray-Darling basin, so that the dynamics of vector and vertebrate hosts during the pre-epidemic period can be assessed.

During the epidemic, mosquitoes should be collected in the river towns as well as in the riverine ecosystem to determine if *Cx. quinquefasciatus* plays a role in epidemics. Passerines and mammals should be serologically surveyed as well as water birds. If waterbird breeding is in progress, the mosquito infection rate both in and remote from rookeries should be determined. If possible, feeding grounds of nankeen night herons should be located and mosquitoes trapped in the vicinity.

IX. PREVENTION AND CONTROL

A. Vector Control

The 1974 epidemic exposed the fact that there was an almost complete lack of suitable infrastructure, contingency plans, trained personnel, equipment, and insecticide stockpiles to combat a mosquito-borne epidemic in the Murray Valley, let alone the other areas of Australia which were to become involved. The Commonwealth Government, through the Health Department, shortly afterward established an Australian Encephalitis Control Program to be funded annually by direct budget vote, with the funds to be allocated to the states on a matching grant basis to develop surveillance and control programs. Subsequently, this was broadened to the National Diseases Control Program to include the ubiquitous Ross River virus/epidemic polyarthritis and the resurgence of dengue in Q.

Final responsibility for mosquito control in most states now rests with local government ("shires" in most states) and they are the final recipients of funds. The State Health Departments retain a coordinating role and are generally responsible for overall surveillance including organizing and testing sentinel chicken flocks.

To train health inspectors and newly recruited staff in the biology, methods of larval and adult surveys, the rudiments of the many aspects of mosquito control, and practical decision making in selecting the appropriate control technique to solve actual specific problems, the first of what became an annual Australian National Mosquito Vector Control Course was held in Mildura in late spring 1975. This intensive 5-day course has provided basic training not only to local government officers throughout Australia but also malaria control and hygiene personnel from the Navy, Army, and Air Force, district medical officers, health departments, and World Health Organization-sponsored mosquito control officers from countries of the South Pacific region. Similar locally organized courses are occasionally held in W.A. and Q. A very active and effective community-based group has flourished in the Mildura district; the Sunraysia Mosquito Advisory Committee, which uses promotional advertising techniques on free time and space slots on local television, radio, and newspapers to educate and create an awareness in the local population on mosquito life cycles and problems, and how they can be solved. Mosquito biology has been introduced into the general biology course in the high schools, with abundant resource material provided by S.M.A.C. More recently, Q., N.T., and W.A. state governments have also embarked on extensive publicity campaigns.

Mosquito populations are monitored in the Murray Valley throughout the summer by light or CO_2 traps, aspiration, and dipping of breeding water. Most riverine shires larvacide problem areas every summer and attempt to maintain a buffer zone to 5 km around towns. As a result of recent flight range studies, this will probably be extended to 7 km.[97] Adulticides are used infrequently and usually in defined areas before an outdoor social event or sporting activity. Contingency plans drawn up by the three Murray Valley states require early aerial adulticiding with ULV malathion and increased larvaciding. The reduction or elimination of virus transmission by reducing the adult populations with one or two well-timed aerial applications is purely theoretical under Australian conditions. During a massive *Cx. annulirostris*-vectored Ross River virus epidemic in N.S.W. in 1984, two groups combined with the intention of carrying out a closely monitored aerial application of ULV malathion at Griffith, a town in the Murrumbidgee irrigation scheme. Unfortunately, this had to be aborted because of pressure from local orchardists who purchase and release parasitic wasps to control citrus scale. This was unfortunate because, although the epidemic was waning, a closely controlled pilot project of this sort would allow better contingency planning.

Darwin and other towns in the N.T. are monitored and controlled to the limit of available facilities, and entomologists are consulted or included in the planning team in new residential developments. Proper design and maintenance of drains and control of tidal swamps are coping with some problem areas. MVE virus has not been isolated from mosquitoes close to Darwin, and no cases have been confirmed there yet.

The irrigated area associated with the Ord River scheme at Kununurra is so heavily polluted with agricultural insecticides that there is no breeding problem.[86] The swamps associated with the Diversion Dam are too extensive to completely control.

Although it is impossible to predict what impression could be made on a *Cx. annulirostris* population as active, enormous, and so widely dispersed as that in the Murray Valley in 1974, if another epidemic does occur, there are human and physical resources available now to at least intelligently attempt to ameliorate the outbreak.

B. Control of Vertebrate Hosts

If water birds are a major vertebrate host of MVE/KUN there is no way acceptable to the community that control could be undertaken, although there might be inadvertent control through draining or otherwise mismanaging wetland habitats. The red and the gray kangaroo are currently being legally slaughtered at the rate of 2 to 3 $\times$ 10^6/year in favor of cattle and sheep, but the residual population is probably more than adequate to continue hosting the viruses. Epidemics of encephalitis are so infrequent that it would be difficult to justify controlling any proven or potential native wildlife host with abatement of MVE/KUN as the only reason.

Feral or exotic species excite fewer benevolent passions. The pig might play a role in virus ecology, but is defying control attempts; many station properties in northern N.S.W. are resigned to spending $20,000 to $30,000/year for a week's slaughter by helicopter "gunships", and others allow as much or more for more conventional control methods. There is no prospect of eradication or even adequate control. Feral goats have not been sufficiently investigated. The role of the rabbit in virus ecology is paradoxical; myxoma virus survives in a single vertebrate host and does not even replicate in its vectors, and yet has now been enzootic in the Murray-Darling basin for 35 years. The high titered, sustained MVE viremias reported in laboratory-infected rabbits[75] should result in an enzootic situation, or at least a major amplifying contribution to epidemics in rabbit-infested regions, but the mostly circumstantial field evidence does not support either role. The rabbit or other pest species would probably be more effectively controlled if it hosted a virus that is life-threatening to humans, but on present evidence it would be rash to claim implication of any of these feral animals in MVE persistence.

A cogent argument against waterbirds as primary hosts is their longevity. Although St. Louis encephalitis has been isolated from herons, the important primary hosts seem to be relatively short-lived, regularly breeding passerines which commune in large flocks and logically should provide a steady supply of young susceptible hosts. In eastern Australia there are enormous flocks of exotic house sparrows *(Passer domesticus)* extending from north Q. to Tasmania, and English starlings *(Sturnus vulgaris)* from southeast Q. through to S.A. and Tasmania, and these are probably stronger candidates as maintenance or amplifying hosts than native passerines. So far they have been ignored by researchers both in the field and in the laboratory. Unlike the rabbit, sparrows and starlings have not succeeded in crossing the Nullarbor Plain to W.A.

Native psittacines are common to all known MVE/KUN habitats, usually in very large flocks. Some of the larger species (sulfur-crested cockatoos, galahs, corellas) have been shown to have moderate antibody prevalence in the field and a viremia response in the laboratory. Their life expectancy is very much longer than even the waterbirds', and if they produce a viremia only once in a lifetime their contribution to arbovirus ecology would be minimal. They are highly destructive of grain crops and even of timber in houses. They are one of the few groups of native bird species for which protection laws might be moderated in the near future, but it is extremely unlikely that control of any such selected species will have an effect on the incidence of MVE/KUN.

C. Environmental Modification

As mentioned in the previous section, some vertebrate host populations might be incidentally reduced in size by habitat destruction carried out for purposes other than MVE/KUN control. More positively, recommended mosquito abatement practices in Australia emphasize integrated programs including judicious environmental modification by permanently "engineering out" problem mosquito-breeding areas. Frequently this involves no more than the attempted eradication of problems created by earlier environmental modification.

Although the major rivers are now controlled, this control can be lost during times of abnormal climatic conditions such as in 1973—1974. River redgum forests, wetland habitats, and even some grazing land are dependent on periodic flooding, and complete eradication of floods without provision for controlled inundation of such areas would be highly undesirable. *Cx. annulirostris* seems extremely sensitive to drying out of breeding grounds; the reduction in the size of the adult population is more rapid than can be accounted for by lack of replacement by breeding. A modification to the environment which would allow rapid drainage of flood waters during an encephalitis epidemic, even if only temporary, could be more effective than application of insecticides.[61,79]

D. Epidemiological Consideration of Use of Vaccines

There has been considerable debate in appropriate public health committees about the prudence of developing a vaccine to protect populations at risk of infection with MVE and KUN, and the usual soundly based pro and con arguments have been raked over. During these deliberations, the use of commercial, inactivated, mouse-brain Japanese encephalitis vaccine was incidentally and gradually sanctioned for Australians taking up temporary residence in Japanese encephalitis endemic areas, and the vaccine was also found effective in mice against challenge with MVE or KUN.[61] However, emergency use of this vaccine in the face of an epidemic of MVE/KUN is thought imprudent. The present consensus is that development and use of a specific MVE/KUN vaccine is not warranted, but current development of novel antigen delivery systems should be monitored, and when appropriate, incorporation of either broadly based flavivirus reactive or MVE/KUN-specific antigens should be considered.

There was a small number of cases among the older patients of the 1974 epidemic where it seemed possible that illness was due to a second infection, probably with the heterologous member of the MVE-KUN pair (Section IV.D).[9] This not only raises the possibility that infection with the arboviral encephalitides during waning immunity might produce a high incidence of clinical response, but also brings into question the susceptibility of the vaccinee, whose immune status, unless repeatedly boosted by vaccination, might reach the predisposing conditions for enhanced clinical response to infection at a much earlier age than those whose immunity is the result of natural infection.

As with the other arbovirus encephalitides, a significant epidemic cycle involving humans appears to be a remote possibility, so vaccination will, at best, protect the individual from overt clinical illness, but will have no effect on the persistence of the virus in the environment.

X. FUTURE RESEARCH

A. Field Studies

Field research directed only at the elucidation of the ecology of MVE and KUN viruses between epidemics cannot be justified, but should be continued as a component of the ecology of Australian mosquito-borne arboviruses in general, and Ross River virus in particular. Field evidence of the involvement of mosquito species in transovarial transmission can be concurrently sought for more than one virus.

The sentinel chicken scheme is directed at flavivirus surveillance, and any evidence of virus activity from this source should be rapidly communicated to research groups. Both surveillance and research groups should be actively aware of the predisposing rainfall patterns postulated by Forbes;[8] if decile 7 rainfall is reached in the defined catchment areas in spring and summer, observations in that year and particularly in the following year should be intensified. Such observations should be as comprehensive and intense as feasible, for it is hard ecological data during pre-epidemic periods that are completely lacking, and on which reductionist approaches, prediction formulas, and modeling should be based.

When another epidemic does occur, the juxtaposition of population densities of *Aedes* species and *Cx. annulirostris* should be evaluated; the role of *Cx. quinquefasciatus* should be investigated by extending mosquito collection to river towns; the serological/virological survey of vertebrate hosts should be extended more comprehensively to passerine birds and to whatever mammals are in the epidemic environment; mosquitoes should be trapped on transects through water-bird breeding rookeries and from the rivers out onto the normally arid plains; and attempts should be made to find feeding grounds of nankeen night herons, their behavior observed, and mosquitoes sampled in the vicinity. Research groups must also collaborate in mosquito control activities. Assessing effectiveness of control measures can often be combined with virus/vector research.

The interepidemic, prospective serological survey of Hawkes et al.[52] should be continued for as long as feasible, or at least until it is reasonably certain that MVE/KUN virus activity is or is not indigenous to the Murray-Darling basin. During this survey, and more particularly in W.A., and during epidemics, syndromes other than encephalitis caused specifically by MVE, KUN, or other flaviviruses should be sought and fully evaluated and documented. Further evidence should be sought for the influence of "original antigenic sin" and waning immunity on the incidence of clinical disease in older age groups.

B. Experimental Studies

The pathogenesis of the arbovirus encephalitides is not well understood, particularly the reasons for the low morbidity but high case mortality/sequelae rates. The peripherally infected young adult mouse is probably a satisfactory model for the human disease, and now that reproducible in vitro assays of flavivirus-activated lymphocytes are becoming available, the role of the various classes of T cells in the pathogenesis of the disease, virus clearance, and immunity can be explored more precisely.

KUN and MVE viruses have been used in investigations of the replication strategies, biochemistry, and molecular biology of flaviviruses.[98,99] As with the T cell responses, technical difficulties have retarded studies of their organization and replication at the molecular level. Westaway and his colleagues used principally KUN virus in characterizing genomic RNA, structural and some nonstructural proteins. Dalgarno et al.[99] have resolved the nucleotide sequence of about half the genome of MVE virus, and this opens up the prospect of constructing cDNA clones which could be useful in pathogenesis studies and in exploring the potential of novel vaccines, perhaps vectored by vaccinia virus, or a multivalent arbovirus vaccine based on and vectored by 17D yellow fever virus.[100] The urgent application of this technology should be with Japanese encephalitis where a low-cost, readily delivered vaccine is immediately required, but it probably matters little whether exploration is carried out with Japanese encephalitis or MVE virus. Although at present a low priority, eventually a multivalent vaccine including MVE/KUN, Ross River, and possible dengue viruses seems a logical package for use in Australia.

If transovarial transmission of MVE or KUN virus occurs in *Cx. annulirostris* in the field, it does not seem to effectively contribute to the maintenance of the viruses. Few of the candidate *Aedes* species in Australia are tree-hole or receptacle breeders, so demonstration of transovarial transmission in the field will be difficult, although ultimately essential before this persistence mechanism can be accepted as proven. Laboratory screening of carefully selected candidate species for competence in the transovarial transmission of MVE and KUN viruses might indicate which species will be worth studying in the field.

REFERENCES

1. **Anderson, S. G.,** Murray Valley encephalitis and Australian X disease, *J. Hyg.,* 52, 447, 1954.
2. **Miles, J. A. R., Fowler, M. C., and Howes, D. W.,** Isolation of a virus from encephalitis in South Australia: a preliminary report, *Med. J. Aust.,* 1, 799, 1951.
3. **Garven, A. K., Margolis, J., and French, E. L.,** A fatal case of Murray Valley encephalitis occurring at Narrabri in New South Wales, *Med. J. Aust.,* 2, 621, 1952.
4. **Fenner, F. and Ratcliffe, F. N.,** *Myxomatosis,* Cambridge University Press, Cambridge, 1965, 281.
5. **Anderson, S. G., Dobrotworsky, N. V., and Stevenson, W. J.,** Murray Valley encephalitis in the Murray Valley, 1956 and 1957, *Med. J. Aust.,* 2, 15, 1958.
6. **Doherty, R. L., Carley, J. G., Cremer, M. R., Rendle-Short, J., Hopkins, I., Herbert, D. H., Caro, A. S., and Stephens, W. B.,** Murray Valley encephalitis in eastern Australia, 1971, *Med. J. Aust.,* 2, 1170, 1972.
7. **Gard, G. P., Marshall, I. D., Walker, K. H., Acland, H. M., and De Sarem, W. G.,** Association of Australian arboviruses with nervous disease in horses, *Aust. Vet. J.,* 53, 61, 1977.
8. **Forbes, J. A.,** *Murray Valley Encephalitis 1974, Also the Epidemic Variance Since 1914 and Predisposing Rainfall Patterns,* Australasian Medical Publishing, Sydney, 1978.
9. **Doherty, R. L., Carley, J. G., Filippich, C., White, J., and Gust, I. D.,** Murray Valley encephalitis in Australia, 1974: antibody response in cases and community, *Aust. N.Z. J. Med.,* 6, 446, 1976.
10. **Muller, D., McDonald, M., Stallman, N., and King, J.,** Kunjin virus encephalomyelitis, *Med. J. Aust.,* 144, 41, 1986.
11. **Stanley, N. F.,** Human arbovirus infections in Australia, in *Proc. 3rd Symp. Arbovirus Research in Australia,* St. George, T. D. and Kay, B. H., Eds., Commonwealth Scientific and Industrial Research Organisation and Queensland Institute for Medical Research, Brisbane, 1982, 216.
12. **French, E. L.,** Murray Valley encephalitis: isolation and characterization of the aetiological agent, *Med. J. Aust.,* 1, 100, 1952.
13. **Anderson, S. G. and Eagle, M.,** Murray Valley encephalitis: the contrasting epidemiological picture in 1951 and 1952, *Med. J. Aust.,* 1, 478, 1953.
14. **Reeves, W. C., French, E. L., Marks, E. N., and Kent, N. E.,** Murray Valley encaphalitis: a survey of suspected mosquito vectors, *Am. J. Trop. Med. Hyg.,* 3, 147, 1953.
15. **McLean, D. M.,** Transmission of Murray Valley encephalitis virus by mosquitoes, *Aust. J. Exp. Biol. Med. Sci.,* 31, 481, 1953.
16. **McLean, D. M.,** Vectors of Murray Valley encephalitis, *J. Infect. Dis.,* 100, 223, 1957.
17. **Miles, J. A. R. and Howes, D. W.,** Observations on virus encephalitis in South Australia, *Med. J. Aust.,* 1, 7, 1953.
18. **Doherty, R. L., Carley, J. G., Mackerras, M. J., and Marks, E. N.,** Studies of arthropod-borne virus infections in Queensland. III. Isolation and characterization of virus strains from wild-caught mosquitoes in north Queensland, *Aust. J. Exp. Biol. Med. Sci.,* 4, 17, 1963.
19. **Marshall, I. D., Thibos, E., and Clarke, K.,** Species composition of mosquitoes collected in the Murray Valley of south-eastern Australia during an epidemic of arboviral encephalitis, *Aust. J. Exp. Biol. Med. Sci.,* 60, 447, 1982.
20. **Marshall, I. D., Woodroofe, G. M., and Hirsch, S.,** Viruses recovered from mosquitoes and wildlife serum collected in the Murray Valley of south-eastern Australia, February 1974, during an epidemic of encephalitis, *Aust. J. Exp. Biol. Med. Sci.,* 60, 457, 1982.
21. **Doherty, R. L., Carley, J. G., Kay, B. H., Filippich, C., and Marks, E. N.,** Murray Valley encephalitis virus infection in mosquitoes and domestic fowls in Queensland, 1974, *Aust. J. Exp. Biol. Med. Sci.,* 54, 237, 1976.
22. **Westaway, E. G.,** The neutralization of arboviruses. II. Neutralization in heterologous virus-serum mixtures with four group B arboviruses, *Virology,* 26, 528, 1965.
23. **Westaway, E. G.,** Assessment and application of a cell line from pig kidney for plaque assay and neutralization tests with 12 group B arboviruses, *Am. J. Epidemiol.,* 84, 439, 1966.
24. **De Madrid, A. T. and Porterfield, J. S.,** The flavi-viruses (Group B arboviruses): a cross-neutralization study, *J. Gen. Virol.,* 23, 91, 1974.
25. **Barrett, E. J.,** Intratypic antigenic variation among strains of the group B arboviruses Murray Valley encephalitis and Kunjin, *Am. J. Epidemiol.,* 93, 212, 1971.
26. **Boyle, D. B.,** Comparative Studies of Two Related Australian Flaviviruses, Murray Valley Encephalitis and Kunjin, Ph.D. thesis, Australian National University, Canberra, 1979.
27. **Faragher, S. G., Hutchison, C. A., and Dalgarno, L.,** Analysis of Ross River virus genomic RNA using HAE III digests of single-stranded cDNA to infected-cell RNA and virion RNA, *Virology,* 141, 248, 1985.
28. **Faragher, S. G., Marshall, I. D., and Dalgarno, L.,** Ross River virus genetic variants in Australia and the Pacific islands, *Aust. J. Exp. Biol. Med. Sci.,* 63, 473, 1985.

29. **Lobigs, M., Weir, R. C., and Dalgarno, L.,** Genetic analysis of Kunjin virus isolates using *HAE* III and *TAQ* I restriction digests of single-stranded cDNA to virion RNA, *Aust. J. Exp. Biol. Med. Sci.,* 64, 185, 1986.
30. **Lobigs, M., Weir, R. C., and Dalgarno, L.,** personal communication, 1986.
31. **Trent, D. W., Grant, J. A., and Vorndam, A. V., and Monath, T. P.,** Genetic heterogeneity among St. Louis encephalitis virus isolates of different geographic origin, *Virology,* 114, 319, 1981.
32. **Lehman, N. I., Gust, I. D., and Doherty, R.,** Isolation of Murray Valley encephalitis virus from the brains of three patients with encephalitis, *Med. J. Aust.,* 2, 450, 1976.
33. **Robertson, E. G. and McLorinan, H.,** Murray Valley encephalitis: clinical aspects, *Med. J. Aust.,* 1, 103, 1952.
34. **Bennett, N. McK.,** Murray Valley encephalitis, 1974. Clinical features, *Med. J. Aust.,* 2, 446, 1976.
35. **Robertson, E. G.,** Murray Valley encephalitis: pathological aspects, *Med. J. Aust.,* 1, 107, 1952.
36. **Cook, I., Allan, B. C., Horsfall, W. R., and Flanagan, J. E.,** A fatal case of Murray Valley encephalitis, *Med. J. Aust.,* 1, 1110, 1970.
37. **Allan, B. C., Doherty, R. L., and Whitehead, R. H.,** Laboratory infections with arboviruses including reports of two infections with Kunjin virus, *Med. J. Aust.,* 2, 844, 1966.
38. **Badman, R. T., Campbell, J., and Aldred, J.,** Arbovirus infection in horses — Victoria 1984, *Commun. Dis. Intelligence,* 84/17, 5, 1984.
39. **Wiemers, M. A. and Stallman, N. D.,** Immunoglobulin M and Murray Valley encephalitis, *Pathology,* 7, 187, 1975.
40. **Kanamitsu, M., Taniguchi, K., Urasawa, S., Ogata, T., Wada, Y., Wada, Y., and Saroso, J. S.,** Geographic distribution of arbovirus antibodies in indigenous human populations in the Indo-Australian Archipelago, *Am. J. Trop. Med. Hyg.,* 28, 351, 1979.
41. **Bowen, E. T. W., Simpson, D. I. H., Platt, G. S., Way, H. J., Smith, C. E. G., Ching, C. Y., and Casals, J.,** Arbovirus infections in Sarawak: the isolation of Kunjin virus from mosquitoes of the *Culex pseudovishnui* group, *Ann. Trop. Med. Parasitol.,* 64, 263, 1970.
42. **Hawkes, R. A.,** Field and Laboratory Studies on the Serology of Arboviruses, Ph.D. thesis, Australian National University, Canberra, 1964.
43. **Marshall, I. D.,** Epidemiology of Murray Valley encephalitis in eastern Australia — patterns of arbovirus activity and strategies of arbovirus survival, in *Proc. 2nd Symp. Arbovirus Research in Australia,* St. George, T. D. and Kay, B. H., Eds., Commonwealth Scientific and Industrial Research Organisation and Queensland Institute for Medical Research, Brisbane, 1979, 47.
44. **McLean, D. M. and Stevenson, W. J.,** Serological studies on the relationship between Australian X disease and the virus of Murray Valley encephalitis, *Med. J. Aust.,* 1, 636, 1954.
45. **Anderson, S. G., Donnelley, M., Stevenson, W. J., Caldwell, N. J., and Eagle, M.,** Murray Valley encephalitis: surveys of human and animal sera, *Med. J. Aust.,* 1, 110, 1952.
46. **Doherty, R. L.,** Surveys of haemagglutination-inhibiting antibody to arboviruses in aborigines and other population groups in northern and eastern Australia, 1966-1971, *Trans. R. Soc. Trop. Med. Hyg.,* 67, 197, 1973.
47. **Doherty, R. L., Whitehead, R. H., Wetters, E. J., and Gorman, B. M.,** Studies of the epidemiology of arthropod-borne virus infections at Mitchell River Mission, Cape York Peninsula, north Queensland. II. Arbovirus infections of mosquitoes, man and domestic fowls, 1963-1966, *Trans. R. Soc. Trop. Med. Hyg.,* 62, 430, 1968.
48. **Anderson, S. G., Price, A. V. G., Nanadai-Koia, A., and Slater, K.,** Murray Valley encephalitis in Papua and New Guinea. II. Serological survey, 1956-1957, *Med. J. Aust.,* 2, 410, 1960.
49. **Wisseman, C. L., Gajdusek, D. C., Schofield, F. D., and Rosenzweig, E. C.,** Arthropod-borne virus infections of aborigines indigenous to Australia, *Bull. WHO,* 30, 211, 1964.
50. **Stanley, N. F. and Choo, S. B.,** Serological epidemiology of arboviruses in Western Australia, *Med. J. Aust.,* 2, 781, 1961.
51. **Stanley, N. F. and Choo, S. B.,** Studies of arboviruses in Western Australia. Serological epidemiology, *Bull. WHO,* 30, 221, 1964.
52. **Hawkes, R. A., Boughton, C. R., Naim, H. M., Wild, J., and Chapman, B.,** Arbovirus infections of humans in New South Wales. Seroepidemiology of the flavivirus group of togaviruses, *Med. J. Aust.,* 143, 555, 1985.
53. **Liehne, P. F. S., Anderson, S., Stanley, N. F., Liehne, C. G., Wright, A. E., Chan, K. H., Leivers, S., Britton, D. K., and Hamilton, N. P.,** Isolation of Murray Valley encephalitis virus and other arboviruses in the Ord River valley 1972-1976, *Aust. J. Exp. Biol. Med. Sci.,* 59, 347, 1981.
54. **Whitehead, R. H., Doherty, R. L, Domrow, R., Standfast, H. A., and Wetters, E. J.,** Studies of the epidemiology of arthropod-borne virus infections at Mitchell River mission, Cape York Peninsula, north Queensland. III. Virus studies of wild birds, 1964-1967, *Trans. R. Soc. Trop. Med. Hyg.,* 62, 439, 1968.
55. **Campbell, J. and Hore, D. E.,** Isolation of Murray Valley encephalitis virus from sentinel chickens, *Aust. Vet. J.,* 51, 1, 1975.

56. **Gard, G. P., Giles, J. R., Dwyer-Gray, R. J., and Woodroofe, G. M.**, Serological evidence of inter-epidemic infection of feral pigs in New South Wales with Murray Valley encephalitis virus, *Aust. J. Exp. Biol. Med. Sci.*, 54, 297, 1976.

57. **Anderson, S. G.**, Murray Valley encephalitis: a survey of avian sera, 1951-1952, *Med. J. Aust.*, 1, 573, 1953.

58. **Liehne, C. G., Stanley, N. F., Alpers, M. P., Paul, S., Liehne, P. F. S., and Chan, K. H.**, Ord River arboviruses — serological epidemiology, *Aust. J. Exp. Biol. Med. Sci.*, 54, 504, 1976.

59. **Marshall, I. D., Brown, B. K., Keith, K., Gard, G. P., and Thibos, E.**, Variation in arbovirus infection rates in species of birds sampled in a serological survey during an encephalitis epidemic in the Murray Valley of south-eastern Australia, February 1974, *Aust. J. Exp. Biol. Med. Sci.*, 60, 471, 1982.

60. **Braithwaite, L. W. and Clayton, M.**, Breeding of the nankeen night heron *Nycticorax caledonicus* while in juvenile plumage, *Ibis*, 118, 584, 1975.

61. **Marshall, I. D.**, unpublished results.

62. **Rozeboom, L. E. and McLean, D. M.**, Transmission of the virus of Murray Valley encephalitis by *Cx. tarsalis* Coquillett, *Ae. polynesiensis* Marks and *Ae. pseudoscutellaris* Theobald, *Am. J. Hyg.*, 63, 136, 1956.

63. **Altman, R. M.**, The behaviour of Murray Valley encephalitis virus in *Culex tritaeniorrhynchus* Giles and *Culex pipiens quinquefasciatus* Say, *Am. J. Trop. Med. Hyg.*, 12, 425, 1963.

64. **Carley, J. G., Standfast, H. A., and Kay, B. H.**, Multiplication of viruses isolated from arthropods and vertebrates in Australia in experimentally infected mosquitoes, *J. Med. Entomol.*, 10, 244, 1973.

65. **Kay, B. H., Fanning, I. D., and Carley, J. G.**, Vector competence of *Culex pipiens quinquefasciatus* for Murray Valley encephalitis, Kunjin, and Ross River viruses from Australia, *Am. J. Trop. Med. Hyg.*, 31, 844, 1982.

66. **Doherty, R. L., Carley, J. G., Kay, B. H., Filippich, C., Marks, E. N., and Frazier, C. L.**, Isolation of virus strains from mosquitoes collected in Queensland, 1972-1976, *Aust. J. Exp. Biol. Med. Sci.*, 57, 509, 1979.

67. **Kay, B. H., Carley, J. G., Fanning, I. D., and Filippich, C.**, Quantitative studies of the vector competence of *Aedes aegypti, Culex annulirostris* and other mosquitoes (Diptera:Culicidae) with Murray Valley encephalitis and other Queensland arboviruses, *J. Med. Entomol.*, 16, 59, 1979.

68. **Kay, B. H., Fanning, I. D., and Carley, J. G.**, The vector competence of Australian *Culex annulirostris* with Murray Valley encephalitis and Kunjin viruses, *Aust. J. Exp. Biol. Med. Sci.*, 62, 641, 1984.

69. **McLean, D. M.**, The behaviour of Murray Valley encephalitis virus in young chickens, *Aust. J. Exp. Biol. Med. Sci.*, 31, 491, 1953.

70. **Maguire, T. and Miles, J. A. R.**, The persistence of arboviruses in domestic chickens, *Arch. Gesamte Virusforsch.*, 15, 441, 1965.

71. **Boyle, D. B., Dickerman, R. W., and Marshall, I. D.**, Primary viraemia responses of herons to experimental infection with Murray Valley encephalitis, Kunjin and Japanese encephalitis viruses, *Aust. J. Exp. Biol. Med. Sci.*, 61, 655, 1983.

72. **Boyle, D. B., Marshall, I. D., and Dickerman, R. W.**, Primary antibody responses of herons to experimental infection with Murray Valley encephalitis and Kunjin viruses, *Aust. J. Exp. Biol. Med. Sci.*, 61, 665, 1983.

73. **Gresser, I., Hardy, J. L., Hu, S. M. K., and Scherer, W. F.**, Factors influencing transmission of Japanese B encephalitis virus by a colonized strain of *Culex tritaeniorrhynchus* Giles, from infected pigs and chicks to susceptible pigs and birds, *Am. J. Trop. Med. Hyg.*, 7, 365, 1958.

74. **McDonald, G., Smith, I. R., and Shelden, G. P.**, Laboratory rearing of *Culex annulirostris* Skuse (Diptera:Culicidae), *J. Aust. Entomol. Soc.*, 16, 353, 1977.

75. **Kay, B. H., Young, P. L., Hall, R. A., and Fanning, I. D.**, Experimental infection with Murray Valley encephalitis virus. Pigs, cattle, sheep, dogs, rabbits, macropods and chickens, *Aust. J. Exp. Biol. Med. Sci.*, 63, 109, 1985.

76. **Kay, B. H., Hall, R. A., Fanning, I. D., and Young, P. L.**, Experimental infection with Murray Valley encephalitis virus: galahs, sulphur-crested cockatoos, corellas, black ducks and wild mice, *Aust. J. Exp. Biol. Med. Sci.*, 63, 599, 1985.

77. **Doherty, R. L., Standfast, H. A., Domrow, R., Wetters, E. J., Whitehead, R. H., and Carley, J. G.**, Studies of the epidemiology of arthropod-borne virus infections of Mitchell River mission Cape York Peninsula, north Queensland, *Trans. R. Soc. Trop. Med. Hyg.*, 65, 504, 1971.

78. **McDonald, G.**, Population studies of *Culex annulirostris* Skuse and other mosquitoes (Diptera:Culicidae) at Mildura in the Murray Valley of southern Australia, *J. Aust. Entomol. Soc.*, 19, 37, 1980.

79. **Russell, R. C.**, Seasonal activity and abundance of the arbovirus vector *Culex annulirostris* Skuse near Echuca, Victoria, in the Murray Valley of southeastern Australia, 1979—1985, *Aust. J. Exp. Biol. Med. Sci.*, 64, 97, 1986.

80. **McDonald, G.,** Factors influencing the growth of mosquito populations and their significance to the transmission of Murray Valley encephalitis virus, in *Proc. 2nd Symp. Arbovirus Research in Australia,* St. George, T. D. and Kay, B. H., Eds., CSIRO and QIMR, Brisbane, 1979, 88.

81. **Reeves, W. C.,** Overwintering of arboviruses, *Prog. Med. Virol.,* 17, 193, 1974.

82. **Kay, B. H. and Carley, J. G.,** Transovarial transmission of Murray Valley encephalitis virus by *Aedes aegypti* (L.), *Aust. J. Exp. Biol. Med. Sci.,* 58, 501, 1980.

83. **Tesh, R. B.,** Experimental studies on the transovarial transmission of Kunjin and San Angelo viruses in mosquitoes, *Am. J. Trop. Med. Hyg.,* 29, 657, 1980.

84. **Kay, B. H.,** Seasonal abundance of *Culex annulirostris* and other mosquitoes at Kowanyama, north Queensland and Charleville, southwest Queensland, *Aust. J. Exp. Biol. Med. Sci.,* 57, 497, 1979.

85. **Standfast, H. A. and Barrow, G. J.,** Studies of the epidemiology of arthropod-borne virus infections at Mitchell River Mission, Cape York Peninsula, north Queensland. I. Mosquito collections, 1963—1966, *Trans. R. Soc. Trop. Med. Hyg.,* 62, 418, 1968.

86. **Liehne, P., Stanley, N., Alpers, M., and Liehne, C.,** Ord River arboviruses — the study site and mosquitoes, *Aust. J. Exp. Biol. Med. Sci.,* 54, 487, 1976.

87. **Kay, B. H.,** Towards prediction and surveillance of Murray Valley encephalitis activity in Australia, *Aust. J. Exp. Biol. Med. Sci.,* 58, 67, 1980.

88. **Kay, B. H.,** Age structure of populations of *Culex annulirostris* (Diptera:Culicidae) at Kowanyama and Charleville, Queensland, *J. Med. Entomol.,* 16, 309, 1979.

89. **Russell, R. C.,** Seasonal abundance and age composition of two populations of *Culex annulirostris* (Diptera:Culicidae) at Darwin, Northern Territory, Australia, *J. Med. Entomol.,* 23, 279, 1986.

90. **Russell, R. C.,** *Culex annulirostris* Skuse (Diptera:Culicidae) at Appin, N.S.W. — bionomics and behaviour, *J. Aust. Entomol. Soc.,* 25, 103, 1986.

91. **Lee, D. J., Clinton, K. J., and O'Gower, A. K.,** The blood sources of some Australian mosquitoes, *Aust. J. Biol. Sci.,* 7, 282, 1954.

92. **Kay, B. H., Boreham, P. F. L., and Williams, G. M.,** Host preferences and feeding patterns of mosquitoes (Diptera:Culicidae) at Kowanyama, Cape York Peninsula, northern Queensland, *Bull. Entomol. Res.,* 69, 441, 1979.

93. **Russell, R. C.,** The efficiency of various collection techniques for sampling *Culex annulirostris* in southeastern Australia, *J. Am. Mosquito Control Assoc.,* 1, 502, 1985.

94. **Edman, J. D., Webber, L. A., and Kale, H. W.,** Effect of mosquito density on the interrelationship of host behaviour and mosquito feeding success, *Am. J. Trop. Med. Hyg.,* 21, 487, 1972.

95. **Russell, R. C.,** Seasonal abundance of mosquitoes in a native forest of the Murray Valley of Victoria, 1979—1985, *J. Aust. Entomol. Soc.,* 25, 235, 1986.

96. **Sellers, R. F.,** Weather, host and vector — their interplay in the spread of insect-borne animal virus diseases, *J. Hyg.,* 85, 65, 1980.

97. **Russell, R. C.,** Dispersal of the arbovirus vector *Culex annulirostris* Skuse (Diptera:Culicidae) in the Murray Valley of Victoria, Australia, *Gen. Appl. Entomol.,* 18, 5, 1986.

98. **Westaway, E. G.,** Replication of flaviviruses, in *The Togaviruses,* Schlesinger, R. W., Ed., Academic Press, New York, 1980, 19.

99. **Dalgarno, L., Trent, D. W., Strauss, J. H., and Rice, C. M.,** Partial nucleotide sequence of the Murray Valley encephalitis virus genome. Comparison of the encoded polypeptides with yellow fever virus structural and nonstructural proteins, *J. Mol. Biol.,* 187, 309, 1986.

100. **Monath, T. P.,** Glad tidings from yellow fever research, *Science,* 229, 734, 1985.

Chapter 33

NAIROBI SHEEP DISEASE

F. Glyn Davies

TABLE OF CONTENTS

I. HISTORICAL BACKGROUND

A. Discovery of Agent and Vector

An apparently tick-borne gastroenteritis of sheep and goats was identified by a Kenya veterinary officer among trade sheep at the Nairobi quarantine area in 1910.[1] The animals had recently been moved from the drier pastoral areas to the north and southeast, and 2000 had died in 1 month while awaiting slaughter. A classical investigation of the disease was carried out by Dr. R. E. Montgomery,[2] who showed that it was caused by a filterable agent, transmitted transstadially and transovarially by the ixodid tick *Rhipicephalus appendiculatus*. This was the first demonstration of the transovarial transmission of a virus by an arthropod. He showed that the indigenous hair sheep were highly susceptible to the disease, and goats and imported wool sheep less so.[2] Subsequently, the disease has been recognized in Uganda,[3-5] Ruanda,[6] and Tanzania,[7] and in the highlands of western Somalia[8-10] and eastern Ethiopia.[10]

Disease outbreaks were frequently observed to follow the movement of sheep for trade or other purposes when mortality rates were very high. One farmer in Kenya, for example, lost 179 sheep out of 200 in 16 days. Daubney and Hudson[11] described outbreaks where the losses were of 600 animals a month, and similar mortality has been encountered on many subsequent occasions. These heavy losses have often been associated with apparent extensions in the range of the tick which have followed prolonged and heavy rainfall with vegetational and microclimatic changes favoring the tick vector.

B. Economic Significance

NSD (Nairobi sheep disease) is the most pathogenic virus disease known for sheep in East Africa and probably Somalia. Losses in the Nairobi quarantine areas to which NSD-susceptible sheep were brought approached 90%.[1,2] Such losses have regularly been encountered following the movement of nonimmune sheep populations into endemic areas. Similar mortality rates were reported from northern Somalia.[8,12]

Abortion is a further consequence of NSD infection in pregnant sheep, which adds to the consequential losses.[2] Shepherds in Kenya, Masailand, and Somalia lose from one half to two thirds of their flocks following NSD epizootics.[8,13] This has considerable social and economic significance.

II. THE VIRUS

A. Antigenic Relationships

NSDV (Nairobi sheep disease virus) is a member of the family Bunyaviridae and of the genus *Nairovirus*.[14] Many members of this genus are transmitted by ixodid ticks, and some transovarially. The Nairobi sheep disease group of this genus includes Ganjam and Dugbe viruses.[15]

NSDV appears to be serologically identical with Ganjam virus.[15] Ganjam virus has been isolated in India from *Haemaphysalis intermedia* and causes a clinical disease in sheep resembling NSD.[16] NSD also has serological relationships with members of the Congo-Crimean hemorrhagic fever group of the *Nairovirus* genus.[15,17]

B. Strain Variation

No antigenic differences have been detected by complement fixation (CF) or indirect fluorescent antibody tests between isolates of NSDV from different parts of Kenya, several Ugandan strains, and one from Tanzania. Cross-protection experiments with the many Kenyan strains have not revealed any which are immunologically distinct.[18]

C. Isolation and Assay of NSDV

The initial virus studies made by Montgomery[2] and by Daubney and Hudson[11,19] were made by sheep-to-sheep passage of the virus. The authors quickly identified the existence of endemic areas for NSDV where sheep were immune, and others where susceptible animals could be obtained for experiments. Sheep have been extensively used for the primary isolation of virus from field outbreaks. The observation of Weinbren et al.[20] that infant mice were susceptible to the virus provided a convenient laboratory animal host. The intracerebral inoculation of suckling mice is the most widely used method for virus isolation and assay, although mice are also susceptible when inoculated intraperitoneally. Adult mice are susceptible to field strains of NSDV when inoculated by the intracerebral route.

Cell cultures may be successfully used for the primary isolation of virus from sheep tissues.[21,22] The BHK-21 C13 cell line is especially suitable for this purpose, and the cytopathic effects may become evident 2 to 4 days after inoculation.[21,22] The viral antigen may be recognized within 12 to 24 hours by immunofluorescent staining.[22,23] Primary isolation may also be possible in lamb kidney and testis cells.[21] VERO cells have given variable results for primary isolation and appear to be less sensitive than BHK cells.[22] They may be used, however, to culture virus which has been isolated in suckling mouse brain. A pig kidney cell line, PS, has also been used for virus assay.[24]

Virus may be isolated from the plasma of infected animals during the febrile reaction or from spleen and mesenteric lymph nodes after death. Ticks may be triturated in a transport medium and centrifuged to provide a supernatant fluid for virus isolation. Thus far, only the intracerebral inoculation of infant mouse brain has been used for the isolation of NSDV from ixodid ticks.[20,25,26] Davies[13] has successfully used BHK cell cultures for primary isolation of the virus from laboratory-reared infected ticks. No tests were made of the comparative sensitivity of the system.

D. Host Range

Sheep are the principal disease hosts for NSDV, and goats, although susceptible, are generally less so. The East African hair sheep varieties are highly susceptible to tick-transmitted infection, and 75 to 95% mortality rates have regularly been observed in non-immune experimental and field populations.[1,2,5,6,9,11,26] The wool sheep which have been imported into Kenya, such as the Romney or Corriedale strains, are less susceptible both in the laboratory and field, with mortality rates of 30 to 40%.[2,26,27] Montgomery[2] reported a 10% mortality in 20 experimentally infected Angora goats; Terpstra[5] reported an 88% mortality in goats in West Nile Province of Uganda; Davies[13] has found a 10 to 40% mortality in the Kenyan goats of Masai type.

Cattle are insusceptible to NSDV, as are all other domesticated animals, including horses, donkeys, pigs, poultry, and dogs.[2] NSDV is lethal for 1- to 4-day-old Swiss white mice inoculated by the intracerebral (i.c.) or intraperitoneal (i.p.) routes.[20] Older mice remain susceptible to i.c. inoculations, and mouse brain-adapted virus is lethal by i.c. and i.p. routes for adult mice.[5,20] Rabbits, guinea pigs, rats, and hamsters were found to be refractory to i.p. and subcutaneous (s.c.) inoculation with NSDV.[2] The wild rodent *Arvicanthus niloticus* was shown to develop a viremia of 10^5 mouse $LD_{50}/m\ell$ after i.p. inoculation.

III. DISEASE ASSOCIATIONS

A. Human Disease

An apparently naturally acquired clinical case of NSDV infection in a human was reported from Entebbe.[28] A 16-year-old boy living near Lunyo complained of headache, shivering, and vague abdominal and back pains accompanied by a pyrexia. There were no gastrointestinal signs and the pyrexia subsided in 3 days. NSDV was isolated from serum taken early in the course of the disease, and paired serum samples showed a fourfold rise in CF antibody to NSD antigen.

No clinical cases of NSD have been identified among the staff at the Veterinary Research Laboratory at Kabete, many of whom have been regularly exposed to the virus for up to 20 years. The exposure has occurred at post-mortem examinations, handling infected sheep and mouse tissues with the dangers of aerosol formation during homogenization, and handling infected cell cultures. None of the staff has suffered any clinical disease which could be related to NSDV, nor have their sera ever shown any antibody to the virus when tested by the IFAT,[29] a test that has been used extensively at the laboratory for NSD serology.

At the Virus Research Institute, Entebbe, however, 23% of the 186 human sera collected in the Entebbe area were considered positive for NSD antibody by a serum neutralization (N) test. Eight of 18 staff members of the Institute also had neutralizing antibody to NSD.[30] The Entebbe area is heavily infested with *R. appendiculatus*. The mouse serum N test was used in these studies. Davies et al.[29] were unable to distinguish between pre- and post-infection sheep sera with N tests (SNT) assayed in infant mice or cell cultures. Other members of the *Nairovirus* group, such as Congo virus, have given equivocal results in the mouse N test.[31,32] Casals[31] reported that sera from many animals, including humans, may exhibit nonspecific antiviral effects in mice. Human infection has also been reported with Ganjam virus.[33]

There is strong evidence to support the conclusion that rare natural infections with NSDV can occur; however, the risk of laboratory infections would appear to be low.

B. Domestic Animals

The disease is similar in sheep and goats.[2,5,20] Initially there is a febrile reaction usually greater than 41°C at the peak, which is 2 to 3 days after onset. This may persist for a further 2 to 8 days. Febrile animals are depressed, disinclined to move, and stand with their heads down with a raised respiratory rate. Anorexia may be partial or complete, conjunctivas are injected or cyanosed, and there may be a serosanguineous nasal discharge. Diarrhea appears 1 to 3 days after the onset of fever, at first watery-green in color and later mucoid and blood tinged. Abdominal pain and tenesmus accompany this. Pregnant animals may abort during or after the febrile reaction.

Deaths occur at any stage after the onset of the fever, with most occurring in the first few days of the disease. In the early cases, little is evident at post-mortem examination other than a nonspecific congestion of most organs and tissues with petechial and ecchymotic hemorrhages. Lymph nodes are enlarged and edematous, the spleen is usually slightly enlarged with small hemorrhages beneath the capsule. No pathognomonic lesions are present, and in many cases there will be no suspicion of NSD.

Later, inflammation in the gastrointestinal tract may be seen, with hemorrhagic changes and ulceration in the abomasum, duodenum, cecum, and colon. These more obvious changes are not found regularly; only one of eight animals in a series of tick-transmission studies showed them. The remainder all died during the early febrile stages.

C. Wildlife

No clinical NSD has been identified in the abundant and varied wild ruminant fauna present in many NSD endemic areas. Extensive attempts at virus isolation from the blood of wild ruminants did not result in any isolations of NSD, and no antibody has been demonstrated in their sera.[34,35] No pathogenicity tests have been carried out in wild ruminants. Two blue duikers apparently died in a zoo in Uganda of NSD. The zoo deaths suggest that some species might be susceptible.[36]

D. Diagnostic Procedures

The virus can be isolated from blood during the febrile stage of the disease, but not usually after the temperature has returned to normal. Thereafter, and from dead animals, the best sources of virus are the spleen or mesenteric lymph nodes.[2,5,20,22]

Viral antigen may be detected by agar-gel precipitation tests[5] or the virus isolated in tissue cultures or infant mice.[20,21,23] Viral identification may be made by fluorescent antibody tests[22,23] for antigen in tissue cultures or mouse brain, or by CF or other serological tests.[22]

IV. EPIDEMIOLOGY

A. Geographic Distribution

Following the initial identification of NSD near Nairobi, it was recognized throughout much of the highland plateau of Kenya, and at the coast. The disease was clinically diagnosed at Entebbe in 1931,[34] and was subsequently found over much of the country, excluding the northern drier areas.[5] Bugyaki[6] described a tick-transmitted disease of viral etiology near Lake Kivu in Ruanda which was clinically identical with NSD. A very high mortality (90%) occurred in sheep moved from high altitude to an *R. appendiculatus* endemic area. This tick species was identified as the vector and the outbreak can be reasonably assumed to have been due to NSD.

Pellegrini[9] described clinical NSD in Somalia and made reference to an earlier record by Pavaglia in 1938. Edelsten[8] also encountered NSD in the western highlands of the Ogaden in Somalia. Sera IFAT-positive for NSDV were found in sheep suffering an NSD-like disease in the contiguous Harar Province of Ethiopia.[10]

A clinical diagnosis of NSD was made in an *R. appendiculatus* endemic area of northern Tanzania in 1968.[7,37] Virus was isolated from clinical cases in sheep in 1977,[38] and a serological survey for antibody to NSD showed a high prevalence in areas where *R. appendiculatus* was common.[39] Throughout Kenya, Tanzania, and Uganda, the Ecoclimatic Zones[40,41] II and III, together with parts of Zone IV, are generally enzootic for NSD (Figure 1). Epizootic NSD is seen in this latter zone after vegetational changes induced by periodic heavy rains.

Increased surveillance in India following the recognition of the NSD-Ganjam relationship, resulted in the recognition of sporadic cases of an NSD-like disease.[16] This was in exotic and exotic-cross sheep in Chittoor District, Andhra Pradesh. The clinical signs were similar to NSD, and Ganjam virus was isolated from sera taken at the acute stage of the disease. A limited serological survey by SNT in the area showed that all the indigenous local sheep had antibody to Ganjam virus; 64% of sheep exotic to the region were positive. NSD antibody is not found in the wild ruminant populations of East Africa, only in the disease hosts.[34] The wild ruminants do have antibody to bluetongue,[42] ephemeral fever,[43] and Akabane virus,[44] and they are apparently involved in these virus-maintenance cycles. NSD may not be an African virus, but may have been imported with sheep or goats in dhows which have traveled from India to East Africa for hundreds of years.

Weinbren[45] reported antibody to NSDV in sheep in Botswana and the Kalahari Desert, along the coastal plain in Mozambique, and in north Zululand. Davies[13] examined some 450 sheep sera from Botswana collected in areas infested with *R. appendiculatus* or the closely related *R. zambeziensis*. Antibodies at low titer were detected by IFAT,[10-20] but none were positive at 100*. It was considered likely that these were cross-reactions with antibody to other members of the Nairovirus group. True positive sera often have titers of 2500.[29] Some 300 sheep sera from near Lusaka in Zambia were also tested, and all were negative. The existence of NSDV south of Tanzania must remain questionable on the evidence thus far available.

---

* Reciprocal of the serum titer.

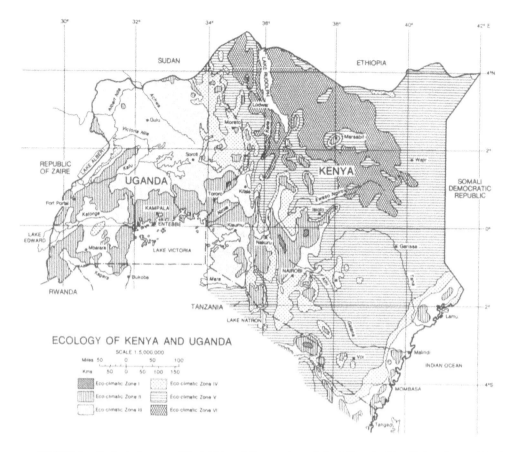

FIGURE 1.   Ecological zones of East Africa. (MI = moisture index.) *Zone I:* Afro Alpine climate. Moorland and grassland at high altitude above the forest lines. *Zone II:* Equatorial climate. Humid to subhumid (MI not less than − 10.) Forests and derived grasslands with or without natural glades. *Zone III:* Dry subhumid to semiarid (MI − 10 to − 30). Land not of forest potential with variable vegetation cover (moist woodland, bushland, or savannah). The trees are typically broad-leaved (e.g., Combretum) and the larger shrubs are mostly evergreen. *Zone IV:* Semiarid (MI − 30 to − 42). Marginal dry woodland or savannah of acacia, Themada type, or derived semievergreen or deciduous bushland. *Zone V:* Arid (MI − 42 to − 51). Very dry Commiphora or acacia, bushed grasslands with other shrubby vegetational types. *Zone VI:* Very arid (MI − 51 to − 57). Dwarf shrub grassland or very dry bushed grassland.

B. Incidence

1. Enzootic Areas in East Africa

Little clinical disease is encountered throughout those parts of Kenya, Tanzania, and Uganda where *R. appendiculatus* is common, and NSDV antibody is found in the small ruminant populations.[34,39]

Young animals rapidly became immune,[34] probably after challenge by infected ticks while they are protected by maternal antibody, as happens with louping-ill.[46] High antibody prevalence to NSD was found in the Taita-Taveta area of Kenya where the disease had not been reported for more than 50 years.[2,34] Many thousands of sheep had died when they were moved from nonendemic areas to feed troops stationed there in 1914.[2]

Clinical disease follows the movement of susceptible hosts into the endemic areas. Attempts at tick control in small ruminants within the endemic area are invariably accompanied by some mortality as the level of control falls below that necessary to ensure complete elimination.[25] The maintenance of a nonimmune population in NSD endemic areas is difficult to accomplish; tick control is expensive and not recommended.

2. Epizootic NSD in East Africa

Lewis[47] recorded the opinion of the Masai that NSD occurred in epizootics "every 7 years", killed most of the sheep, and left the remainder immune. They recognized other areas where the disease was present, together with East Coast fever, and avoided grazing them. Lewis could not support this report with personal experience of such an epizootic.

Davies[25] and Davies and Mwakima[26] saw epizootics of NSD in ecological zone IV where the small or nonexistent populations of *R. appendiculatus* expanded after heavy and prolonged rains. These rains altered the microclimate in favor of this tick species. After a period of 4 to 6 months, the populations had increased dramatically in numbers, either from small local nidi where the tick had maintained itself, or when ticks dropped off animals from the contiguous *R. appendiculatus* endemic areas. The vegetation changes over the period 1977 to 1979 allowed large populations of ticks to become established where previously they had been scarce. Several hundred could be found on a single animal, in areas where previously few *R. appendiculatus* were present on cattle. Epizootics of NSD followed in the totally susceptible animals. East Coast fever also appeared in cattle in these areas. The observation made by the nomadic, pastoral Masai appears to be correct: epizootic NSD does occur, not necessarily at 7-year intervals, but whenever the periodic heavy and prolonged rainfall temporarily alters the vegetational characteristics in areas otherwise marginal for the vector tick.

3. NSD in Somalia

Edelsten[8] described an epizootic of a tick-borne disease in Somalia which was probably NSD; the virus was not positively identified. This outbreak persisted for 2 to 3 years, with morbidity rates in different flocks of 30 to 70% and a case-fatality rate of 50 to 80%. The figures suggest that totally susceptible flocks of sheep were introduced into areas where the virus was endemic in tick populations. Hartley[58] has suggested that the susceptible sheep population was indirectly produced by the preceding drought conditions. The drought reduced the tick populations virtually to extinction and the animals moved several hundred miles away from the endemic areas in search of grazing. These animals may thus have remained away from the infected areas for 2 years or more, and returned when heavy rains once more produced good grazing. Circumstantial evidence supports a view that *R. pulchellus* was the tick vector.

C. Genetic Predisposition

Clinical observations of the losses caused by NSD show that there are distinct variations in the susceptibility of different breeds and strains of sheep and goats to the disease (see Section II.D). A limited amount of data in support of this has been acquired from laboratory experiments. No experiments have been carried out to study the susceptibility of various strains of laboratory mice to NSD.

D. Seasonal Distribution

In some parts of East Africa, as at Iringa, South Tanzania, *R. appendiculatus* follows a strictly seasonal breeding pattern.[48] NSD has not been studied in such areas. Elsewhere, there are population fluctuations associated with the seasonal rains, but the ticks are present throughout the year.

E. Serologic Epidemiology

Serological investigations of NSD epidemiology have been carried out in Kenya,[34] Tanzania,[39] and Uganda.[5] These studies have related NSD distribution with that of possible tick vectors. Davies[34] also determined the host range for NSD by screening wild ruminants and rodent species by IFAT for NSD antibody. His results showed that neither appeared to be involved in the natural history of NSD in Kenya. Some low-titer positives in giraffes and game animals were considered to be cross-reactions with other members of the *Nairovirus* group.[15] No rodent sera from enzootic areas were found to have antibody to NSD.

V. TRANSMISSION CYCLES

The classical study by Montgomery[2] of this disease clearly demonstrated that the virus could be transmitted transstadially and transovarially by the Kenyan race of the tick *R. appendiculatus*. These observations were confirmed by later workers at Kabete. The suggestion that feeding infected ticks upon immune sheep rendered them unable to transmit the virus at the next instar was made by Daubney and Hudson.[19] Experiments to reexamine this possibility were carried out by Davies and Mwakima.[26] Their results showed that feeding a vertically infected generation of *R. appendiculatus* ticks upon immune sheep or insusceptible hosts did not result in their sterilization, and they were able to transmit the virus to susceptible sheep at the next or later instars. Lewis[49] showed that vertically infected *R. appendiculatus* larvae were infective 245 days after hatching, nymphs were infective 348 days after moulting, and an adult tick transmitted NSD after 871 days of incubation. These studies confirm field observations of the persistence of NSDV in infected tick populations over very long periods of time.

Davies and Mwakima[26] showed that a vertically infected tick population of *R. appendiculatus* readily transmitted NSD to susceptible sheep despite feeding at earlier instars upon immune or insusceptible hosts. This is important, for the tick feeds principally upon cattle and to a much lesser extent upon sheep and goats.[50,51] Many wild ruminant species are also suitable feeding hosts, especially the waterbuck *(Kobus ellipsiprymnus)* and buffalo *(Syncerus caffer)*. None of these animals has been found seropositive to NSDV,[34] and it is presumed that viremias do not occur.[35] It is possible, however, on the basis of the virus survival studies of Lewis,[49] to understand how the virus may persist in a tick population when the disease hosts are unavailable for some lengthy periods. Rodents are not preferred feeding hosts for any stage of *R. appendiculatus*. Virus has not been isolated from organs of wild-caught specimens in endemic areas,[25] nor has specific antibody been demonstrated in rodent sera from such areas.[34] Rodents are not considered to be involved in a maintenance cycle for NSDV in Kenya, as suggested by Daubney and Hudson.[11] They described apparent outbreaks of NSD in parts of Kenya where *R. appendiculatus* was not commonly found. These areas were, however, in the zones into which this tick species extends and greatly increases in numbers when the vegetational characteristics are altered by prolonged rains. There are many such zones in Ecoclimatic Zone IV.[40] Such situations were also described by Davies and Mwakima.[25,26] Daubney and Hudson[11] suggested that *Amblyomma variegatum* may have been transmitting the virus during these outbreaks, and it is likely that some populations of *A. variegatum* can transmit NSDV. The distribution of this tick and that of *R. appendiculatus* in Kenya largely coincide.[34] The distribution of antibody to NSD in Kenya, Uganda, and Tanzania more closely correlates, however, with the distribution of *R. appendiculatus* than *A. variegatum*.[34,35,39]

Experimental studies with *A. variegatum* in Kenya showed that the tick was capable of transmitting NSDV transstadially, but was a far less efficient vector than *R. appendiculatus*. NSDV has been isolated from *A. variegatum* in the lake basin area of western Kenya, but not from the high-altitude, high-rainfall endemic areas.[25,52]

Experimental work presented by Pellegrini[9] and circumstantial evidence obtained by Edelsten[8] suggest that *R. pulchellus* may be the vector of NSD in Somalia. *R. appendiculatus* was not present in the areas where the disease occurred in epidemic form and the distribution of the disease was wider than that of the *Amblyomma* spp. in Somalia. The Kenyan races of *R. pulchellus* were reported to be unable to transmit NSD by Daubney and Hudson,[11] and were found to be unable to transmit the virus transstadially or transovarially by Davies and Mwakima.[59] The Somali populations of this tick can, however, do so. Lewis[53] had also suggested that it could on occasion transmit NSD. These observations suggest and support the view that different populations of the same tick species differ in their ability to transmit NSDV. Some may be highly effective vectors and others incapable of transmitting the virus. Daubney and Hudson[11] and Lewis[53] found that some populations of *R. appendiculatus* were inefficient transmitters of NSDV.

VI. ECOLOGICAL DYNAMICS

A. Macro- and Microenvironment

The physical determinants of the range of *R. appendiculatus* have been studied under laboratory[54] and field conditions.[55] Temperatures within the range 19 to 30°C are necessary for normal maturation of the different instars of the tick. No development occurs below 15°C although such temperatures are not lethal. A relative humidity of less than 30% is lethal, 30 to 45% unfavorable, and greater than this necessary for the normal development of the tick, particularly of the egg and larval stages which are most susceptible to desiccation. Such conditions are found throughout the endemic areas of East Africa and probably Ruanda and Zaire, although this is not true of some parts of southern Tanzania where the ticks prevail for only part of the year. A seasonal pattern of behavior occurs, and this is also true for most of the *R. appendiculatus* populations of countries farther south. No study has been made of the ecology of NSD in an area where the tick is only seasonally active.

R. appendiculatus is found in East Africa in the moist forest, derived or natural grasslands of Ecoclimatic Zone II,[40] and throughout most of Zone III[40] which is of drier bushed and wooded grasslands. The grasslands of acacia type in Zone IV[40] are generally too dry for this tick species, other than in the exceptional conditions when there are changes in the vegetational characteristics of the zone produced by prolonged and heavy rainfall. The tick is, however, found along the acacia-wooded river margins which are a feature of this zone.

A. variegatum shares a similar range of habitat with *R. appendiculatus*,[50] but it is able to survive in hotter and drier environments.[56] A seasonal breeding season commonly occurs in the drier range of the tick, where the larval stages develop during the rainy seasons when the humidity is higher. The more resistant nymphal and adult instars occur in the drier seasons, which they are better able to withstand. *R. pulchellus* is more of a dry-country tick than either of the above. Its range in Kenya extends over most of the country east of the Great Rift Valley.[50] Typically, its habitat is the dry, bushed, and wooded grasslands of Ecological Zones IV and V. It occurs in the semidesert of Zone VI, and also in the coast hinterland. The range of this tick does not correlate either with that of clinical disease[25] or antibody[34] to NSDV in Kenya or Tanzania.[39] Its range correlates with the range of the disease in Somalia,[8] and the tick is present in Harar Province of Ethiopia, where NSD antibody has been found in sheep.[27]

B. Climatic Effects

The influence of climate upon NSD is to determine the extensions and contractions in the range of the tick, which occur in concert with the periods of prolonged heavy rainfall in East Africa.[57] Such rainfall is seen in 5- to 15-year cycles, and marked changes are induced in the vegetation, particularly of Zone IV. These changes are favorable to *R. appendiculatus* and allow the range of the tick to extend. The climatic changes are a function of the characteristics of the Intertropical Convergence Zone (ITCZ)[57] which in periods of heavy rainfall also increase cloud cover and humidity, both of which favor development of the NSD vector tick. The changes may persist for 1 to 3 years before a predominantly dry period prevails again.

C. Epidemiological Significance of Environmental Changes

The steady state whereby an immune, small ruminant population exists in the presence of infected tick vectors is desirable for successful animal production in NSD endemic areas. This situation exists over most of the NSD areas in East Africa. A particular problem occurs where there are extensions in the range of the tick and new infected tick populations are established. This is seen in East Africa in Ecoclimatic Zone IV[41] following the periodic cycles of heavy rainfall in these normally semiarid rangelands.[57] Such cycles may be predicted by active surveillance of rainfall patterns and vegetational changes by satellite monitoring.[57] A knowledge of the distribution of Ecoclimatic Zone IV and the changes which are observed by satellite pictures may allow the accurate definition of extension areas. The lag period before the establishment of large, infected tick populations may give an opportunity for preventive measures to be instituted. These could be movement of animals out of the danger zones, or their vaccination. Together with the movement of nonimmunes into endemic areas, this constitutes the only situation in which it is envisaged that vaccination against NSD might be recommended.

D. Biting Activity, Tick Density, and Host Preferences

R. appendiculatus, originally a tick of wild Bovidae, has adapted well to domestic livestock and is now a major pest of cattle.[50,51] It also feeds upon sheep and goats and may be found when these animals are kept together with cattle. All stages of the tick will readily feed upon domestic ruminants,[50,51] although sheep and goats are hosts of secondary importance when compared with cattle. This tick is also found in large numbers upon buffalo and waterbuck, but lower infestations occur on other wild animals.[50] The density of *R. appendiculatus* in different endemic areas is related to numbers and availability of feeding hosts and to the level of suitability of the microenvironment. The tick populations may be very high indeed where humidity and temperature are optimal and there are many hosts, especially if no tick control if practiced. Lower populations of all instars, especially of larvae, will be found where conditions are less suitable and the hosts are resistant or less common. The host-resistance factor is of some importance in determining these tick populations. Naive hosts may carry many thousands of ticks, but following the acquisition of resistance (immunity), the numbers may decline to a few hundred on *Bos indicus* cattle or approach 1000 on *B. taurus. A. variegatum* has not been encountered as a possible vector of NSD in recent outbreaks in Kenya, and has very rarely been found on affected sheep. Sheep are not favored hosts for this tick species, unless the tick populations are high, for cattle are the principal and preferred hosts.[50,51]

The distribution of *R. pulchellus* in Kenya does not correlate with that of the disease or antibody to the virus.[25,34] This tick does feed readily upon sheep and goats, and these hosts are almost as favored as cattle.[50,51] Zebra and other wild animal species are fed upon extensively by this tick species, and in many areas it is known as the zebra tick.[50] There is evidence to support a conclusion that it is the vector of NDS in Somalia.[8,12] The populations of *R. pulchellus* in the ''haud'' of the western highlands of Somalia were found to increase 40-fold after rains.[12] This peak in population coincides with the heavy utilization of these pastures by sheep, goats, and cattle.

VII. INVESTIGATION OF EPIDEMICS

The investigation of endemics in Kenya has revealed that they are due to one of the following causes:[25]

1. Introduction of nonimmune populations into endemic areas
2. Breakdowns in attempted tick control

3. Extension of NSD into areas with nonimmune disease-host populations following ecological changes which favor the establishment of the tick vector
4. Introduction of infected tick populations into new receptive areas by movement of the disease or other hosts

The pattern of NSD in Somalia or Ethiopia is not known. The nomadic nature of the population suggests that nonimmune or partially immune flocks are moved into endemic areas for only a part of the year for grazing. This will give rise to losses in the younger animals which have not previously been exposed to the virus. Investigations are necessary to examine this possibility.

VIII. PREVENTION AND CONTROL

The control of the principal tick vector of NSD in East Africa, *R. appendiculatus*, has been widely practiced to protect cattle from infection with East Coast fever. Success has been achieved by the use of acaricides of different types, arsenicals, organochlorides, and organophosphorous compounds. Degrees of resistance to most acaricides have been observed, and the cost-effectiveness of the dipping or spraying procedures has questioned the validity of this policy. The toxic and growth rate depressing effects of acaricides and problems of residues in meat destined for human consumption are further disadvantages. Traditionally, small ruminants have not been dipped to any significant extent, and the practice is not to be recommended.

The indications for vector control exist only when a state of enzootic stability, with immune disease hosts in the presence of a regular infected-tick challenge, is altered. The introduction of totally susceptible disease hosts into such an environment may justify tick-control measures. The presence of a single infected tick on an animal is all that is required to initiate infection, so absolute tick control is rarely possible. Vector control procedures are not considered to have any significant, economic long-term application for the control of NSD.

IX. FUTURE RESEARCH

The relationship between NSDV and Ganjam virus requires further investigation. The viruses are identical serologically, and the report of a clinical disease from India which resembles NSD further supports the view that they may be the same virus. There is a need to carry out cross-pathogenicity and immunity experiments with the two viruses to confirm this relationship.

Basic research is required to investigate the biology of NSD virus and the tick vectors about which little is known. Some qualitative work on transmission has been carried out.[2,5,6,11,19,26,49,53] Quantitative studies of the replication of the virus in different tick populations, species, and genera are necessary to evaluate their vector potential. The inability of the Kenya strains of *R. pulchellus* to transmit NSDV must be compared with strains of this tick from Somalia which do. The role of *A. variegatum* in epizootics must be investigated. Certain populations of *R. appendiculatus* in Kenya were found to be less efficient vectors of NSDV than others. The determination of the infecting virus dose for different tick species and populations of diverse ecological and geographical origins is relevant to a greater understanding of the natural history of NSD.

A fascinating problem remains to determine the true range of NSD. No clinical disease has been reported south of Tanzania, and it is unlikely that such a highly pathogenic virus disease would have escaped detection for such a long time. There may be a biological barrier to effective transmission of NSDV. The populations of *R. appendiculatus* in southern Tanzania and those of Zambia, Mozambique, Botswana, and South Africa exhibit a marked annual seasonality. This may be the determining factor restricting the range of NSD.

REFERENCES

1. Kenya Veterinary Department Annual Report, 1910.
2. **Montgomery, R. E.,** On a tick-borne gastroenteritis of sheep and goats occurring in East Africa, *Br. J. Comp. Pathol., 30,* 28, 1917.
3. Uganda Veterinary Department Annual Report, 1931.
4. Uganda Veterinary Department Annual Report, 1933.
5. **Terpstra, C.,** Nairobi Sheep Disease: Studies on Virus Properties, Epizootiology and Vaccination in Uganda, Ph.D. thesis, University of Utrecht, Holland, 1969.
6. **Bugyaki, L.,** La maladie de Kisenyi du mouton, due a un virus filtrable et transmise par des tiques, *Bull. Agric. Congo Belge,* 46, 1455, 1969.
7. Uganda Veterinary Department Annual Report, 1969.
8. **Edelsten, R. M.,** The distribution and prevalence of Nairobi sheep disease and other tick-borne infections of sheep and goats in northern Somalia, *Trop. Anim. Health Prod.,* 7, 29, 1975.
9. **Pellegrini, D.,** La gastro-enterite emorragica delle pecore. Experimenti di transmissione con *Rhicephalus pulchellus, Boll. Soc. Ital. Med. Igiene Trop.,* 10, 164, 1950.
10. **Davies, F. G., Mungai, J. N., and Shaw, T.,** A Nairobi sheep disease vaccine, *Vet. Rec.,* 94, 128, 1974.
11. **Daubney, R. and Hudson, R. N.,** Nairobi disease: natural and experimental transmission by ticks other than *Rhipicephalus appendiculatus, Parasitology,* 26, 496, 1934.
12. **Pegrum, R. G.,** Ticks (Acarina, Ixodidea) of the northern regions of the Somali Democratic Republic, *Bull. Entomol. Res.,* 66, 345, 1976.
13. **Davies, F. G.,** unpublished data.
14. **Mathews, R. E. F.,** Classification and nomenclature of viruses, in *Intervirology: 4th Report of the International Committee on Taxonomy of Viruses,* Vol. 17, Melnick, J. L., Eds., S. Karger, Basel, 1982, 115.
15. **Davies, F. G., Casals, J., Jessett, D. M., and Ochieng, P.,** The serological relationships of Nairobi sheep disease, *J. Comp. Pathol.,* 88, 519, 1978.
16. **Ghalsasi, G. R., Rodrigues, F. M., Dandawate, N. P., Gupta, C. G., Khasnis, B. D., Pinto, B. D., George, S., and Pargaonkar, V. N.,** in Arthropod-Borne Virus, Information Exchange, 1979, 80.
17. **Casals, J. and Tignor, G. H.,** The *Nairovirus* genus: serological relationships, *Intervirology,* 14, 144, 1980.
18. **Davies, F. G. and Mungai, J. N.,** unpublished data.
19. **Daubney, R. and Hudson, J. R.,** Nairobi sheep disease, *Parasitology,* 203, 507, 1931.
20. **Weinbren, M. P., Gourlay, R. N., Lumsden, W. H. R., and Weinbren, B. M.,** An epizootic of Nairobi sheep disease in Uganda, *J. Comp. Pathol.,* 68, 174, 1958.
21. **Coakley, W. and Pini, A.,** The effect of Nairobi sheep disease upon tissue culture systems, *J. Pathol. Bacteriol.,* 90, 672, 1965.
22. **Davies, F. G., Mungai, J. N., and Taylor, M.,** The diagnosis of Nairobi sheep disease, *Trop. Anim. Health Prod.,* 9, 75, 1977.
23. **Ohder, H., Lund, L. J., and Whiteland, A. P.,** Observations on the growth and development of blue-tongue, Nairobi sheep disease and Rift Valley Fever viruses by fluorescent antibody technique and titration in a tissue culture system, *Arch. Ges. Virusforsch.,* 29, 127, 1970.
24. **David West, T. S. and Porterfield, J. S.,** Dugbe virus: a tick-borne Arbovirus from Nigeria, *J. Gen. Virol.,* 23, 297, 1974.
25. **Davies, F. G.,** Nairobi sheep disease in Kenya. The isolation of virus from sheep, goats, ticks and possible maintenance hosts, *J. Hyg.,* 81, 259, 1978.
26. **Davies, F. G. and Mwakima, F.,** Qualitative studies of the transmission of Nairobi sheep disease virus by *Rhicephalus appendiculatus* (Ixodidea, Ixodidae), *J. Comp. Pathol.,* 92, 15, 1982.
27. **Davies, F. G., Otieno, S., and Jessett, D. M.,** The antibody response in sheep vaccinated with experimental Nairobi disease vaccines, *Trop. Anim. Health Prod.,* 9, 181, 1977.
28. **Kirya, G. B., Tukei, P. M., Lule, M., and Mujomba, E.,** Nairobi sheep disease in man, East African Virus Research Institute Rep. 284,
29. **Davies, F. G., Jessett, D. M., and Otieno, S.,** The antibody response to Nairobi sheep disease virus, *J. Comp. Pathol.,* 86, 497, 1976.
30. Annual Reports, East African Virus Research Institute, Entebbe, 1957, 1958, and 1959.
31. **Casals, J.,** Crimean-Congo haemorrhagic fever, in *Proc. Colloq. Ebola Virus and Other Haemorrhagic Fevers,* Pattyn, S. R., Ed., Elsevier, Amsterdam, 1978, 301.
32. **Zavodava, L. L., Butenko, A. M., Tkachenko, E. A., and Chumakov, M. P.,** Properties of the neutralisation test in Crimean haemorrhagic fever, in *Viral Haemorrhagic Fevers,* Chumakov, M. P., Ed., Tr. Inst. Polio Virus *n.* Entsefolitov Akad. Med. Nauka, SSSR, 1971, 61.
33. **Dandawate, C. N., Work, T. H., Webb, J. K., and Shah, K. V.,** Isolation of Ganjam virus from a human case of febrile illness: a report of a laboratory infection and serological survey of human sera from three different States of India, *Indian J. Med. Res.,* 57, 975, 1969.

34. **Davies, F. G.,** A survey of Nairobi sheep disease antibody in sheep, goats and wild ruminants within Kenya, *J. Hyg.,* 81, 251, 1978.
35. **Davies, F. G.,** Attempted isolation of arboviruses from wild ruminants in Kenya, *Trop. Anim. Health Prod.,* 13, 116, 1981.
36. Annual Report East African Virus Research Institute, Entebbe, 1957.
37. **Groocock, C. M.,** personal communication.
38. **Davies, F. G. and Shakas, S.,** unpublished data.
39. **Jessett, D. M.,** Serological evidence of Nairobi sheep disease in Tanzania, *Trop. Anim. Health Prod.,* 10, 99, 1978.
40. **Pratt, D. J., Greenway, P. J., and Gwynne, M. D.,** A classification of East African range land with an appendix on terminology, *J. Appl. Ecol.,* 3, 369, 1966.
41. **Pratt, D. J. and Gwynne, M. D.,** *Rangeland Management and Ecology in East Africa,* Hodder & Stoughton, London, 1977, 310.
42. **Davies, F. G. and Walker, A. R.,** The distribution of bluetongue virus, antibody and the Culicoides vector in Kenya, *J. Hyg.,* 72, 265, 1974.
43. **Davies, F. G., Shaw, T., and Ochieng, P.,** Observations on the epidemiology of ephemeral fever in Kenya, *J. Hyg.,* 75, 231, 1975.
44. **Davies, F. G. and Jesset, D. M.,** A study of the host range and distribution of antibody to Akabane virus (genus *Bunyavirus* family Bunyaviridae) in Kenya, *J. Hyg.,* 95, 191, 1985.
45. **Weinbren, M. P.,** in *International Catalogue of Arboviruses,* Berge, T. O., Ed., Washington, D.C., 1975, 508.
46. **Reid, H. W. and Boyce, J. B.,** The effect of colostrum derived antibody on louping ill virus infection in lambs, *J. Hyg.,* 77, 349, 1976.
47. **Lewis, E. A.,** A Study of the Ticks in Kenya Colony. The Influence of Natural Conditions, and Other Factors on their Distribution and the Incidence of Tick-Borne Disease. Part III, Bull. No. 7, Government Printer, Nairobi, 1934.
48. **Tatchell, R. E.,** Technical Report of the Improvement of Tick Control in Tanzania, Rep. No. AG DP/URT/72/009, Food and Agriculture Organization, Rome, 1977.
49. **Lewis, E. A.,** Nairobi sheep disease: the survival of the virus in the tick *Rhipicephalus appendiculatus,* *Parasitology,* 37, 55, 1946.
50. **Walker, J. B.,** The Ixodid Ticks of Kenya. A Review of Present Knowledge of Their Hosts and Distribution, Commonwealth Institute of Entomology, London, 1967.
51. **Yeoman, G. H. and Walker, J. B.,** The Ixodid Ticks of Tanzania. A Study of the Zoogeography of the Ixodidae of an East African Country, Commonwealth Institute of Entomology, London, 1967.
52. **Johnson, B. K., Chanas, A. C., Squires, E. J., Shockley, P., and Simpson, D. M., Parsons, J., Smith, D. H., and Casals, J.,** Arbovirus isolations from ixodid ticks infesting livestock in the Kano plain, Kenya, *Trans. R. Soc. Trop. Med. Hyg.,* 74, 732, 1980.
53. **Lewis, E. A.,** in Kenya Veterinary Department Annual Report, Nairobi, 1949.
54. **Branagan, D.,** The development of the ixodid tick *Rhipicephalus appendiculatus* Neumann under laboratory conditions, *Bull. Entomol. Res.,* 63, 155, 1973.
55. **Branagan, D.,** Observations on the development and survival of the ixodid tick *Rhipicephalus appendiculatus* Neumann under quasi-natural conditions in Kenya, *Trop. Anim. Health Prod.,* 5, 153, 1973.
56. **McCleod, T.,** Tick infestation patterns in the southern Province of Zambia, *Bull. Entomol. Res.,* 60, 253, 1970.
57. **Davies, F. G., Linthicum, K. J., and James, A. D.,** Rainfall and Rift Valley fever, *Bull. WHO,* 63, 941, 1985.
58. **Hartley, M.,** personal communication.
59. **Davies, F. G. and Mwakima, F.,** unpublished data.

Chapter 34

OMSK HEMORRHAGIC FEVER

Dimitri K. Lvov

TABLE OF CONTENTS

I. HISTORICAL BACKGROUND

Study of the etiology of the disease, which first occurred in the spring of 1945—1946 in one of the West Siberian subregions, was carried out by M. P. Chumakov and colleagues in 1947 and 1948. The virus (Kubrin prototype strain) was first isolated from the acute-phase blood specimen of a patient with typical symptoms in the Sarghat region, Omsk district (latitude 156° north, longitude 73° east) on July 14, 1947. The strain, isolated by inoculation of white mice, proved to have antigenic and biologic characteristics close to those of tick-borne encephalitis (TBE) virus,[1-3] and electron microscopic examination subsequently confirmed this virus to be a member of the Flaviviridae.[4] In spite of the close antigenic relationship between the agent and TBE virus, Omsk hemorrhagic fever (OHF) appears to be a unique nosological entity with respect to its clinical features, epidemiology, and ecology.

OHF was initially diagnosed by the clinicians as an atypical form of typhus or generalized tularemia.[5] Based on the pecularities of its clinical, epidemiologic, and pathoanatomic characteristics,[6-12] however, it was classified as a unique disease by the end of 1946. Between 1945 and 1949, high morbidity was recorded in the endemic regions of West Siberia. In 1949, a drastic decline in incidence was noted, followed by the virtual disappearance of the disease.[13-15] In subsequent years, only sporadic, individual cases have been registered during autumn and winter among persons hunting for muskrats[16-18] and members of the hunters' families. Cases of laboratory infection have been described.[19-21]

II. THE VIRUS

A. Antigenic Relationships and Strain Variation

Omsk hemorrhagic fever virus is a member of the TBE complex. No difference can be determined between OHF and TBE using hyperimmune sera.[22-24] Nevertheless, these viruses may be differentiated by immune precipitation in agar with cross-adsorbed monospecific sera.[25-29] The existence of at least two antigenic varieties of the virus has been shown, represented by (1) the Kubrin prototype strain and another strain isolated from a patient, and (2) the Bogolubovka strain from *Dermacentor marginatus* and the Guriev strain from a human patient.[25,26] Comparative study of seven strains (two from muskrats, one from *D. pictus*,* two from *D. marginatus,* and two from humans, one of them being isolated from the blood of a patient with fever and another from the brain of a fatal case) revealed significant variability of the biological and antigenic characteristics.[30,31] These data have been confirmed by comparative study of 15 strains isolated from various sources.[32]

B. Host Range and Assay Methods

The virus is pathogenic for a wide range of laboratory and wild animals. Muskrats *(Ondatra zibethica)* are highly susceptible by all routes of infection, developing fatal hemorrhagic disease within 5 to 20 days.[33] Inoculation of white mice and Syrian hamsters of all age groups leads to lethal encephalitis.[34-38] OHF virus pathogenicity for white mice is remarkably greater than that of different RSSE viral strains. Inbred BALB and C57 mice are extremely sensitive to the virus. Among wild rodents, *Microtus gregalis* also proved to be highly susceptible by all routes of infection.[31] Other rodents, such as *M. oeconomus*[39] and *Arvicola terrestris,*[35] also have been shown to be susceptible. Guinea pigs inoculated by the subcutaneous route develop febrile reactions with scattered deaths. Infection of rabbits leads to antibody production only.[30] Intracerebral inoculations of rhesus monkeys leads to lethal encephalitis, whereas subcutaneous inoculation results in inapparent infection. Subcutaneous inoculation of 2-month-old calves results in fever. The virus has been isolated from frogs

* *Dermacentor pictus* is a junior synonym of the currently accepted *D. reticulatus;* the latter is used in subsequent references in this chapter.

and lizards.[40] Viremia has been determined after experimental infection of poikilothermal species and of birds.[42,43]

All viral strains tested replicate to high titer in a variety of primary and continuous cell lines. Cytopathic effect is produced in several cell lines, including swine embryo kidney, HeLa, and Detroit-6.[23,24,31,36,44] The highest virus titers were obtained on days 2 to 3 in culture fluids of primary chick embryo cells, HeLa, and pig embryo kidney cells.

Suckling mice have been generally used for primary isolation of OHF virus.

III. DISEASE ASSOCIATIONS

A. Humans

The disease in humans is characterized primarily by dysfunction of the microvasculature.[9,19,45-47] The incubation period is 2 to 4 days long. Typical OHF has an abrupt onset, accompanied by fever, headache, myalgia, facial congestion, injection of the sclerae, and leukopenia. The fever lasts for 3 to 4 days (39 to 40°C) and is followed by a gradual temperature decrease, with abrupt defervescence on the 7th to 10th day. From the first days of illness, the patient suffers from diapedesic bleeding, especially nasal. Bronchopneumonia may complicate the disease. Convalescence is usually uneventful, without residual effects; fatal cases are also registered, but rarely (0.5 to 3%). Infection which develops after direct contact with sick muskrats and from bites of the infected animals is characterized by a similar clinical picture.[16,18]

B. Domestic Animals

No morbidity of domestic animals in nature has been described. Nevertheless, intracerebral and subcutaneous inoculation of calves resulted in the development of a diphasic febrile reaction, loss of appetite with drastic weight loss, motor disorders, conjunctivitis, lacrimation, and development of viremia.[5] Antibody prevalence in cows in the OHF endemic focus reaches 6.3%.[48]

C. Wild Animals

Epizootics of disease in muskrats, which were introduced into West Siberia in 1928, have a 10-year cycle. The morbidity has no connection with Errington disease,[49] the hemorrhagic disease of muskrats in North America.[50] The muskrats passed a long step-wise quarantine in the northern regions of the country — Arkhangelsk, Vologda, and Tomsk — to where they have been introduced from Great Britain, Czechoslovakia, and Finland. This may be regarded as the reason of absence of Errington disease within the western Siberian population of muskrats. Epizootics in West Siberia are often of complex etiology, including tularemia plus OHF[51-53] and leptospirosis plus OHF.[54] OHF infection of muskrats begins after a 10- to 12-day-long incubation period and is severe, with a high fatality rate (about 80%) and symptoms of central nervous system disease. Folitarek[55] explained this high sensitivity of muskrats to OHF virus by the "effect of contact of strangers". Other inhabitants of the foci — rodents, insectivores, mustelids, birds, and poikilotherms — are not involved in epizootics, in spite of having close contact with muskrats. Indigenous species, including *A. terrestris, M. oeconomus,* and other rodents, birds of prey, frogs, and lizards, develop inapparent infections, accompanied by prolonged viremia and dissemination of the virus in inner organs.[56]

D. Diagnostic Methods

The clinical syndrome is not pathognomonic, and laboratory verification is necessary. The virus may be isolated from blood during the acute phase of illness using intracerebral inoculation of suckling mice. Paired acute and convalescent sera may be tested by enzyme-linked immunosorbent assay (ELISA), complement-fixation (CF), or neutralization (N) test for serologic confirmation.

IV. EPIDEMIOLOGY

A. Geographic Distribution

Omsk hemorrhagic fever was first recognized in two forest-steppe regions of Omsk district and was attributed solely to transmission by ixodid ticks, *D. reticulatus* and *D. marginatus*. The disease foci were later shown to include other forest-steppe areas of western Siberia as well, where muskrats played a role in dissemination of the agent. Typical landscapes associated with OHF are composed of northern and, in some areas, southern forest-steppe with numerous swamps and a wide network of small and large lakes covered with dense reed. Thus, the distribution of OHF crosses the entire forest-steppe landscape zone of western Siberia within the borders of Omsk, Novosibirsk, Kurgan, and Tumen regions and is adjacent to TBE foci.[13,15,22,57-63] Data indicating OHF activity in adjacent regions of the southern taiga landscape zone need further investigation and verification.[64]

B. Incidence

Between 1945 and 1949, the disease incidence was high, reaching 500 to 1400/100,000 population. Subsequently, a decrease of morbidity occurred, and only single cases have been registered, followed after 1958 by a complete disappearance of typical forms resulting from tick transmission.[13,15] A total of 1488 cases have been registered between 1945 and 1958 in the Omsk region; these data include results of retrospective analyses.[62] The morbidity rate between 1945 and 1958 in the transitional zone of the northern forest-steppe to the southern taiga zone of the Omsk region varied from 0 to 30/100,000; in the lake region of the northern forest-steppe, it varied from 0 to 25/100,000; in the lake belt of the northern forest-steppe, it was 33 to 356/100,000; and in the southern forest-steppe, it was 0 to 108/100,000.[62,65] The vast majority (96.8%) of all cases have been registered within the lake belt of the northern forest-steppe. This zone occupies 14.5% of the total endemic area and is inhabited by only 15.3% of the rural population of the Omsk district. Of the hyperendemic territory, 30% is covered by meadows and over 20% by grassy swamps and lakes.[62] It is important to recognize that the northern forest-steppe represents one of the youngest landscape zones in western Siberia, having replaced the southern taiga landscape.

C. Seasonal Distribution

Single cases occur in April, followed by peak incidence in May and declining morbidity in June and July. In some areas, a less-well-defined increased incidence has been registered from May to September.[8] The seasonal pattern definitely correlates with the dynamics of activity of *D. reticulatus* in the northern forest-steppe and of *Ix. marginatus* activity in the southern forest-steppe. Human cases resulting from direct contact with muskrats occur during the hunting season, i.e., autumn and winter (October to January).[16,17,32]

D. Risk Factors

During the spring and summer, the rural population engaged in field work is primarily affected. The patients' ages range from 5 to 70 years, with peak incidence in the 40- to 50-year age group. Muskrat hunting is associated with a high risk of infection; during the autumn and winter, 60% of cases are in hunters, 28% in adult members of their families, and 12% in children.[16,17,33]

E. Serologic Epidemiology

Seroprevalence in the human population of the endemic region, determined by N or hemagglutination-inhibition (HI) test, varies from 0 to 32%. Antibody prevalence has been determined mainly in elderly persons and not in children.[23,61] Among hunters engaged in muskrat hunting, antibodies are detected only in persons who have a history of clinical illness. These data suggest that all cases of infection acquired from muskrats lead to clinically overt disease.

Analysis of the data on human population immunity and disease incidence has defined ecologic correlations between zones of TBE and OHF activity. The following zones have been determined: (1) TBE hyperendemic zone (southern taiga), (2) interface zone (junction of southern taiga and northern forest-steppe), (3) OHF hyperendemic zone (northern and southern forest-steppe), and (4) zone of sporadic cases (part of southern forest-steppe).[61]

In zone 1, 92.6% of the total number of TBE cases have been registered, and no cases of OHF occur. In zone 1, 1.1% of both TBE and OHF have been registered. In zone 3, the proportion of TBE and OHF cases are 4.3 and 96%, respectively. In zone 4, 2% of the total number of TBE cases has occurred, OHF cases being absent.[63]

V. TRANSMISSION CYCLES

A. Evidence from Field Studies

1. Vectors

As noted above, between 1945 and 1949, OHF transmission was detected only in the landscape subzone of the northern forest-steppe in the lake region of Omsk district. A high density of *D. reticulatus* ticks was found in this region, with up to 216 specimens per kilometer;[2] 100% of cattle harbored ticks.[10] Preimagal forms of *D. reticulatus* were mainly found on voles, among which *M. gregalis* was the dominant species; 67 to 90% of larvae and nymphs were associated with this rodent. In the period 1959 to 1962, the density of *D. reticulatus* was reduced, fluctuating between 26 and 48 specimens per kilometer.[2] Virus infection rates in ticks in the center of endemic zone of forest-steppe was shown to be low during this period; rates varied between 0.1 ± 0.09% to 0.9 ± 0.2% during a nonepidemic period of low tick population density[32] and 6% during epidemic outbreaks.[57,58,66] A long-term depression of *D. reticulatus*, accompanied by a very low OHF virus infection rate in this species, which has an optimum of development in the northern forest-steppe, explains the absence of virus transmission to humans. Absence of mutual hosts for *D. reticulatus* imagos on one hand, and larvae and nymphs on the other hand (except for hares), implies that maintenance of OHF virus in natural foci is due only to transovarial virus transmission. Wide dissemination of the virus by *D. reticulatus* is possible only during massive population increases of the vole, *M. gregalis*. The latter situation occurred during the large epidemic of OHF in 1945 to 1948, and decline of this species in 1947 led to the gradual disappearance of natural meadow foci of OHF in 1949 and to the absence of tick-borne infection of humans. Under these conditions, the virus may have been restored in wetland areas due to transmission by other blood-sucking arthropods.

Relatively high densities of *Ix. apronophorus* are typical in grassy marshes of the western Siberian lowland. In some places, these ticks comprise 18 to 68% of collections from small vertebrates.[62,65-68] All stages of *Ix. apronophorus* feed on small vertebrates during the warm seasons. Under these conditions, virus infection of adults, larvae, and nymphs and transstadial transmission occur without the need for transovarial infection. *Arvicola terrestris* is the main vertebrate host for *Ix. apronophorus;* these rodents migrate in search of food in July and August from wet areas to meadows, where *D. reticulatus* larvae and nymphs reach peak activity during the same period. In turn, transovarial transmission of OHF virus in *D. reticulatus* may contribute to maintenance of gradually disappearing meadow foci. In swampy areas, *Ix. apronophorus* may also be responsible for infection of muskrats. Unfortunately, these ticks have not yet been studied by virological methods.

D. marginatus, principally active in the steppe landscape zone, plays a certain role in maintenance of OHF natural foci in the lake region of the southern forest-steppe subzone.[70] This role is not a critical one, however. Isolation of OHF virus from *Ix. persulcatus*[32] ticks needs further verification.

The participation of gamasid mites in OHF virus transmission has been proposed by Tagiltsev and Tarasevitch,[71-73] but no virus isolations from field-collected ticks have been reported.

Discovery of significant immunity rates among water birds in the natural foci of OHF led to the study of mosquitoes. Collections were made in July and August of 1966 and 1968 in northern and southern forest-steppe regions. During this period, *Mansonia richiardii* dominated, and *Aedes flavescens*, *Ae. excrucians*, and *Culex modestus* were minor species in the northern forest-steppe; *Ae. flavescens*, *Ae. cinereus*, and *Cx. modestus* were minor species in the southern forest-steppe.[74]

Infection of *Ma. richiardii*, *Ae. flavescens*, and *Ae. excrucians* was established. In the northern forest-steppe,[9] OHF strains have been recovered from each species. In the southern forest-steppe,[1] a single isolate has been recovered from *Ma. richiardii*. The infection rates in mosquitoes in the northern and southern forest-steppe was 0.3 to 2.0 and 0.09 to 0.4%, respectively. Study of 140 sentinel mice exposed in the northern forest-steppe revealed seroconversions in 9% of mice and, in one case, the virus has been isolated.[75] These data indicate the possibility that mosquitoes participate in OHF virus circulation. Nevertheless, no humans cases have occurred within the areas of mosquito collection. Under natural conditions, the source of mosquito infection is not clear. Mosquitoes could become infected from ill muskrats. *Arvicola terrestris*, *M. oeconomus*, and several bird species may play a role as hosts.[76] We believe that mosquitoes in natural foci may be involved in virus dissemination and may contribute to inapparent infections in animals and humans.

2. Vertebrate Hosts

In wetland areas of the northern forest-steppe of western Siberia, the muskrat is a significant component of the fauna. Having been introduced into this biocenose in 1928, muskrats spread widely due to the abundance of aquatic habitat. By the beginning of the 1940s, the density of the muskrat population had become high. Close interspecies associations occurred between muskrats and indigenous species, such as *A. terrestris*. For *A. terrestris*, cyclic population explosions followed by intensive epizootics were typical. Muskrats thus became involved in the concurrent epizootics of tularemia or leptospirosis and OHF.[51-54] These epizootics occurred with a 10-year periodicity. One of these epizootics in 1946 to 1950 coincided with the human OHF epidemic. OHF virus was isolated from muskrats in almost all sites within the endemic area.[33] In total, over 70 strains have been isolated from muskrats, the rate of infection determined by virological study reaching 14.6%, whereas the prevalence of OHF antibodies was as high as 30%. High infection rates were determined in both summer and winter, explaining exposure of hunters in autumn and winter. However, the mechanism of vertebrate infections during these seasons remains somewhat unclear. Ticks play a role in transmission, but horizontal spread from rodent to rodent and rodent to human by the alimentary and respiratory routes may also occur. During epizootics, the virus has been successfully isolated from the urine and feces of sick animals.[77] The introduction of muskrats must have significantly enhanced the epizootic and epidemic situation, was a factor in maintenance of OHF natural foci, and was principally responsible for direct contact exposure of humans during hunting.

Arvicola terrestris is another widely distributed species within the endemic area. Infection of this species by OHF virus has been documented in numerous investigations.[76-78] The role of *A. terrestris* and *Ix. apronophorus* ticks in muskrat epizootics was mentioned earlier. During dry seasons, *A. terrestris* populations are connected with lake reservoirs and river banks (the main habitat of muskrats), thereby favoring the development of concurrent tularemia and OHF epizootics in muskrat populations.[53] Nevertheless, epizootics of OHF alone have never been detected among *A. terrestris* or rodent species other than muskrats. During massive population increases of *A. terrestris,* it becomes a leading component of the lake-reservoir fauna, whereupon the density of some other species decreases drastically. Among these, *M. oeconomus,* which possesses similar ecological features, should be mentioned.[79] Conversely, when the population density of *A. terrestris* is low, a marked increased of *M. oeconomus* has been found; the kidney of this species appears to be the primary source for isolation of OHF virus.[56,79,82] Immunity rates in this species reach 36.6%. The virus has also been isolated from other rodents, including *Sicista betulina*[14,80,82] and *Citellus erythrogenus* with seroprevalences of up to 27%,[56,83] and also from *M. gregalis, Clethrionomys rutilis, Microtus minutus, Apodemus agrarius,* and *Rattus norvegicus.*[65] Four virus strains have been isolated from *Sorex araneus*[56] shrews, and a high antibody prevalence (35.5%) has been found in this species. The virus has also been isolated from the carnivore *Mustela eversmanni.*[65] Based on serological surveys, *Clethrionomys rutilus* ($\geq$48% seroprevalence), *Lagurus lagurus* ($\geq$17%), *Mus musculus* ($\geq$10%),[56] and *C. erythrogenus*[84] may be involved in the virus circulation among rodents.

The role of birds in virus transmission has been intensively studied. In OHF foci, birds appeared to be free from tick infestation. Nevertheless, 13 to 16% of Passeriformes (ten species), Anseriformes (five species), Limicolae (five species), Laridae (two species), Podicipediformes (one species), Rallidae (two species), and Accipitridae (one species) have been found with antibodies against the virus.[84,85] Antibodies were found in young birds more often (>17.6%) than in adults (<7.2%). This difference was more notable in birds of the water-marsh complex, with antibody prevalences of 24.0 and 3.6%, respectively. The virus has been isolated from the brain of a coot *(Fulica atra)* and a gadwall *(Anas strepera)* collected in natural foci.[86] As mentioned above, birds may be infected by mosquito bite. Antibody prevalence in August (29.4%) is four times higher than in July (7.2%). These data provide evidence for infection of birds, but no information on their role in virus maintenance. Serologic study of fish (four species) and frogs in OHF foci has been conducted.[87] The prevalence of antibodies in these poikilotherms reached 21.5 and 8.1%, respectively. Infection of these vertebrates may have occurred due to contamination of water by infected rodents, including *A. terrestris, M. oeconomus,* and sick muskrats. Poikilothermal animals are regarded as dead ends in virus circulation.

B. Evidence from Experimental Infection Studies

1. Vectors

Transovarial transmission of OHF virus in *D. reticulatus* has been documented in experimental studies. Transovarial transmission is, however, inefficient, and the imperfection of this mechanism was a factor in the disappearance of the virus after 1949, as discussed earlier.[65] The possible role of *Ix. apronophorus* in virus maintenance has been mentioned; unfortunately, this species has not been studied by virologists either in field experiments or under laboratory conditions. In studies of gamasid mites, infection of *Laelaps multispinosus* engorged on sick muskrats was followed by transmission of the virus to healthy animals (400 to 600 mites per animal). After the infection, the animals fell ill and died on days 6 to 8 of the experiments.[33,71,73] This may be a way in which virus is transmitted from *A. terrestris,* which has been shown to maintain the virus for long periods, to muskrats. No transmission of the virus by bird fleas *(Ceratophyllus styx)* collected in the nests of the sand martin, *Riparia riparia,* has been shown.[88]

Isolation of OHF virus from field-collected mosquitoes was followed up by experimental studies with *Ma. richiardii,* the predominant species in the OHF foci. Mosquitoes were fed on infected white mice and virus recovered on days 5, 7, 8, and 10 after the infection. Nevertheless, the mosquito infection rates were very low (0.5 to 0.6% on the 5th to 7th day and 1.7% on days 8 to 9 after the infection).[89] Biological transmission of the virus to white mice by the mosquitoes has been described; however, viremia and clinical manifestations of the disease have rarely been determined, and infection of most animals was determined by the appearance of antibodies. These data suggested that transmission of the virus by mosquitoes is inefficient and leads solely to the development of inapparent infection, even in highly susceptible animals. The mosquitoes may be responsible for the infection of birds, but other routes of infection are also possible.

2. Vertebrate Hosts

Muskrats proved to be highly susceptible to the infection, the virus titer in blood being as high as 10^5 to 10^6 $LD_{50}/0.03$ mℓ.[33] In all cases, infection with 10 to 100 $LD_{50}/0.03$ mℓ usually leads to the death of 80 to 90% of these animals. Survivors show no viremia.[56] Experimental infection of *A. terrestris* with 100 to 100,000 LD_{50} does not lead to clinical manifestations. The virus has been found between days 7 to 14 and on day 20 after the infection in blood, brain, liver, lungs, and kidneys. The long-term viremia is a significant feature of the role of this species in virus maintenance. Infection of *M. oeconomus* resulted in death in 25% of these animals on days 6 to 20 after inoculation. Virus was isolated from blood and viscera of animals which survived up to day 30 (termination of the experiment). Infection of *C. erythrogenus* was followed by the development of clinical manifestations and by death of some animals. No long-term viremia was determined.[48,83]

Alimentary and subcutaneous infection of birds of prey (March harrier [*Circus aeruginosus*], kestrel [*Falco tinnunculus*], long-eared owl [*Asio otus*]) sometimes leads to death, but the rest of the birds produced antibodies.[48] Subcutaneous infection of the rook *(Corvus frugilegus)* led to death of all the birds within 1 month. The birds were shown to be viremic (up to 10^5 $LD_{50}/0.03$ mℓ) for as long as 18 days. No clinical manifestations were found in the Pochard *(Aythya ferina)*, but virus was present in blood on days 2 to 11 (up to 10^5 $LD_{50}/0.03$ mℓ) and in feces on days 6 to 60. These data evidence the possibility of the participipation of some birds in virus maintenance and dissemination during their migrations.

VI. PREVENTION AND CONTROL

Risk of infection may be reduced by measures of personal protection against tick bite and exposure to muskrats.[90] Since OHF and TBE viruses proved to be antigenically related, the existing TBE vaccines may be successfully used;[90] under the present circumstances of low disease incidence, immunization would appear not to be a practical public health measure.

VII. FURTHER INVESTIGATIONS

Surveillance of the dynamics of the main components of the OHF biocenosis — muskrats, *A. terrestris, M. oeconomus,* and ticks infesting them — should be continued. Virological study of *Ix. appronophorus* populations is needed to define its role in virus transmission/maintenance cycles. Surveillance of muskrat populations at the borders of the present endemic territory should also be carried out.

REFERENCES

1. **Chumakov, M. P.,** Results of field work of Institute of Neurology on the study of Omsk haemorrhagic fever (OHF), *Vestn. AMS U.S.S.R.,* 2, 19, 1948.
2. **Chumakov, M. P.,** Materials of Institute of Neurology AMS U.S.S.R. on the study of Omsk haemorrhagic fever, *Vestn. AMS U.S.S.R.,* 3, 21, 1949.

3. **Chumakov, M. P.**, On the classification and nomenclature of viruses of antigenic subgroup of tick-borne encephalitis, *Vopr. Virusol.*, 3, 376, 1965.

4. **Shestopalova, N. M., Reingold, V. N., Gavrilovskaya, I. N., Belyaeva, A. P., and Chumakov, M. P.**, Electron microscopic study of the morphology and localization of Omsk haemorrhagic fever virus in infected tissue culture, *Vopr. Virusol.*, 4, 425, 1965.

5. **Smorodintsev, A. A., Kazbintsev, L. I., and Chudakov, V. G.**, Virus Haemorrhagic Fevers, Medical Literature, Leningrad, 1963, 292.

6. **Akhrem-Akhremovich, R. M.**, Spring-summer fever in the Omsk region, in *Proc. Omsk Med. Inst.*, Vol. 13, Omsk, 1948, 3.

7. **Akhrem-Akhremovich, R. M.**, Final study of Omsk haemorrhagic fever, *Proc. Omsk Med. Inst.*, Vol. 18, Omsk, 1952, 211.

8. **Gavrilovskaya, A. A.**, Materials on the etiology and epidemiology of Omsk hemorrhagic fever, in *Proc. 5th Conf. Omsk Inst.*, Omsk, 1949, 219.

9. **Tatarintsev, N. M., Bisyarina, V. P., and Kverel, R. M.**, On clinical characterisation of Omsk haemorrhagic fever among children, *Pediatria*, 6, 49, 1952.

10. **Fedyushin, A. V. and Netskii, G. I.**, Results of field work of Omsk Department of Health and Omsk Medical Institute on the study of spring-summer fever in the Sargat region, 1946, *Proc. Omsk Med. Inst.*, Vol. 13, Omsk, 1948, 59.

11. **Konstantinov, V. P., Sizemova, G. A., and Veselov, Yu. V.**, Spring-summer fever in Omsk oblast, *Proc. Omsk Med. Inst.*, Vol. 13, Omsk, 1948, 81.

12. **Veselov, Yu. V., Konstantinov, V. P., and Egorova, L. S.**, On clinico-epidemiological characterization of Omsk haemorrhagic fever, in *Proc. 3rd Session Altai Medical Institute*, Barnaul, 1959, 71.

13. **Kucheruk, V. V., Ivanova, L. M., and Neronov, V. M.**, Tick-borne encephalitis. Omsk haemorrhagic fever, in *Geography of Natural-Foci Infections of Man in Connection with Their Prophylaxis*, Moscow, 1969, 171.

14. **Lebedev, E. P., Dunaev, N. B., Busygin, F. F., and Matyukhina, L. V.**, Virological value of activity of Omsk haemorrhagic fever, natural foci and their epidemiological manifestation in the Omsk oblast, in *Questions of Medical Virology*, Moscow, 1975, 313.

15. **Vasyuta, Yu. S.**, Hemorrhagic fevers in the RSFUR. Tick-borne encephalitis and virus haemorrhagic fevers, Omsk, 1963, 355.

16. **Fedorova, T. N. and Sizemova, G. A.**, Cases of Omsk haemorrhagic fever of people and muskrats in a winter period, *J. Microbiol. Epidemiol. Immunobiol. (U.S.S.R.)*, 11, 134, 1964.

17. **Fedorova, T. N., Chudinov, P. I., Sizemova, G. A., Fedorov, V. G., and Tofanyuk, E. V.**, Winter cases of Omsk haemorrhagic fever in the zone of muskrat epizooty in West Siberia, in *Muskrat of West Siberia*, Novosibirsk, 1966, 162.

18. **Chudinov, P. I., Fedorova, T. N., Melenteva, L. A., Konstantinov, V. P., Sizemova, G. A., Tofanyuk, E. F., and Pospelov, E. S.**, Winter outbreaks of Omsk haemorrhagic fever in the Novosibirsk oblast, in *Natural Foci Diseases of West Siberia*, Novosibirsk, 1965, 127.

19. **Sizemova, G. A.**, Clinical features of Omsk haemorrhagic fever after laboratory infection, *Sov. Meditsina*, 5, 105, 1962.

20. **Belyaeva, A. P. and Chumakov, M. P.**, Study of cases of laboratory infections of people by Omsk haemorrhagic fever virus, in *Epidemic Virus Infections*, Nauka, Moscow, 1965, 396.

21. **Konstantinov, V. P., Kotova, N. S., and Sizemova, G. A.**, On cases of Omsk haemorrhagic fever by experimental infection, in *Proc. 11th Session Inst. Poliom. and Viral Enceph. AMS U.S.S.R.*, Moscow, 1964, 316.

22. **Gagarina, A. V.**, New foci of OHF: relationships between OHF virus and tick-borne encephalitis virus, in *Proc. 7th Session D. I. Ivanovskii Inst. Virology in Tomsk*, Tomsk, 1954, 46.

23. **Gavrilovskaya, I. N. and Chumakov, M. P.**, Application of the neutralization reaction in tissue culture for serological investigations on Omsk hemorrhagic fever, in *Proc. 11th Session Inst. Poliom. Viral Enceph. AMS U.S.S.R.*, Moscow, 1964, 314.

24. **Levkovich, E. N., Pogodina, V. V., Zasukhina, G. D., and Karpovich, G.**, *Viruses of Tickborne Encephalitis Complex*, Meditsina, Leningrad, 1967, 24.

25. **Clarke, D. H.**, Antigenic relationships among viruses of the tick-borne encephalitis complex as studied by antibody absorption and agar gel precipitin techniques, in *Proc. Symp. Biol. Viruses Tick-Borne Enceph. Complex*, 1962, 67.

26. **Clarke, D. H.**, Further studies on antigenic relationships among the viruses of the group B tick-borne complex, *Bull. WHO*, 31, 45, 1964.

27. **Rzhakhova, O. E.**, Typing of tick-borne encephalitis viruses by means of precipitation test in agar, in *Proc. 13th Session Inst. Poliom. and Viral Enceph. AMS U.S.S.R.*, Moscow, 1967, 113.

28. **Solovei, E. A. and Chumakov, M. P.**, Diffuse precipitation reaction in agar (DPRA), by studying Omsk haemorrhagic fever (OHF) natural focus, in *Proc. 15th Session Inst. Poliom. and Viral Enceph. AMS U.S.S.R.*, Moscow, 1969, 189.

29. **Fedorova, T. N. and Rubin, S. G.,** Antigenic characterization of viruses isolated in foci of Omsk haemorrhagic fever, in *Questions of Medical Virology,* Moscow, 1975, 365.

30. **Kornilova, E. A.,** Study of biological properties of strains of Omsk haemorrhagic fever virus in laboratory animals and tissue culture, *Arboviruses,* Moscow, 1967, 2, 183.

31. **Kornilova, E. A., Gagarina, A. V., and Chumakov, M. P.,** Comparative characterization of strains of Omsk haemorrhagic fever virus isolated from different sources in natural foci, *Vopr. Virusol.,* 2, 232, 1970.

32. **Fedorova, T. N.,** Comparative characterization of strains of viruses of Tick-borne Encephalitis antigenic complex isolated in natural foci of Omsk haemorrhagic fever, in *Characterization of Foci of Omsk Hae-morrhagic Fever in West Siberia,* Nauka, Novosibirsk, 1974, 160.

33. **Fedorova, T. N.,** The role of muskrats in epidemiology of Omsk haemorrhagic fever, in *Characterization of Foci of Omsk Haemorrhagic Fever in West Siberia,* Nauka, Novosibirsk, 1974, 81.

34. **Barkova, E. A. and Melenteva, L. A.,** Materials of virological study of epizootic among muskrats in the Krutinsk region of the Omsk oblast, *Proc. Omsk. Inst. Epidemiol. Microbiol. Hyg.,* 6, 23, 1959.

35. **Kharitonova, N. N.,** Susceptibility of muskrats and water rats to Omsk haemorrhagic fever virus in an experiment. News of Siberian Branch AS U.S.S.R., *Ser. Biol. Sci.,* 15, 106, 1967.

36. **Chumakov, M. P. and Belyaeva, A. P.,** Experimental data on relationships between Omsk haemorrhagic fever and Tick-borne Encephalitis viruses, *Endemic. Virus Infect.,* 7, 356, 1965.

37. **Kosova, T. V.,** The role of muskrat in natural foci of Omsk haemorrhagic fever, in *Proc. Problem Commission AMS U.S.S.R. (Arboviruses),* Moscow, 1967, 2, 184.

38. **Gagarina, A. V., Zimina, V. E., and Ravdonikas, O. V.,** Natural infection of muskrats with Omsk haemorrhagic fever, *Proc. Omsk Inst. Epidemiol. Microbiol. Hyg.,* 5, 31, 1958.

39. **Leonov, Yu. A. and Kharitonova, N. N.,** The experimental confirmation of the role of *Microtus aeconomus* in epizootiology of Omsk haemorrhagic fever, in *Transcontinental Connections of Migratory Birds and Their Role in the Distribution of Arboviruses,* Novosibirsk, 1972, 355.

40. **Vorobeva, N. N., Kharitonova, N. N., and Khadzhieva, T. M.,** Ecological connections of OHF virus with animals in a natural focus, News of Siberian Branch AMS U.S.S.R., *Ser. Biol. Sci.,* 5, 98, 1970.

41. **Radkova, O. A. and Grigorev, O. V.,** Viremia among poikilothermic animals experimentally infected by Omsk haemorrhagic fever virus, in *Characterization of Foci of Omsk Haemorrhagic Fever in West Siberia,* Nauka, Novosibirsk, 1974, 152.

42. **Fedorova, T. N., Stavskii, A. V., Yurlov, K. T., Voronin, Yu. K., Fedorov, V. G., and Volynets, L. G.,** Experimental study of susceptibility of rooks and wild ducks to Omsk haemorrhagic fever virus, in *Characterization of Foci of Omsk Haemorrhagic Fever in West Siberia,* Nauka, Novosibirsk, 1974, 101.

43. **Kharitonova, N. N., Danilov, O. N., and Leonov, Yu. A.,** The importance of birds of prey in a focus of Omsk haemorrhagic fever, in *Transcontinental Connections of Birds and Their Role in Distribution of Arboviruses,* Novosibirsk, 1972, 353.

44. **Karmysheva, V. Ya., Gavrilovskaya, I. N., and Chumakov, M. P.,** Study of interaction between Omsk haemorrhagic fever virus and the cells of a sensitive tissue culture, *Vopr. Virusol.,* 5, 557, 1965.

45. **Pokrovskii, V. I. and Lvov, D. K.,** Omsk haemorrhagic fever, in *Textbook on Zoonoses,* Pokrovskii, V. I., Ed., Meditsina, Leningrad, 1969, 53.

46. **Zubov, V. A.,** The change of nervous system after Omsk haemorrhagic fever, *Questions Infect. Pathol.,* 2, 237, 1970.

47. **Sizemova, G. A.,** The clinical picture of Omsk haemorrhagic fever (OHF) associated with epizootics among muskrats, *Endemic Virus Infect.,* Moscow, 12, 449, 1968.

48. **Vorobeva, N. N., Kharitonova, N. N., and Radkova, O.,** A latent infection by Omsk haemorrhagic fever among animals in the nautral focus, in *Characterization of Foci of Omsk Haemorrhagic Fever in West Siberia,* Nauka, Novosibirsk, 1974, 73.

49. **Errington, P. L.,** *Muskrat Populations,* Iowa State University Press, Ames, 1963.

50. **Abashkin, S. A.,** On a question about identity muskrats diseases in North America and the Soviet Union with connection of introduction, in *Characterization of Omsk Haemorrhagic Fever Foci in West Siberia,* Nauka, Novosibirsk, 1974, 91.

51. **Egorova, L. S., Korsh, P. V., Ravdonikas, O. V., and Fedorova, T. N.,** Tularemia and OHF among muskrats in West Siberia, in *Tularemia and Other Diseases,* Omsk, 1965, 74.

52. **Gurbo, G. D., Filatov, V. G., Kuzovlev, A. P., Khomutova, N. V., Kazantseva, Z. I., and Khomutova, V. A.,** Natural foci of Omsk haemorrhagic fever connected with other infectious diseases in forest steppe landscapes of West Siberia, in *Proc. 10th Conf. Natural Foci of Diseases,* Dushanbe, 1979, 63.

53. **Maksimov, A. A., Vladimirskii, M. G., and Kiseleva, N. V.,** Geographical distribution and development conditions of epizootics of Omsk haemorrhagic fever and tularemia in the muskrat in the region of Novosibirsk in the years 1957—1964, in *Muskrat of West Siberia,* Novosibirsk, 1966, 101.

54. **Popov, V. V., Getsold, S. G., Zuevskii, A. P., Tsuryateva, F. M., Dubov, A. V., and Bilalova, E. Z.,** On mixed winter tularemia, Omsk haemorrhagic fever, and leptospirosis epizootics in the muskrat in the Tyumen region, in *Muskrat of West Siberia,* Novosibirsk, 1966, 152.

55. **Folitarek, S. S.,** Effect of contact of strangers "Alienconflictus" in an epizootic of muskrat in West Siberia, in *Migratory Birds and Their Role in Distribution of Arboviruses,* Novosibirsk, 1969, 128.

56. **Vorobeva, N. N., Kharitonova, N. N., Leonov, Yu. A., Radkova, O. A., and Blumberg, L. E.,** New data on persistence of Omsk haemorrhagic fever virus among small animals of northern Kulunds, in *Characterization of Foci of Omsk Haemorrhagic Fever in West Siberia,* Nauka, Novosibirsk, 1974, 70.

57. **Netskii, G. I.,** Contemporary state of natural foci and further tasks for studying prophylactic measures against Omsk haemorrhagic fever, in *Epidemiology of Tick-borne Encephalitis: Rickettsiae Tularemia and Leptospiroses,* Omsk, 1961, 3.

58. **Netskii, G. I., Trop, I. E., Fedorova, T. N., Alifanov, V. I., Bogdanov, I. I., Melenteva, L. A., and Fedorov, V. G.,** Characterization of natural foci of tick-borne encephalitis and Omsk haemorrhagic fever in West Siberia, in *Tick-Borne Encephalitis,* Minsk, 1965, 221.

59. **Busygin, F. F.,** Characterization of modern status of foci of Omsk haemorrhagic fever, in *Tick-Borne Encephalitis,* Omsk, 1965, 37.

60. **Busygin, F. F.,** The role of the muskrat in formation of natural foci of Omsk haemorrhagic fever, *Questions Infect. Pathol.,* 2, 34, 1970.

61. **Busygin, F. F. and Prigorodov, V. I.,** Characterization of foci of tick-borne encephalitis and Omsk haemorrhagic fever, in *Epidemiology Omsk Haemorrhagic Fever and Tick-Borne Rikketsiae of Asia and West Siberia,* Omsk, 1973, 65.

62. **Ravdonikas, O. V., Korsh, P. V., and Ivanov, D. I.,** Landscape peculiarities of Omsk haemorrhagic fever (OHF), *Endemic Virus Infect.,* 12, 441, 1968.

63. **Tsaplin, I. S. and Busygin, F. F.,** Epidemiological characterization of tick-borne encephalitis and Omsk haemorrhagic fever in "transitional" zone of West Siberian lowland, in *Questions of Infections and Pathology,* Omsk, 1970, 31.

64. **Dokuchaeva, Yu. I. and Vorobeva, A. M.,** Characterization of foci of tick-borne encephalitis and Omsk haemorrhagic fever in the south taiga of Irtysh region, in *Characterization of Foci of Omsk Haemorrhagic Fever in West Siberia,* Nauka, Novosibirsk, 1974, 168.

65. **Ravdonikas, O. V., Chumakov, M. P., Solovei, E. A., Ivanov, D. I., and Korsh, P. V.,** On epizootiological activity of natural foci of Omsk haemorrhagic fever, in *Natural Foci of Diseases in Ural, Siberia and Far East,* Sverdlovsk, 1969, 122.

66. **Gagarina, A. V.,** Virologic confirmation of participation of the tick *Dermacentor pictus* Herm. on transmission and maintenance of Omsk haemorrhagic fever virus, *Proc. Omsk Inst. Epidemiol. Microbiol.,* 1, 19, 1952.

67. **Alifanov, V. T.,** Ecology and distribution of *Ix. apronophorus* ticks in West Siberia and their role as tularemia vectors, *Zool. J.,* 44, 291, 1965.

68. **Alifanov, V. I., Netskii, Y. I., Ravdonikas, O. V., and Fedorov, V. G.,** Zoological-parasitological characterization of natural foci of Omsk haemorrhagic fever in the Tyukalinsk Region of Omsk oblast, In *Questions of Epidemiology and Prevention of natural foci, enteric and children's diseases,* Omsk, 1961, 9.

69. **Alifanov, V. I., Netskii, G. I., and Bogdanov, I. I.,** Ixodidae ticks of West Siberia: vectors of tick-borne encephalitis and Omsk haemorrhagic fever, in *Characterization of Omsk Haemorrhagic Fever Foci in West Siberia,* Nauka, Novosibirsk, 1974, 106.

70. **Gagarina, A. V.,** Natural infection of Omsk haemorrhagic fever virus by tick *Dermacentor marginatus* Sulz, *Proc. Omsk Inst. Epidemiol. Microbiol. Hyg.,* 4, 15, 1957.

71. **Tagiltsev, A. A. and Tarasevich, L. N.,** *Arthropoda of Shelter Complex in Natural Foci of Arborvirus Infections,* Nauka, Novosibirsk, 1982.

72. **Tarasevich, L. H. and Tagiltsev, A. A.,** Relationship of Gamasidae with Omsk haemorrhagic fever virus, in *Ecology of Viruses Associated with Birds,* Minsk, 1974, 75.

73. **Tarasevich, L. H. and Tagiltsev, A. A.,** Experimental study of relationships of hematophagous Gamasidae with Omsk haemorrhagic fever virus, in *Epidemiology and Prophylaxis of Omsk Haemorrhagic Fever,* Omsk, 1975, 89.

74. **Volynets, L. V.,** Virological characterization of viruses isolated from mosquitoes in forest-steppe foci of Omsk haemorrhagic fever, *Questions Infect. Pathol.,* 2, 225, 1970.

75. **Matyukhina, L. V.,** The role of blood-sucking mosquitoes in foci of Omsk haemorrhagic fever, in *Natural Foci Anthroponoses,* Omsk, 1976, 127.

76. **Bogdanov, I. I. and Volynets, L. V.,** Ecological connection between blood-sucking mosquitoes and birds in foci of Omsk haemorrhagic fever in south forest-steppe of West Siberia, in *Proc. 6th Symp.*

77. **Kharitonova, N. N. and Leonov, Yu. A.,** Possibility of alimentary route of infection by Omsk haemorrhagic fever, in *Characterization of Foci of Omsk Haemorrhagic Fever in West Siberia,* Nauka, Novosibirsk, 1974, 155.

78. **Kharitonova, N. N. and Leonov, Yu. A.,** Ecology of Omsk haemorrhagic fever (OHF) virus, in *Proc. Symp. Transcontinental Connections of Migratory Birds and Their Role in the Distribution of Arboviruses,* Novosibirsk, 1976, 71.

79. **Leonov, Yu. A. and Fedorov, T. N.,** The role of rodents and shrews in Omsk haemorrhagic fever foci in northern Kulunda, in *Role of Migratory Birds in the Distribution of Arboviruses,* Novosibirsk, 1969, 322.

80. **Lebedev, E. P.,** Characterization of Omsk haemorrhagic fever focus in interepidemic periods, in *Natural Foci of Anthroponoses,* Omsk, 1976, 125.

81. **Maksimov, A. A. Lenov, Yu. A., Glotov, I. N., Vladimirskii, L. G., and Barbash, L. A.,** Mouse-like rodents and shrews in lake foci of Omsk haemorrhagic fever in northern Kulunda and Baraba, in *Characterization of Foci of Omsk Haemorrhagic Fever in West Siberia,* Novosibirsk, 1974, 6.

82. **Lebedev, E. P. and Busygin, F. F.,** Perspectives of study of Omsk haemorrhagic fever virus ecology in north forest-steppe of West Siberian lowlands, in *Proc. 10th Conf. Natural Foci Diseases,* Dushanbe, 1979, 123.

83. **Leonov, Yu. A., Kharitonova, N. N., and Sapegina, V. F.,** Redcheeked gopher and its role in the epizootiology of Omsk haemorrhagic fever (OHF), News of Siberian Branch AS U.S.S.R., *Ser. Biol. Sci.,* 10, 126, 1970.

84. **Danilov, O. N., Fedorova, T. N., and Matyukhin, V. N.,** Results of serological investigation of birds in North Kulunda to Omsk haemorrhagic fever virus, in *Migratory Birds and Their Role in the Distribution of Arboviruses,* Novosibirsk, 1964, 333.

85. **Matyukhin, V. N., Fedorova, T. N., Danilov, O. N., Malkov, G. B., and Voronin, Yu. K.,** The role of birds in natural foci of Omsk haemorrhagic fever, in *Characterization of Foci of Omsk Haemorrhagic Fever in West Siberia,* Novosibirsk, 1974, 94.

86. **Fedorova, T. N., Matyukhin, V. N., Malkov, G. B., Bogdanov, I. I., and Volynets, L. V.,** Comparative study of role of birds in natural foci of tick-borne encephalitis and Omsk haemorrhagic fever, in *Transcontinental Connections of Migratory Birds and Their Role in the Distribution of Arboviruses,* Nauka, Novosibirsk, 1972, 336.

87. **Leonov, Yu. A., Kharitonova, N. N., Kotlyarevskaya, V. A., and Volgin, M. V.,** The role of poikilothermal animals in the circulation of Omsk haemorrhagic fever virus, in *Characterization of Foci of Omsk Haemorrhagic Fever in West Siberia,* Nauka, Novosibirsk, 1974, 146.

88. **Sapegina, V. F. and Kharitonova, N. N.,** On the ability of bird fleas to transmit Omsk haemorrhagic fever to white mice under experimental conditions, in *The Role of Migratory Birds in the Distribution of Arboviruses,* Novosibirsk, 1969, 363.

89. **Volynets, L. U. and Bogdanov, I. I.,** Experimental infection of mosquitoes *Mansonia richiardii* by Omsk haemorrhagic fever virus, in *Characterization of Foci of Omsk Haemorrhagic Fever in* West Siberia, Novosibirsk, 1974, 126.

90. **Los, M. V., Shvanbauer, B. Ya, and Shaiman, O. T.,** Materials on the specific prophylaxis of Omsk haemorrhagic fever, in Proc. Siberian Conf. Inst. Epidemiol. Microbiol., Tomsk, 1949, 57.

Chapter 35

O'NYONG-NYONG VIRUS DISEASE

Bruce K. Johnson

TABLE OF CONTENTS

I. HISTORICAL BACKGROUND

A. Discovery of Agent

In early 1959, large numbers of people in the Acholi region of northwestern Uganda were noted to be falling ill with disease characterized by fever, severe back and joint pain, lymphadenopathy, and an itchy rash. Locally, the disease was known as o'nyong-nyong, meaning very painful and weak. Workers at the East African Virus Research Institute at Entebbe recovered a previously unknown alphavirus from the blood of infected patients and from the mosquitoes *Anopheles funestus* Giles and *An. gambiae* Giles.[1]

B. History of Epidemic

The epidemic was detected soon after it began and was closely followed by investigators from that institute who documented a huge and unique outbreak. By the time the epidemic died out more than 3 years later, it is estimated that more than 2 million people had been infected with o'nyong-nyong virus, with evidence of infection being found as far south as Mozambique and as far west as Senegal.[2-4] The epidemic spread in a southeasterly direction through Uganda, moving into western Kenya by December 1959. The disease was calculated to be moving through Uganda at a rate of 1.7 mi/day. The spread also occurred around the southern end of Lake Victoria and continued through Tanzania to the coast of the Indian Ocean. By June 1962, it was estimated that 50% of the population south of Mombasa on the Kenya coast had been infected.[5,6]

Serological evidence of exposure to o'nyong-nyong virus continued to be detected in human populations until the late 1960s, after which transmission appears to have declined.[7,8] A single strain of o'nyong-nyong virus was recovered from a pool of *An. funestus* collected in 1978 in western Kenya, following a period of several years during which no evidence of virus activity could be detected.[9,10] No indication of o'nyong-nyong fever in epidemic form has been found since the original outbreak.

C. Social and Economic Impact

O'nyong-nyong fever was of sudden onset and was extremely debilitating. This and the explosive nature of the spread in areas of high anopheline mosquito density led to large numbers of people being ill over a short period. Labor-intensive production or agricultural industries were affected during the epidemic with 25% or more of the labor force being off for 5 days or more, sometimes 10% at one time.[5]

Fortunately, the disease was associated with rapid and full recovery, and no mortality could be attributed directly to o'nyong-nyong fever.

II. THE VIRUS

A. Antigenic Relationships

The *Alphavirus* genus of the family Togaviridae, of which o'nyong-nyong virus is a member, consists of 26 serologically related members, all but one of which are transmitted by mosquitoes. O'nyong-nyong falls within the Venezuelan equine encephalitis virus subgroup in the genus and is very closely related to chikungunya virus and less closely related to Semliki Forest virus.[11,12] O'nyong-nyong virus is now considered to be a subtype of chikungunya virus.[13] The very close relationship between o'nyong-nyong and chikungunya viruses makes the two difficult to differentiate except by reciprocal testing with antigens and antisera against both agents.[14] It is not known whether chikungunya virus may represent the progenitor from which o'nyong-nyong was derived, but the former expresses antigens common to both, while o'nyong-nyong virus fails to express some of the antigens of chikungunya. This results in a one-way relationship, in which chikungunya antiserum reacts strongly against o'nyong-nyong virus in specific assays such as neutralization tests, but o'nyong-nyong antiserum reacts only weakly or not at all against chikungunya virus. This relationship is more pronounced in early immune sera than in hyperimmune sera where the difference is a matter of degree.[15,16] More recently, the nucleotide sequence homology between the prototype strains of o'nyong-nyong and chikungunya viruses have been compared, and o'nyong-nyong virus was found, by RNA-RNA hybridization, to share a 13% base sequence homology with chikungunya.[17]

B. Host Range

In mosquitoes, o'nyong-nyong virus has been recovered only from *An. funestus* and *An. gambiae*. It is only the arthropod-borne virus known to have occurred in epidemic form transmitted apparently exclusively by anophelines.[18,19] The majority of mosquito isolates came from *An. funestus*. This and limited experimental transmission studies suggested that this species was a more efficient vector than *An. gambiae*. It must be borne in mind that, at the time of these studies, *An. gambiae* was considered to be a single species; subsequently it has been shown to be a complex of species with different habits and possibly different vector capabilities. Attempts to demonstrate o'nyong-nyong virus replication in culicine species or even cell cultures derived from culicine species have been successful.[15,20] O'nyong-nyong virus has been recovered in nature only from these mosquitoes, man, and sentinel mice. Attempts to infect monkeys by subcutaneous (s.c.) virus inoculation were unsuccessful, although antibodies could be detected.[19,21,22]

C. Strain Variation

Many strains of o'nyong-nyong virus were recovered during the course of the original outbreak. Virus was isolated at Entebbe from Ugandan, Kenyan, and Tanzanian serum samples and mosquitoes. The only strain which appeared to differ from the others was an isolate obtained from sentinel mice in Senegal, over 3000 miles to the northwest of the main epidemic area. This variant, recovered in late 1962 by workers at the Institute Pasteur in Dakar, appeared to be even more closely related to chikungunya virus than o'nyong-nyong virus isolates from eastern Africa, in that antiserum produced in mice against the Senegal virus was capable of neutralizing chikungunya virus.[4]

In 1966, a new virus was isolated from blood samples of two febrile patients in Nigeria; a third isolate was obtained in 1969 from a Peace Corps volunteer with fever, rash, and joint pains.[23] The virus, named Igbo-Ora, was biologically similar to o'nyong-nyong in that it was weakly pathogenic for infant mice and required repeated passage. In cross-complement-fixation (CF) test, Igbo-Ora virus showed a one-way difference from o'nyong-nyong virus, but was distinct in both directions from chikungunya. Igbo-ora virus thus appears to represent a West African variant of o'nyong nyong, and has been associated with human illness.

III. METHODS OF ASSAY

A. Virus Isolation

O'nyong-nyong virus is strikingly avirulent in intracerebrally inoculated suckling mice when compared to other alphaviruses, which usually kill inoculated mice rapidly on primary isolation from field material or adapt within a few mouse passages to cause uniformly fatal disease. On primary inoculation, o'nyong-nyong virus may cause no symptoms in suckling mice, requiring repeated blind brain passage. Signs of illness in mice may never develop further than runting, failure to thrive, macular rash, and/or alopecia, which normally appears on or after the 5th day postinoculation.[24] The 16th suckling mouse brain passage of an o'nyong-nyong virus strain isolated in Kenya in 1978 failed to consistently kill or even cause illness in inoculated mice.[9] The need to rely on this method of isolation during the 1959—1962 epidemic probably resulted in isolates being missed and failure to fully characterize others owing to the large numbers of mice required. Although cell cultures have not been successful for primary isolation, early mouse brain passage material causes cytopathic changes or plaque formation in primary chick or mouse embryo cells, VERO cells, and BHK-21 cells.[15,16] The ideal isolation method might be the inoculation of 7-day blind-passaged mouse brain of field material into cell culture. In this way, toxicity problems arising from mosquito, blood, or organ pool inoculation might be avoided and amplification of the virus from the original material might be achieved.

B. Virus Identification

Identification of o'nyong-nyong virus can only be carried out by comparative reciprocal neutralization tests against chikungunya virus, unless monoclonal antibodies are used.[5,14] Antiserum is raised in adult mice immunized with a single intraperitoneal (i.p.) inoculation of $\geq 10^6$ virus particles 14 days earlier. It may be necessary to use formalin-inactivated virus in the case of chikungunya, as this more virulent virus may kill adult mice. Earlier workers found that antiserum raised in rabbits against o'nyong-nyong virus reacted more strongly against chikungunya virus than did similar antiserum produced in mice.[25] Guinea pigs were said to be relatively poor antibody producers.[5]

Immunofluorescence or ELISA (enzyme-linked immunosorbent assay) techniques which are IgM specific might also differentiate between chikungunya and o'nyong-nyong viruses.

The close serological relationship between the two agents makes seroepidemiological investigations difficult. Sera reacting against both viruses by neutralization test may actually contain antibodies only against chikungunya. On the other hand, those immunized against o'nyong-nyong by repeated challenge may react to some degree against chikungunya virus. As a result, the two have been combined in such studies despite their ecological differences.

IV. DISEASE ASSOCIATIONS

A. Humans

The onset of o'nyong-nyong fever is sudden after an incubation period that is believed to be 8 days or more.[26] The disease is characterized by rigors, fever in about 75% of the cases (seldom found to exceed 38.3°C), headache, retro-orbital pain in about 60% of the cases, injected conjunctivas, eyeballs painful to pressure, photophobia, and edema of the eyelids. A universal symptom is severe joint pain which usually affects most or all of the joints. This pain is often very severe, completely disabling the patient. Lumbar back pain often accompanied the onset of fever. A pruritic maculopapular rash occurs between the 4th and 7th day in about 60% of the cases. This begins on the face and descends to the extremities and trunk. The rash may become confluent, but is sometimes confined to discrete papules on the neck, chest, back, sides, or inner aspects of the arms and thighs. The palms are sometimes involved. The irritating nature of the rash is characteristic. A firm, rubbery enlargment of the lymph nodes of the neck, postoccipital, axillary, and inguinal areas is so pronounced that it may be visible from a distance. Epistaxis may occur early in the illness and coryza and a dry cough later. Examination of blood shows a characteristic leukopenia with a relative lymphocytosis. The fever usually lasts 5 days but joint pain, weakness, and mental depression may last longer. All known patients made full recovery. Subclinical seroconversions were common. Malaria, measles, rubella, and other arbovirus diseases must be considered in the differential diagnosis. Diagnosis can be confirmed only by virus isolation from, or serological studies on, serum samples from patients.

B. Other Animals

Other than man and the laboratory mouse, disease caused by o'nyong-nyong virus has not been observed in any vertebrate.

V. EPIDEMIOLOGY

A. Geographic Distribution

O'nyong nyong virus has been isolated in Kenya, Tanzania, Uganda, Malawi, and Mozambique in East Africa[27]; in Cameroon[28] and the Central African Republic[29] in Central Africa; and in Senegal[4] in West Africa. The closely related Igbo-Ora virus has been recovered in Nigeria[23] and in the Central African Republic,[30] on both occasions from humans with febrile illnesses. In contrast, chikungunya virus is very widely distributed, occurring both in areas of Africa in which o'nyong nyong is unknown and in Asia.

B. Seasonal Distribution

O'nyong-nyong virus appears to be limited to areas which contain suitable habitat for the vectors, *An. funestus* and *An. gambiae*. Therefore, the areas of virus endemicity would coincide with areas of malaria endemicity in East Africa. Being dependent upon mosquitoes to maintain the chain of transmission, more effective spreading can occur with higher densities of anopheline mosquitoes. In areas with sufficient water, these species are abundant all year. In areas that do not have continuous water, the vectors are present in large numbers following seasonal rains, during which an epidemic bridge would exist allowing the virus to spread to new areas.

C. Risk Factors

During the period of epidemic spread, the entire susceptible human population is at risk if exposed to the bite of infected anopheline mosquitoes. Antibody prevalence rates of 90% were recorded in some endemic areas.

D. Serologic Epidemiology

In western Kenya, the high attack rates noted during the epidemic were followed by a period of endemicity during which the annual infection rate was estimated at 10%. Activity appeared to then decline, and by 1975 no evidence of exposure to the virus could be detected in children under 3 years of age.[7,10,31] However, the virus reappeared in the area in 1978 when a strain was recovered from *An. funestus*.[9]

Antibodies were detected by hemagglutination-inhibition (HI) test in a low percentage of cattle and goats during the period when virus activity was occurring but declining in western Kenya.[32] During this study antibodies were also detected in the sera of rodents *Arvicanthus niloticus* (Desmeret), *Mastomys natalensis* (Smith), and *Pelomys isseli* (de Beaux) at rates of 1% or less. Significantly higher percentages of positive sera were found in the rats *Aethomys kaiseri* (Noack) (77/250 or 30%) and *Dasymys imcomtus* (Sundevall) (92/575 or 16%). These findings could not be confirmed by neutralization (N) tests against either o'nyong-nyong or chikungunya viruses, and thus their meaning remains unresolved. No antibodies were found in the sera of more than 1200 birds, but two of eight monitor lizards *Varanus niloticus* (L.), were found antibody positive.[32]

VI. TRANSMISSION CYCLES

A. Field and Experimental Infection Studies

In the epidemic setting the cycle of transmission was almost certainly man-*Anopheles*-man as the disease spread through various ecological areas of which the common features were humans and anopheline mosquitoes. Wild-caught *Anopheles* mosquitoes held alive for up to 20 days were found infected with virus, and experimentally infected *An. funestus* transmitted o'nyong-nyong to suckling mice.[19,21]

No detailed studies have been carried out on extrinsic incubation periods, threshold infectivity, or population densities required to sustain the chain of transmission. Since significant numbers of *Anopheles* mosquitoes survive long enough to transmit malaria, and given the climatological conditions under which o'nyong-nyong virus occurs, transmission of the agent would be expected after infection.

B. Maintenance Mechanisms

The o'nyong-nyong virus epidemic of 1959 to 1962 appeared to continue unabated until areas of susceptible humans or *Anopheles* habitat were exhausted. The fact that the virus has not since been shown to occur in epidemic form suggests that any maintenance cycle is rather isolated from human intrusion. This cycle remains unelucidated.

VII. SURVEILLANCE, PREVENTION, AND CONTROL

A. Surveillance

In areas where o'nyong-nyong fever is potentially epidemic, a surveillance system would probably best be based on clinical diagnosis or, if feasible, periodic serological screening. If cases were detected in an area, intensive mosquito control measures would have to be taken.

B. Prevention and Control

An. funestus and *An. gambiae* are anthropophilic species which will readily enter dwellings to feed. They are difficult species to control economically, especially in rural or semirural settings. Malariologists have learned that vector control is a never-ending problem which, if relaxed, results in the return of the resilient *Anopheles* mosquito. As o'nyong-nyong fever appears to be benign and relatively rare, it would seem that the best control measures would be focal-intense anopheline mosquito control such as wall spraying if and when the disease was diagnosed clinically with laboratory virological or serological backup.

REFERENCES

1. **Haddow, A. J., Davies, C. W., and Walker, A. J.**, O'nyong-nyong fever: an epidemic virus disease in East Africa I. Introduction, *Trans. R. Soc. Trop. Med. Hyg.*, 54, 517, 1960.
2. **Haddow, A. J.**, Introduction, in East Afr. Virus Res. Inst. Rep. No. 12, Entebbe, 1962, 1.
3. **Theiler, M. and Downs, W.**, *The Arthropod-borne Viruses of Vertebrates*, Yale University Press, New Haven, 1973, 281.
4. **Brés, P., Chambon, L., Pape, Y., and Michel, R.**, Les arbovirus au Senegal. II. Isolelment de plusieurs souches, *Bull. Soc. Med. Afr. Noire Lang. Fr.*, 8, 710, 1963.
5. **Williams, M. C., Woodall, J. P., and Gillett, J. D.**, O'nyong-nyong fever: an epidemic virus disease in East Africa. VII. Virus isolations from man and serological studies up to July, 1961, *Trans. R. Soc. Trop. Med. Hyg.*, 59, 186, 1965.
6. **Woodall, J. P., Williams, M. C., and Hewitt, L. E.**, Virology: studies on epidemics and outbreaks, East Afr. Virus Res. Inst. Rep. No. 12, Entebbe, 1962, 15.
7. **Geser, A., Henderson, B. E., and Christensen, S.**, A multipurpose serological survey in Kenya. II. Results of arbovirus serological tests, *Bull. WHO*, 43, 539, 1970.
8. **Bowen, E. T. W., Simpson, D. I. H., Platt, G. S., Way, H., Bright, W. F., and Day, J.**, Large scale irrigation and arbovirus epidemiology, Kano Plain, Kenya. II. Preliminary serological survey, *Trans. R. Soc. Trop. Med. Hyg.*, 67, 702, 1973.
9. **Johnson, B. K., Gichogo, A., Gitau, G., Patel, N., Ademba, G., Kirui, R., Highton, R. B., and Smith, D. H.**, Recovery of o'nyong-nyong virus from *Anopheles funestus* in western Kenya, *Trans. R. Soc. Trop. Med. Hyg.*, 75, 239, 1981.
10. **Marshall, T. F., DeC., Keenlyside, R. A., Johnson, B. K., Chanas, A. C., and Smith, D. H.**, The epidemiology of o'nyong-nyong in the Kano Plain, Kenya, *Ann. Trop. Med. Parasitol.*, 76, 153, 1981.

11. **Karabatsos, N.,** Antigenic relationships of group A arboviruses by plaque-reduction neutralization testing, *Am. J. Trop. Med. Hyg.,* 24, 527, 1975.

12. **Chanas, A. C., Johnson, B. K., and Simpson, D. I. H.,** Antigenic relationships of alphaviruses by a simple micro-culture cross-neutralization method, *J. Gen. Virol.,* 32, 295, 1976.

13. **Calisher, C. H., Shope, R. E., Brandt, W., Casals, J., Karabatsos, N., Murphy, F. A., Tesh, R. B., and Wiebe, M. E.,** Proposed antigenic classification of registered arboviruses. I. Togaviridae, Alphavirus, *Intervirology,* 14, 229, 1980.

14. **Williams, M. C. and Woodall, J. P.,** O'nyong-nyong fever: an epidemic virus disease in East Africa. II. Isolation and some properties of the virus, *Trans. R. Soc. Trop. Med. Hyg.,* 55, 135, 1961.

15. **Williams, M. C., Woodall, J. P., and Porterfield, J. S.,** O'nyong-nyong fever: an epidemic virus disease in East Africa. V. Human antibody studies by plaque inhibition and other serological tests, *Trans. R. Soc. Trop. Med. Hyg.,* 56, 166, 1962.

16. **Chanas, A. C., Hubalek, Z., Johnson, B. K., and Simpson, D. I. H.,** A comparative study of o'nyong-nyong virus with chikungunya virus plaque variants, *Arch. Virol.,* 59, 231, 1979.

17. **Wengler, G., Wengler, G., and Filipe, A. R.,** A study of nucleotide sequence homology between nucleic acids of different alphaviruses, *Virology,* 78, 124, 1977.

18. **Corbet, P. S., Williams, M. C., and Gillett, J. D.,** O'nyong-nyong fever: an epidemic virus disease in East Africa. IV. Vector studies at epidemic sites, *Trans. R. Soc. Trop. Med. Hyg.,* 55, 463, 1961.

19. **Williams, M. C., Woodall, J. P., Corbet, P. S., and Gillett, J. D.,** O'nyong-nyong fever: an epidemic virus disease in East Africa. VIII. Virus isolations from *Anopheles* mosquitoes, *Trans. R. Soc. Trop. Med. Hyg.,* 59, 300, 1965.

20. **Buckley, S. M.,** Multiplication of chikungunya and o'nyong-nyong viruses in Singh's Aedes cell lines, *Curr. Top. Microbiol. Immunol.,* 55, 133, 1971.

21. **Gillett, J. D., Woodall, J. P., Williams, M. C., and Corbet, P. S.,** Transmission of the virus by laboratory reared *Anopheles funestus,* East Afr. Virus Res. Inst. Rep. No. 10, Entebbe, 1960, 20.

22. **Binn, L. N., Harrison, V. R., and Randall, R.,** Patterns of viremia and antibody observed in rhesus monkeys inoculated with chikungunya and other serologically related group A arboviruses, *Am. J. Trop. Med. Hyg.,* 16, 782, 1967.

23. Arbovirus Research Project, Annu. Rep. University of Ibadan, Nigeria, 1970.

24. **Walker, G. M., Woodall, J. P., Haddow, A. J., and Williams, M. C.,** O'nyong-nyong fever: an epidemic virus disease in East Africa. VI. Alopecia in mice experimentally infected with o'nyong-nyong virus, *Trans. R. Soc. Trop. Med. Hyg.,* 57, 496, 1962.

25. **Porterfield J. S.,** Cross-neutralization studies with group A arthropod-borne viruses, *Bull. W.H.O.,* 24, 735, 1961.

26. **Shore, H.,** O'nyong-nyong fever: an epidemic virus disease in East Africa. III. Some clinical and epidemiological observations in the Northern Province of Uganda, *Trans. R. Soc. Trop. Med. Hyg.,* 55, 361, 1961.

27. **Woodall, J. P., Williams, M. C., and Lule, M.,** O'nyong-nyong, East Afr. Virus Res. Inst. Rep. No. 13, Entebbe, 1963, 31.

28. **Salaun, J. J. and Brottes, H.,** Les arbovirus au Cameroun, *Bull. W.H.O.,* 37, 343, 1967.

29. **Chippaux, A. and Chippaux-Hyppolite, C.,** Stigmates de la diffusion du virus o'nyong nyong en Republique Centafricaine, *Med. Trop. (Marseilles),* 28, 345, 1968.

30. **Georges, A. J., Saluzzo, J. F., Gonzalez, J. P., and Dussarat, G. V.,** Arboviroses en Centafrique: incidence et aspects diagnostiques chez l'homme, *Med. Trop. (Marseilles),* 40, 561, 1980.

31. **Simpson, D. I. H.,** The Kisumu study, in *Man-made Lakes and Human Health,* Stanley, W. F. and Alpers, M. P., Eds., Academic Press, London, 1975, 193.

32. **Johnson, B. K., Chanas, A. C., Shockley, P., Squires, E. J., Gardner, P., Wallace, C., Simpson, D. I. H., Bowen, E. T. W., Platt, G. S., Way, H., Parsons, J., and Grainger, W. E.,** Arbovirus isolations from, and serological studies on, wild and domestic vertebrates from Kano Plain, Kenya, *Trans. R. Soc. Trop. Med. Hyg.,* 71, 512, 1977.

INDEX

H